Plants and the Chemical Elements

Biochemistry, Uptake, Tolerance and Toxicity

Edited by Margaret E. Farago

VCH

Distribution:

VCH, P.O. Box 101161, D-69451 Weinheim (Federal Republic of Germany)

Switzerland: VCH P.P. Box, CH-4020 Basel (Switzerland)

United Kingdom and Ireland: VCH (UK) Ltd., 8 Wellington Court, Cambridge CB1 1HZ (England)

USA and Canada: VCH, 220 East 23rd Street, New York, NY 10010-4606 (USA)

Japan: VCH, Eikow Building, 10-9 Hongo 1-chome, Bunkyo-ku, Tokyo 113 (Japan)

ISBN 3-527-28269-6 (VCH, Weinheim) ISBN 1-56081-135-8 (VCH, New York)

Plants and the Chemical Elements

Biochemistry, Uptake, Tolerance and Toxicity

Edited by Margaret E. Farago

VCH Weinheim · New York · Basel · Cambridge · Tokyo

Dr. Margaret E. Farago
40 The Fairway
Northwood
Middlesex HA6 3DY
United Kingdom

Published jointly by
VCH Verlagsgesellschaft, Weinheim (Federal Republic of Germany)
VCH Publishers, New York, NY (USA)

Editorial Director: Dr. Hans-Joachim Kraus
Production Manager: Dipl.-Wirt.-Ing. (FH) Bernd Riedel

Library of Congress Card No. applied for.

A catalogue record for this book is available from the British Library

Die Deutsche Bibliothek – CIP Einheitsaufnahme

Plants and the chemical elements : biochemistry, uptake, tolerance and toxicity / ed. by Margaret E. Farago. – Weinheim ; New York ; Basel ; Cambridge ; Tokyo : VCH, 1994
ISBN 3-527-28269-6
NE: Farago, Margaret E. [Hrsg.]

Printed on acid- and chlorine-free paper

Composition: Lichtsatz Glaese GmbH, D-69502 Hemsbach
Printing: Strauss Offsetdruck, D-69509 Mörlenbach
Bookbinding: J. Schäffer GmbH & Co. KG, D-67269 Grünstadt

Preface

Twenty years ago in his preface to his book on plant mineral nutrition E. J. Hewitt wrote: *"The study of plant mineral nutrition is a vast and complex subject . . ."*, and over those twenty years the whole subject has become considerably more complex and of ever widening scope as the boundaries of the subject have been pushed back to encompass many multidisciplinary areas. The subject was rooted in the agricultural aspects of the effects of mineral nutrients, the effects of their deficiencies and excesses, largely in the production of food. More recently, the wider environmental aspects of plants in relation to chemical elements have, in addition, come to the fore. This area includes the study of toxicity and tolerance in particular where anthropogenic inputs have resulted in high concentrations of toxic elements, the use of plants as biomonitors of contamination. In order to arrive at some understanding of the mechanisms and processes involved in some 92 elements and their compounds moving from a number of differing soil types, under different environmental conditions, and their fate in various plant species, taking into account the dynamic aspects including plant responses to environmental change, then workers in a number of disciplines must attack this mammoth task. These will include: soil scientists, plant physiologists, microbiologists, geneticists, chemists, analytical chemists, environmental scientists, statisticians and modellers, and probably many more.

This book summarises many of the important aspects of this very large topic. In Chapter 1, Brian Davies sets the scene with a review of soil chemistry and in particular bioavailability of trace elements to plants, and discusses how we assess this. In the second Chapter Aradhana Mehra and Margaret Farago give a general account of the role of some elements in plant life. In Chapter 3 Mark Macnair and Alan Baker discuss the rapid evolution of adaptation to high concentrations of potentially toxic elements to give metal tolerant plants. This is followed by a review of plants that hyperaccumulate metals by Robert Brooks (Chapter 4), and it is suggested that these plants represent the ultimate in plant tolerance to extremely hostile edaphic environments. Robert Hay discusses the plant enzymes that require enzymes for activity in Chapter 5, and in particular concentrates on their chemistry and structure. In Chapter 6, J. Vangronsveld and H. Clijsters examine some aspects of the toxic effects of metals on plants, in particular those which occur at the cellular level. The pathways and mechanisms involved in the incorporation of man-made nuclides into plants are examined by G. Shaw and J. N. B. Bell in Chapter 7. The use of plants to indicate and biomonitor both natural mineralisation and anthropogenic pollution is discussed in Chapter 8 (Margaret Farago). Finally, the experimental techniques used in the study of elements in plants are discussed; methods of multielement analysis for the determination of concentrations in plant tissues (Chapter 9, Margaret Farago and Aradhana Mehra) and phytochemical methods where the role of the element within the plant is deduced (Chapter 10, Margaret Farago).

I am most grateful to the authors for their hard work and time, and in particular to my coauthor Dr. Aradhana Mehra, for producing work at short notice when some chapters originally agreed were not able to be produced. Thanks are also due to Dr. Don Emerson and Mrs. Christa Schultz of VCH and Mrs. Judith Johnson for the copy editing.

London, April 1994 *Margaret E. Farago*

Contents

List of Contributors

Dr. Alan J. M. Baker
Department of Animal and Plant Sciences
University of Sheffield
P. O. Box 601
Sheffield S10 2UQ
United Kingdom
Chapter 3

Prof. Dr. J. N. B. Bell
Department of Biology
Imperial College at Silwood Park
Ascot, Berkshire SL5 7TE
United Kingdom
Chapter 7

Prof. Dr. Robert R. Brooks
Department of Soil Sciences
Massey University
Palmerston North
New Zealand
Chapter 4

Prof. Dr. Herman Clijsters
Limburgs Universitair Centrum
Department SBG
Universitaire Campus
B-3590 Diepenbeek
Belgium
Chapter 6

Prof. Dr. Brian E. Davies
Department of Environmental Science
University of Bradford,
West Yorkshire BD7 1DP
United Kingdom
Chapter 1

Dr. Margaret E. Farago
Environmental Geochemistry Research
Royal School of Mines
Imperial College of Science, Technology
and Medicine
Prince Consort Road
London SW7 2BP
United Kingdom
Chapters 2, 8, 9 and 10

Prof. Dr. Robert W. Hay
Department of Inorganic Chemistry
University of St. Andrews
St. Andrews KY16 9ST
United Kingdom
Chapter 5

Dr. Marc R. Macnair
Department of Biological Sciences
University of Exeter
Hatherly Laboratories
Prince of Wales Road
Exeter EX4 4PS
United Kingdom
Chapter 3

Dr. Aradhana Mehra
School of Environmental and
Applied Sciences
University of Derby
Kedleston Road
Derby DE22 1GB
United Kingdom
Chapters 2 and 9

Dr. G. Shaw
Centre for Analytical Research in the
Environment
Imperial College at Silwood Park
Ascot, Berkshire SL5 7TE
United Kingdom
Chapter 7

Dr. Jaco Vangronsveld
Limburgs Universitair Centrum
Department SBG
Universitaire Campus
B-3590 Diepenbeek
Belgium
Chapter 6

1 Soil Chemistry and Bioavailability with Special Reference to Trace Elements

Brian E. Davies

Department of Environmental Science, University of Bradford
Bradford, West Yorkshire BD7 1DP, UK

Contents

Introduction

Soils are a negligibly thin veneer at the interface between the lithosphere and the atmosphere. Productive agricultural soils are only about 0.001 km in depth yet below these soils lie 10–70 km of crustal rocks which, in turn, overlie 2900 km of mantle rocks: the equatorial radius of the planet is 6378 km. Human nutrition depends on this thin surface since, while acknowledging the important contribution of fish to human diets, most of our food comes from the land. Sadly, soil is dirt and dirt is despised, but if we do not unterstand the dirt beneath our feet we will not know how to keep this planet healthy for our survival and that of all its other inhabitants. This chapter presents an overview of the composition of soils and its relation to plant nutrition.

Most soils contain approximately 95% of their dry mass as mineral material and 5% as organic material. This chapter is largely concerned with the mineral material, and organic matter is discussed only so far as it influences the fate of the inorganic elements. The mineral material is derived from the break down of rocks. Embryonic soils are little more than a few millimetres of rock debris colonised by lichens while at the other extreme, in the tropics, tens of metres of weathered material may overlie unaltered rock. In all cases, however, the chemical nature of the soil will be influenced by the geochemical nature of the parent rock. In turn, plant contents tend to reflect soil contents, albeit indirectly.

Rock Geochemistry

The Elements and their Abundances

The chemical elements are often listed as the familiar periodic table which can be seen in any standard chemistry text book and will show 92 naturally occurring elements. A useful source of factual data about the elements has been provided by Emsley (1989). However, we are not concerned with all the 92 elements. Five of the 'inert gases' of Group 0 (He, Ne, Ar, Kr and Xe) need no further consideration because they are inert and therefore form no natural compounds. Radon, although it is also a Group 0 element, is of environmental interest because it is a radioactive gas and concentrations can rise to unacceptable levels in closed environments above certain rocks, especially granites (Hewitt and Kelly, 1990). However, this is a specialised topic and not one within the scope of this chapter. Promethium may occur as traces in uranium ores but its longest persisting isotope is ^{145}Pm which has a half life of only 17.7 years so that, effectively, it exists only as an artifact. Astatine also has a short half life (^{210}At = 8.3 h) and it too can be omitted. Geochemists and soil chemists therefore have some 84 naturally occurring elements as their subject matter.

Beyond element 92 (U) lie the transuranic elements of the actinide series. These are all artificial but $^{239}_{94}Pu$ is of interest because it is produced in nuclear reactors from $^{238}_{92}U$ and may be released to the environment from accidents or weapons testing. It has a very long half life (2.4×10^4 years) and is a very dangerous alpha emitter, but, like radon, its geochemistry is too specialised to be included in this chapter. Choppin and Stout (1991) have written an overview of the general chemistry of Pu, to mark the 50th anniversary of its original isolation, and Rai *et al.* (1980) have discussed its soil chemistry.

The naturally occurring elements do not occur everywhere in equal amounts or abundances. Assuming the universe did indeed begin at a finite instant of time (the 'big bang' theory) then in less than four minutes it consisted entirely of light, neutrons, subatomic particles and hydrogen and helium. It took another 700000 years for the universe to be cool enough for the hydrogen to accrete into stars and only after these had formed, evolved and finally exploded did the heavier elements, which are synthesised in the stars, join the two light gases. The consequence is that hydrogen is still the most common element in the universe.

Table 1-1 (from Emsley, 1989) lists those elements with crustal concentrations >0.1% (10000 mg kg^{-1}). The table shows that only 8 elements account for over 99% of the mass of crustal rock and 12 account for 99.9% of crustal rocks. These are the *major elements* and the remaining elements are conveniently referred to as the *trace elements.* On earth, hydrogen and helium are rare because they are too volatile to be retained by terrestrial gravity and when released to the atmosphere they eventually diffuse back to space.

Oxygen is not easily determined directly in rocks or soils and therefore the major elements are usually reported as their oxides and as *percentages*. Trace elements are

Table 1-1. Elements with a crustal abundance >0.1% (data from Emsley, 1989).

Element	Content (mg kg^{-1})
O	455000
Si	272000
Al	83000
Fe	62000
Ca	46600
Mg	27640
Na	22700
K	18400
Ti	6320
H	1520
P	1120
Mn	1060

reported as, loosely, parts per million (ppm) or per thousand million, ie, billion (ppb) but more accurately in units such as mg kg^{-1}. For geochemists the trace elements are often little more than interesting impurities in natural minerals. For biologists, however, the trace elements include essential elements, without which the organism must die or grow deformed, and highly toxic elements.

The Essential Elements

As rocks are transformed to soil so a proportion of each element is usually converted to a form which plant roots can absorb. Consequently, most of the elements in the Periodic Table would probably be detected in any plant sample if sufficiently sensitive analytical methods were used. Uptake of an element is no evidence that it plays any role in the development of the plant since a root has no power to reject any soluble element entirely. One must therefore differentiate between those elements which are needed, the essential nutrients, and the rest. The conventional criteria by which the presence of an element is regarded as essential rather than adventitious are these: the plant cannot complete its life cycle in the absence of the element; the action of the element must be specific in that no other element can wholly substitute for it nor is the element simply involved in beneficially altering the plants root environment; the element must be shown to be a constituent of an essential metabolite, or required for the proper functioning of an essential enzyme system or be uniquely involved in maintaining the overall ionic composition of tissue. Table 1-2 lists the elements which are generally accepted as being essential for plants. Major or macronutrients are separated from micronutrients, the latter being present in tissue concentration $<0.1\%$.

Table 1-2. The essential plant nutrients. The major or macronutrients are divided into (a) the major structural elements and (b) the other macronutrients.

Macronutrients
(a) Carbon, hydrogen, oxygen, nitrogen
(b) Phosphorus, potassium, sulphur, calcium, magnesium
Micronutrients
Boron, chlorine, cobalt, copper, iron, manganese, molybdenum, nickel, silicon, sodium, zinc

There is no reason to suppose that all the essential micronutrients have been identified and careful experimentation will undoubtedly prove the essentiality of other elements in due course. The agricultural significance of the most recently discovered plant micronutrients, sodium and nickel, is uncertain. There is no evidence to suggest that they limit crop production anywhere since they are required in such tiny amounts that most soils can probably satisfy a plant's normal requirements.

Both the micronutrients and most of the non-essential trace elements are known to have undesirable effects on plant and animal growth if present in excess concentration in their tissues. Plants commonly have their growth affected by excess manganese and aluminium in acid soils, and by nickel, cobalt and chromium on acid soils derived from ultramafic igneous rocks. In addition there is growing concern that several toxic elements are accumulating in some soils as a consequence of man's industrial or urban activities which lead to soil pollution or because of the use of metal-contaminated sewage sludge as a substitute for farm yard manures. Whenever an element or its compounds are heated, dissolved or pulverised it becomes environmentally labile and may escape from the working environment and ultimately accumulate in soils or sediments. Much research has been carried out on the 'heavy' metals, especially cadmium, copper, lead, mercury and zinc. The qualifier 'heavy' is not rigorously defined but most authors use it to describe metallic elements having a density greater than 6000 km m^{-3}. The more important heavy metals are listed in Table 1-3 together with

Table 1-3. Some heavy metals with their densities, average crustal rock contents and whether they are essential (E) for life processes and whether they are a known environmental pollutant (P). From Davies (1980).

Element	Density ($\times 10^3$) ($kg\ m^{-3}$)	Mean content crustal rocks ($mg\ kg^{-1}$)	Essential or pollutant
Ag	10.5	0.07	P
Au	19.3	0.05	–
Bi	9.8	0.17	P
Cd	8.7	0.2	P
Cr	7.2	100	EP
Co	8.9	25	EP
Cu	8.9	55	EP
Fe	7.9	60000	EP
Hg	13.6	0.08	P
La	6.2	25	–
Mn	7.4	950	E
Pb	11.3	13	P
Mo	10.2	1.5	EP
Ni	8.9	75	EP
Pt	21.5	0,05	–
Tl	11.9	0.45	P
Th	11.5	9.6	P
Sn	7.3	2	EP
U	19.1	2.7	P
V	6.1	135	E
W	19.3	1.5	EP
Zn	7.1	70	EP
Zr	6.5	165	–

an indication of whether they are also essential and whether they have been identified as important pollutants. Contamination of land and crops by arsenic and fluorine can also present problems locally.

Igneous Rocks

The crust and the upper part of the earth mantle are solid and comprise rigid, interlocking plates which ride over the partially molten mantle (the *asthenosphere*) in which rising convection currents force the crustal plates to move away from or into and under each other. Where the plates are parting, under the oceans, new rock is formed from the upwelling molten material, called *magma*. The new and spreading sub-oceanic crust is rich in iron and magnesium and is basaltic. The terrestrial crust formed early in the earth's history and is granitic in character. However, the rocks are continually being reworked. In particular, where one plate slides under another the consequent heating causes remelting and plumes of magma force their way back up through the crustal rocks and may emerge at the surface as volcanic ejecta or lava.

Whatever the particular mode of formation of the crustal rocks, all have in common the fact that they have originally resulted from crystallisation processes in a cooling magma. These rocks are therefore called *igneous rocks* and they are made up from intergrown and interlocking minerals (crystals) which formed as the magma cooled. Silicate minerals dominate the igneous rocks and, despite their complexity and variety, they form from a very simple building unit, a silicon-oxygen tetrahedron. Silicon has 4 electrons in its outer (M) shell ($3s^2$, $3p^2$) and loss of these gives the inert gas electron arrangement of the Si^{4+} ion. Oxygen atoms ($2s^2$, $2p^4$ electrons) accept these electrons. According to Degens (1989) there are a few thousand different minerals accessible to visible examination but only about twenty make up most rocks.

Atoms and ions of the different elements have different sizes. The theoretical ion Si^{4+} has a radius of 41 pm and that of O^{2-} is 140 pm. For a crystal to form there must be a *stable geometrical arrangement* which results in a coordination polyhedron of anions surrounding a central cation. The geometry can be predicted by deriving the cation/anion radius quotient, assuming undistorted spheres. For Si/O the radius quotient is 0.293 which corresponds to a tetrahedron. The Si–O building block is therefore a central Si^{4+} surrounded by four O^{2-} each at the apex of a regular tetrahedron. This unit is written in conventional form as $(SiO_4)^{4-}$ and each O^{2-} is half saturated by the electrons donated from the silicon. The remaining negative charges can be satisfied in two ways. Firstly, the tetrahedra can and do polymerise through shared oxygens. For example, the next most complex structure would be a double tetrahedron corresponding to $(Si_2O_7)^{6-}$. Doublets can form rings, single chains, double chains, sheets or three-dimensional frameworks. The other way involves the metals of the group of major elements. The silicate minerals consist of threedimensional arrays of silica tetrahedra held together by metal ions that saturate the unshared oxygens. The formation of the tetrahedra can be regarded as an ionic process, but the Si–O bonds are not entirely ionic in character. Using the conventional Pauli methodology and assuming that the electronegativity of Si = 1.8 and of O = 3.5 then the Si–O bond has only 51% ionic character. Thus, the silicate minerals are neither purely ionic nor purely covalent in character.

The first minerals to crystallise, at highest temperatures and pressures, are poor in silicon (<50% SiO_2) and are rich in calcium, iron and magnesium and they form minerals, such as olivine and the pyroxenes, which make up the basic or mafic rocks. The acidic or silicic rocks are the last to form, are rich in silicon (>65% SiO_2) and are dominated by quartz, the potash feldspars and muscovite.

The discussion of crystals and ion geometry leads to a question. If the mineral crystals in the igneous rocks can largely be accounted for by Si–O tetrahedra plus the major elements, how do the trace elements occur? The answer lies in the natural impurity of the minerals. As magma cools and the minerals appropriate to a particular fluid composition and temperature–pressure combination form, then other elements in the magma are incorporated into the growing crystals, but this is not a haphazard affair and involves *isomorphous substitution*, ie, substitution by an ion of similar size.

When two ions replace each other in a crystal lattice the differences between their ionic radii must not exceed 15% and their charge differences must not be greater than one. For ions of the same charge the smaller ion tends to be incorporated preferentially and if their radii are the same but their charges differ then the ion with the greater charge is incorporated. The ionic radii of magnesium and iron are 66 pm and 74 pm, respectively and they can be replaced by cobalt (72 pm), trivalent chromium (63 pm), nickel (69 pm), zinc (74 pm) and to some extent by divalent copper (72 pm). Sometimes the ions are so similar that a complete interchange is possible (solid solution). This happens in the olivine family where the two end members are fayalite (Fe_2SiO_4) and forsterite (Mg_2SiO_4) and the relevant ionic radii are Fe^{2+} = 74 pm and Mg^{2+} = 66 pm.

Since the primary minerals are electrically neutral a compensation must occur when differently charged elements replace each other. In plagioclase feldspars Si^{4+} can be replaced by Al^{3+} but, simultaneously Na^+ is replaced by Ca^{2+}. Large, well formed crystals are prized as gems and their colours depend on these ionic substitutions. Corundum (Al_2O_3) is colourless, yet just a few Cr^{3+}ions turn it into a ruby (Burns, 1983). For a fuller discussion of igneous rocks the reader is referred to Dercourt and Paquet (1985).

Sedimentary and Metamorphic Rocks

The minerals which make up the igneous rocks form at temperatures (typically, 200–1200 °C) higher than those found at the surface of the earth (5–25 °C) (again, see Dercourt and Paquet, 1985). Unsurprisingly, they are generally unstable and break down, albeit slowly, when exposed at the surface, the process called weathering. Two kinds of weathering may be distinguished, chemical and physical.

Physical weathering involves the mechanical disruption of rock fragments to release the individual mineral grains. The two most important processes are ice formation and crystal formation. When water enters pores or cracks in the rock and then freezes its volume expands by 9% and the mineral grains are wedged apart. Similarly, since the same water carries dissolved constituents, should the water evaporate, the residual precipitates will also wedge apart the mineral grains. Mechanical abrasion (in sand storms) or shock, entry of plant roots into cracks and differential expansion and contraction of the mineral grains over heating and cooling cycles will also contribute to physical weathering.

As physical weathering continues, the relative surface area of the rock fragments increases and the rock is increasingly subject to chemical attack through interaction with water and its dissolved constituents. This is chemical weathering. Minerals which form at the highest temperatures decompose most readily and, since these tend to contain the greatest variety and contents of essential trace elements soils which form on basic rocks are generally well endowed with micronutrients and crop deficiencies are rare. Indeed, toxicities may occur where soils form in ultrabasic rocks such as serpentine since they contain high concentrations of chromium and nickel. In contrast, acid igneous rocks are generally poor in some of the micronutrients but are enriched in lithium, tin or tungsten. For example, cobalt deficiencies in stock are common where soils are formed from granite or rhyolite.

Without water chemical weathering cannot take place. Desert soils are thin and barely distinguishable from the underlying rock or sediment whereas soils of the humid tropics may be tens of metres thick. Higher temperatures as well as the ubiquity of water account for deep weathering in the tropics and a familiar rule of thumb is that a change of 10 °C in mean temperatures corresponds to a doubling, or halving, in weathering rates. The polar nature of the water molecule makes it effective in dissolving ionic compounds and therefore all silicates are to some extent, at least, water soluble. However, the major weathering process is hydrolysis where hydrogen and hydroxyl ions react with silicate lattices leading both to breakdown products and to a synthesis of new, secondary minerals. Hydrolysis also involves the dissolved constituents in water such as carbon dioxide or organic acids released from lichens (the first colonisers of rock surfaces) or higher plant roots. The decomposition of forsterite (olivine) exemplifies a simple case of hydrolysis:

$$Mg_2SiO_4 + 4H_2O = 2Mg^{2+} + 4OH^- + H_4SiO_4$$

During the process of weathering, new kinds of minerals, the secondary minerals, are synthesised. As well as precipitates (hydroxides, oxides, sulphates, carbonates and phosphates) there are the clay minerals. For example, potassium feldspar can hydrolyse and lead to the formation of kaolinite:

$$2KAlSi_3O_8 + 9H_2O + 2H^+ = Al_2Si_2O_5(OH)_4 + 2K^+ + 4H_4SiO_4$$

The general structure of clay minerals involves the silicon–oxygen tetrahedra, as described above, and aluminium at the centre of an octahedral arrangement of hydroxyl ions. The octahedral and tetrahedral units link through shared oxygens and the minerals are formed as sheet structures of polymerised tetrahedral units bonded to sheet of polymerised octahedral units. The two broad classes of clay minerals are the 1:1 minerals (chiefly kaolinite) comprising one tetrahedral and one octahedral sheet and the 2:1 minerals (illite, momtmorillonite) where the octahedral sheet is sandwiched between two tetrahedral sheets. Isomorphous replacement of Si and Al occurs, typically, Si is substituted by aluminium or Al by Mg or Fe(II). In each case the consequence is the development of a negative charge on the surface of the mineral which is neutralised by the adsorption of cations from the weathering or soil solution. These cations are held only by electrostatic forces and therefore readily exchange with other cations. The cation exchange process is of great importance in the chemistry of soils and the reader is referred to Sposito (1989) or Talibudeen (1981) for a detailed account.

Erosion processes ultimately redistribute the weathering products and new sediments are formed. When buried they undergo diagenesis to form the sedimentary rocks. Ultimately, these sedimentary rocks will be exposed and will themselves undergo weathering and form the parent materials of new soils.

More extreme subterranean processes (especially heat) cause substantial changes in the sediments and lead to the formation of metamorphic rock. Thus, marble is heat metamorphosed limestone and slate is pressure metamorphosed shale.

From an agricultural viewpoint the sedimentary rocks are far more important than the igneous rocks since, although they represent only 15% of the crustal volume they occur spread over the igneous basement and amount to approximately 75% of the earth's surface. The most common sedimentary rocks are shales and mudstones (80%); sandstones, conglomerates and limestones account for the remaining 20%.

Generally for sedimentary rocks substitution in silicate lattices of primary minerals is of lesser importance. As the primary minerals decompose during weathering the alkaline earths and alkali metals tend to remain in solution and some of the metallic micronutrients pass into the lattices of the secondary or clay minerals. Others become adsorbed onto clay sized particles, are incorporated into humified organic matter by complexation or separate as precipitates following changes in redox potential.

Ore and Sulphide Minerals

Ores are metalliferous minerals which can be extracted from their host rocks to yield their metals economically, ie, the costs of mining and extraction must be compared with the revenue from selling the ore concentrates. Ores are therefore localised concentration of metal compounds. These minerals are not usually the parent materials for soils but they are of importance because mining, ore processing and smelting and the later use of metal compounds can cause soil contamination as metals are lost to the environment during processing. It is helpful to know a little about certain ore minerals in order to understand better the chemistry of contaminated soils.

The greatest interest centres on sulphide ore minerals. Metals react as electron-pair acceptors with electron-pair donors and both donors and acceptors may be classified as hard or soft (Pearson, 1967). Soft donors preferentially bind to soft acceptors such as SH^-, S^-. Thus, zinc or lead, being soft acceptors, bind to sulphide and geochemists classify them as chalcophilic metals. They occur as localised concentrations within the host rock from which they are liberated during mining. Some rock formations may contain sulphide minerals in too low a content to make economic extraction feasible but their presence will modify the total content of metals in the rock. When ore minerals form contemporaneously with the host rock they are called syngenetic. Often hot seawater or groundwater will react with igneous rocks and the elements dissolved from the primary silicates may then be reprecipitated in faults in the overlying sedimentary rocks. These are the epigenetic minerals.

Isomorphous substitution can occur both for the metal ion in the mineral or for the sulphide. The minor elemental constituents of the ore minerals may have considerable economic importance. For example, galena (PbS) can contain several percent Ag, and Cd is a prominent 'guest' in sphalerite (ZnS).

Alterations to the bulk rock content arising from the presence of dispersed sulphide phases are not confined to igneous rocks. The black shales are intermediary between shales and organic deposits (coal, lignite). They are deposited in still, marine basins and incorporate organic matter under reducing conditions. Pyrite (FeS_2) forms and they are enriched in the chalcophilic metals. Altschuler *et al.* (1983) have studied sulphur diagenesis in peat of the Florida (USA) Everglades. They concluded that pyrite forms directly from the reduction of organic oxysulphur compounds which generate HS^- or organic sulphides which react with ferrous iron in degrading tissues. These mechanisms can probably be applied to the formation of coals and organic sediments.

Thomson *et al.* (1973) have reported that clinical copper deficiency in farm animals is associated with black shales in England as a consequence of them, and their superjacent soils, being enriched in molybdenum. The resulting excess molybdenum in the diet of the grazing animals reduces the utilisation of dietary copper.

Ore metals and their crystal 'guests' may accumulate in soils and then move up the food chain. In organisms they will bind to the SH-groups of proteins (including enzymes) and thereby denature them. Thus they are toxic to organisms, and metals like Cd are well known for their pronounced toxicity. The trace constituents of the sulphide ore minerals are therefore of more than geochemical interest.

Table 1-4 provides some data selected from El Shazly *et al.* (1956) for epigenetic sphalerite. The high concentrations of Cd should be noted. During smelting the ZnS is reduced to Zn metal and the Cd is then an impurity in the ingot metal. Subsequent

Table 1-4. Concentrations ($\mu g\ g^{-1}$) of 'guest' trace elements in selected samples of epigenetic sphalerite. Data from El Shazly *et al.* (1956).

Location	Ga	Ge	Sb	Cd
Wales	20	100	200	1%
	30	200	100	1%
	200	200	100	2000
	200	250	100	1000
Salop	200	100	–	5000
Devon	100	10	–	3000

refining may isolate the Cd as a valuable coproduct or it may be passed on to the Zn compounds used in industry. Thus, environmental contamination by Zn often implies simultaneous contamination by Cd. Galena can contain appreciable amounts of Ag. El Shazly *et al.* (1956) reported as much as 5000 μg Ag g^{-1} in samples of epigenetic galena from Wales and the West of England. Jones *et al.* (1986) reported that while uncontaminated soils in West Wales contained <0.1 μg Ag g^{-1} those affected by past lead mining could contain as much as 9 μg Ag^{-1}.

Soil Geochemistry

The Nature of Soils

Soil formation is more than a simple chemical and physical attack on rocks since it also involves the participation of living organisms, such as plants or microbes, in the weathering process. Biological and non-biological weathering are not always distinguishable but it is the biological factor which differentiates soil from any other loose surface material which is referred to as regolith. It should be noted that this restrictive understanding of the word soil is not that of the engineer for whom soil includes regolith.

Soil must not be thought of as simply a residual layer which supplies nutrients for plant, and hence animal, growth. Soils are dynamic bodies in the landscape which change and evolve as environmental conditions change. Soils are the consequence of the interaction of climate and organisms on their parent material over time and modified by local topography. Human influences can be considered as a particularly intensive subset of the general effects of organisms.

The characteristic feature of soils and the way in which they are classified and recognised is the presence of distinct layers or horizons formed by pedogenesis. The smallest volume of soil which can be recognised as having all the distinctive features of that particular soil is called a pedon and a profile is the vertical section through the pedon to reveal the arrangement of horizons. A pedon is not as distinctive as, say, a plant species. It is more like the unit cell of a mineral crystal. The soil landscape can be visualised as a mosaic of pedons. For practical mapping and classificatory purposes the *soil series* is used as the fundamental soil unit. Soils are grouped together within a series on the basis of similar assemblages of horizons, a common parent material and formation under similar environmental conditions.

Influence of the Parent Material

The soil inherits from the parent material a stock of trace elements which, in turn, is controlled by the geochemical nature of the parent material (West, 1981). In global terms this relationship can be illustrated by comparing the commonly accepted mean contents of soil elemental constituents with crustal rock abundances. Reasonably reliable data are available for 67 elements.

Figure 1-1 illustrates the relationship between soil and rock composition and is based on soil data derived from Ure and Berrow (1982) and rock data from Emsley

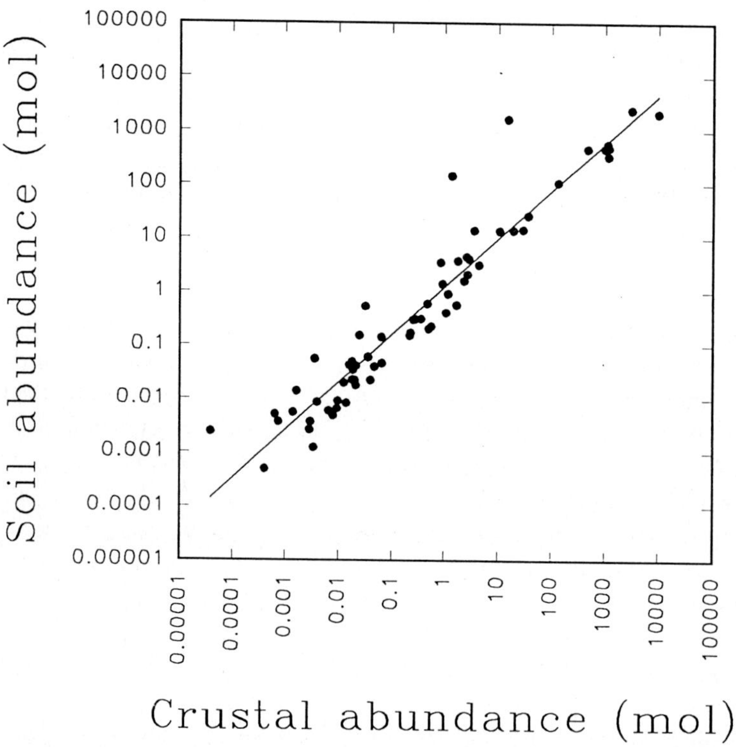

Fig. 1-1. A log–log plot of elemental abundances in soil and rock to demonstrate the close relationship of the former on the latter (data from Emsley, 1989 and Ure and Berrow, 1982).

Table 1-5. Chromium and nickel contents ($\mu g\ g^{-1}$) of soil in relation to parent material. Data taken from Berrow and Reaves (1986).

Rock type	Cr	Ni
Granitic	49	16
Intermediate igneous	100	42
Basic and ultrabasic	280	89
Metamorphic	54	26
Sands, gravels	30	16
Shales	190	53

(1989). For most elements there is a close relationship between rock and soil. The graph indicates an overall enrichment in soil compared with rock and this can be explained by a tendency to lose Si (silicic acid) to the drainage water during weathering. Among the elements showing a strong enrichment in soils (mol ratio > 1.5) are carbon and nitrogen which is accounted for by the accumulation of humus in surface horizons.

Detailed, localised studies have broadly confirmed the close relationship between rock and soil elemental contents. Table 1-5 illustrates the relationship between soil contents and rock type for chromium and nickel in Scottish soil (Berrow and Reaves, 1986). The two elements are more abundant in basic than in granitic rocks and the data confirm the control of parent material. There have been few multi-element studies of different soil series. Pierce *et al.* (1982) analysed soil samples for Cd, Cr, Cu, Ni, Pb and Zn from two dominant soil series from each of six parent materials and four series from a seventh parent material in Minnesota (USA). Differences were attributed to their differing parent materials and they concluded that the soil series is a basic sampling unit for assessing the metal content of soils. In contrast, Khan and Nortcliff (1982) working in England evaluted the micronutrient variability of a singlc soil series, the Sonning series, which has developed in terrace sands and gravels in the Thames valley. They concluded that relatively homogeneous units delineated by soil surveyors were not homogeneous for micronutrients when a considerably finer subdivision was needed. More studies are needed to settle the question whether the soil series is useful in classifying trace element data.

The weathering pathways of trace elements are complex and diagenesis causes further changes in the composition of sediments so that only general guidance is possible concerning the composition of sedimentary rocks. Trace element concentrations in sandstones are generally low since these rocks are frequently dominated by quartz particles, although the matrix may be formed from other resistant minerals, and it is the matrix cement which carries the micronutrients which eventually become available to plants. Consequently, micronutrient deficiencies in plants or animals are often more frequent in areas where soils have developed in arenaceous sediments. Purves and Ragg (1962) found that the distribution of 'wither tip', the classical symptom of copper deficiency in cereals, and low soil copper contents were directly related to the presence of one soil series, the Eckford series developed in fluvioglacial sands. Similar problems are encountered on other light textured parent materials such as aeolian sands, fluvioglacial sands, cover sands and coarse glacial outwash deposits. Carbonate sediments may also present micronutrient fertility problems. Where the sediments are very pure the only significant source of trace elements in the soil may be from extraneous materials. In southern England copper deficiencies occur in cereals growing in shallow rendzinas except where the soil is influenced by Pleistocene loess or other incorporations of non-carbonate materials. Isomorphous substitution for calcium is possible by barium, strontium and lead and limestones are often solid solutions of calcium and magnesium carbonates. However many carbonate rocks are impure and their trace element contents may be boosted by iron/manganese oxides and clays. In between the two extremes of sandstone and limestone are the shales which generally have a satisfactory trace element composition but they are so variable in composition that broad guidelines may be of little help. Special mention must be made of the black shales which are enriched in sulphide minerals and organic matter and grade into bituminous shales and coals. They tend to contain very high concentrations of chalcophilic micronutrients and in England and Wales they have been associated with molybdenum induced hypocuprosis in sheep and cattle. Tinker (1986) has written a general overview of trace element problems in arable agriculture in Britain.

These geochemical generalisations are useful when considering young soils where soil composition is broadly similar to that of the parent material, but little general guidance is available for old, deeply weathered soils rich in iron, manganese and aluminium oxides and the published literature tends of be contradictory. In Ghana, a survey of 19 profiles representing most of the major soils of the forest zone led to the conclusion that total trace element content was related to the underlying geological formations (Burridge and Ahn, 1965). On the other hand, Oertel (1961) concluded from studies of Australian soils that the soil/rock relationship is often not close enough for the satisfactory prediction of soil contents from parent material values because of the influence of pedogenetic factors.

Profile Development: Influence of Pedogenesis

Nearly all the changes which occur during pedogenesis require liquid water: in dry or frozen soils the rate of weathering or pedogenesis is negligible. In general, water moves vertically up or down in a soil carrying with it (leaching) the soluble or colloidal products of weathering. Some lateral movement does occur but the overall trend of movements is downwards. The matrix constituents of soils are not all equally soluble, quartz can be regarded as essentially insoluble.

The profile development process is therefore marked by differential movement of constituents: some, such as salts of potassium or sodium may be leached completely from the soil into the drainage water, others will accumulate relatively because they are left behind while others will accumulate in subsoil horizons because of changes in the chemical conditions within the soil. Figure 1-2 illustrates this for Cu, Ni and Cr in a podsol and in a brown earth using data from Finch *et al.* (1964) and Swaine and Mitchell (1960). Bache (1983) has described how the composition of surface waters reflects that of the soil through which water has passed but when the residence time in soil is low then surface waters more nearly resemble local precipitation in composition.

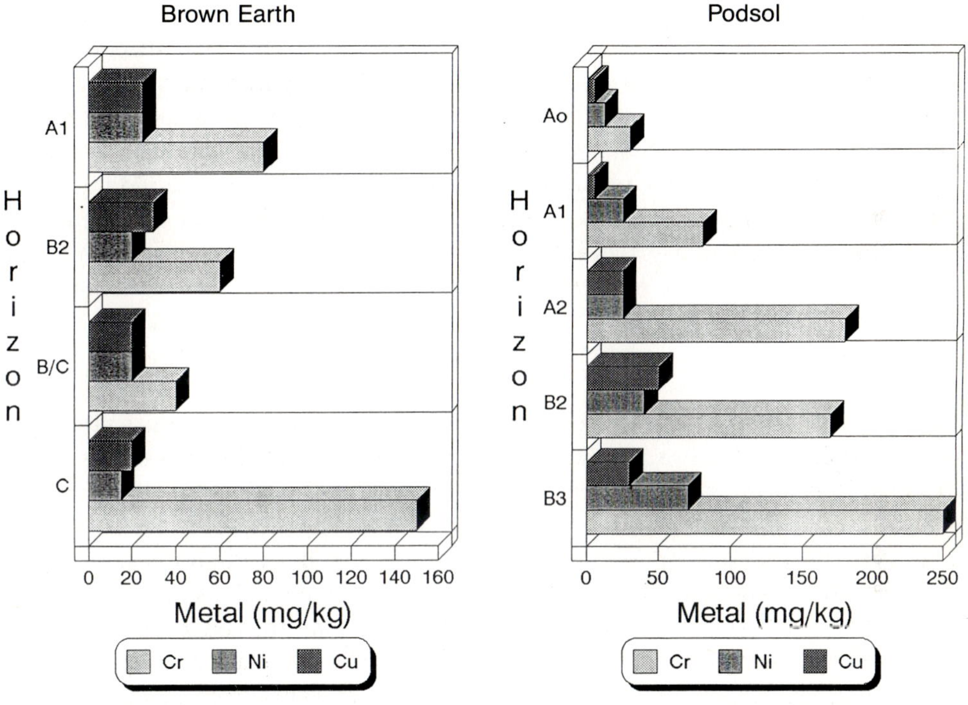

Fig. 1-2. Chromium, copper and nickel contents of genetic soil horizons of a brown earth and a podsol. In the podsol there is a marked depletion of metals in A horizons compared with B horizons whereas differences are less pronounced in brown earths (data from Finch *et al.*, 1964 and Swaine and Mitchell, 1960).

As silicate minerals decompose, trace elements are released and enter the soil solution. Thereafter their fate depends on a number of factors. Immediately, they may precipitate or remain in solution and the likelihood of this happening can be predicted by considering the ion's charge/radius (in nm) ratio, which is often called the *ionic potential* (IP). Values of IP >95 refer to soluble anions of the elements B, Cr(VI), Mo(VI), Si and W. Soluble cations are found in a group where IP<30, namely Cd, Co, Cu, Fe(II), Pb, Mn(II), Hg, Ni, Ag, Sr, Sn and Zn.

Elements may be trapped in those soil fractions formed from compounds of elements where IP ranges from 30 to 95. This is a group of elements which tend to accumulate in weathering residues and includes the elements Cr(III), Fe(III), Mn(III), Mo(IV), V(III) and V(V). Iron and manganese dominate the group and in most soils these elements precipitate as hydrous oxides either in concretionary forms or as coatings on mineral and clay surfaces. They exert a chemical control on the activities of other ions far greater than might be supposed from their concentrations (Jenne, 1968). The hydrous oxides act as sinks for other trace elements through several mechanisms (Hem, 1977; Loganathan *et al.*, 1977; Okazaki *et al.*, 1986). As they form, other trace metals can be occluded in the oxide precipitate and cobalt is notable for its association with manganese oxides. They tend to have high adsorption affinities which increase with pH and the trace cations are readily sorbed onto the oxide surfaces after which they may enter the precipitate through solid state diffusion. Superimposed on these mechanisms is the possibility that the hydrous oxides can dissolve and then precipitate again in response to changes in the soil's redox potential or pH.

Two important chemical parameters control the nature of the leaching regime. The first is acidity, ie, pH. Desert and semidesert soils may have pH values up to about 10 because of the presence of sodium bicarbonate in the soil solution but usually a pH value of about 8.3 represents the upper limit. Soils with pH values in the range 7.0–8.3 are those formed on limestone and the soil solution is dominated by carbonate/bicarbonate. Many metal carbonates have low solubilities and leaching losses are therefore minimal. With increasing acidity leaching losses become more pronounced and at pH = ca. 4.0 the clay minerals start to degrade and aluminium dominates the soil solution. Agricultural soils are usually maintained at pH values above 6 and enhanced solubility of metal ions under the more extreme acid conditions is not important. However, in recent years concern has mounted over the effects of 'acid rain', ie, increased wet and dry deposition of sulphates or nitrates from industrial sources. Berggren *et al.* (1990) have reported that forest soils in southern Scandinavia have become, on average, 0.5–1.0 pH units lower during the course of recent decades. The solubility of Al, major cations, Mn, Cd and Zn has increased and large areas of forest soils now have acidity conditions such that any additional input of strong mineral acids causes a rapid increase in the release rates of these elements.

The second control is the reduction–oxidation potential, *Eh.* In most soils the dissolved oxygen content of the soil solution is high enough to ensure that oxidising conditions prevail. Where water percolates slowly through the profile or where water is stagnant in the soil, reducing conditions are found and the process of *gleisation* occurs. Sulphate reduces to sulphide and most metal sulphides are very insoluble. For agricultural soils marked reductions in soil *Eh* do not occur but the control of redox potential on metal solubility is an important consideration for padi rice and in wetland soils which may be drained and aerated during reclamation. De Laune *et al.* (1981) maintained sediment suspensions at pH values of 5.0, 6.5 and 8.0 and *Eh* values of −200 mV, 0 mV, 250 mV and 500 mV. Iron and Mn concentrations decreased significantly as pH and *Eh* were raised. Calmano *et al.* (1986) have reported how changes from neutral, reducing environments to moderately acid, oxic environments lead to increased bioavailability of Cd.

Weathering and leaching are not the only processes that cause the development of horizons. Plants draw their moisture and nutrients from a relatively large volume of soil but, assuming they are not harvested, will return their inorganic constituents to the *surface* through leaf fall or generally to the upper layers through root death. The fresh organic material is food for many organisms, which, in turn, live on each other. Thus, organic material changes until the end product, a black or brown organic material with a lignin-like structure, called humus which enriches the surface layers. Humus is a store of trace metals through adsorption, since it carries a negative charge like the clays, or through complexation (chelation). Metals may be released as humus oxidises (mineralises) or by desorption or decomplexation (Shuman, 1988; Livens, 1991).

Natural or Baseline Contents and Pollution

Little has yet been done to put the presentation of soil data on a firm statistical footing and most compilations merely summarise the elemental values recorded in the literature. Such summaries may well be biased by an inherent research interest in unusually high or unusually low concentrations. In recent years there has been a movement towards reporting trace element concentrations in terms of probability ranges for carefully defined parent materials.

There is no simple, unequivocal way of recognising when a soil has been contaminated or polluted by trace elements since all the naturally occurring elements are present in all soils, albeit at very low concentrations. The problem of recognising whether contamination by metals has taken place becomes one of deciding whether the measured concentration of a particular element is within the range of what could occur naturally for that soil or whether the measured concentration is *anomalous*.

It is not clear what is the best model to describe the variability of soil metal concentrations. Ahrens (1954, 1966) has proposed that the distribution of elements in igneous rocks approximates to a log-normal distribution. This model does not necessarily apply to soils but the available evidence suggests it may. Its applicability underlies the interpretation of geochemical data in mineral exploration.

Rose *et al.* (1979) discuss the concept of *threshold*, the upper limit of normal background fluctuations. Values above background are considered anomalous. This approach is directly applicable to contamination studies since a contaminated soil is an anomalous soil. The simplest way of identifying threshold concentrations is by collecting samples from apparently uncontaminated areas (eg, those remote from urban or industrial influences). After analysis the geometric means and deviations are calculated. The threshold is then the value lying two or more standard deviations from the mean, depending on the probability level required. An anomalous value is one which lies above the threshold. Negative anomalies may also be recognised and are applicable to nutrient deficiency studies.

Very often it is not possible *a priori* to separate contaminated and uncontaminated soils at the time of sampling. The best that can be done in this situation is to assume the data comprise several overlapping log-normal populations. A plot of percent cumulative frequency versus concentration (either arithmetic or log-transformed values) on probability paper produces a straight line for a normal or log-normal population. Overlapping populations plot as intersecting lines. These are called broken line plots and Tennant and White (1959) and Sinclair (1974) have explained how these composite curves may be partitioned so as to separate out the background population and then estimate its mean and standard deviation. Davies (1983) applied the technique to soils in England and Wales and thereby estimated the upper limits for lead content in uncontaminated soils.

It should not be assumed that anomalous concentrations necessarily indicate contamination. Bolviken and Låg (1977) have described areas in Norway where the absence of vegetation is due to the toxic effects of high concentrations of metals in soils as a result of weathering of sulphide ores close to the surface. This is a natural process having nothing to do with contamination.

Identification of a geochemical anomaly should, in the first instance, be considered as only that, an anomaly. Other evidence must be taken into account to decide whether the anomaly is natural or is a *neoanomaly*, one caused by anthropogenic contamination.

Chemical Forms and Fractions

Soil is a mechanically and chemically complex and heterogeneous material. A casual glance at a spadeful will show that it is composed of particles of very different sizes, some large (gravel, stones, boulders) and other small (sand, silt, clay). Soil scientists classify soil into *texture classes* based on the proportions of different particle diameters. The particles are usually stuck together to give aggregates of different sizes and shapes, collectively described as soil *structure.* Closer examination reveals even greater complexity. A microscopic examination of thin slices of intact soil will reveal its microstructure a characteristic of which is the formation of *cutans*. These are distinct deposits of manganese or iron oxides on surfaces or orientated clay or precipitate fillings of pores and channels. Soils contain humified organic matter. Some soils have visible nodules of iron oxides or calcium carbonate or sulphate. This chemical heterogeneity has led to the concept that bulk soil can be thought to contain different chemical fractions and that trace elements are distributed non-uniformly among these fractions. Various dilute solutions have been used which are believed to dissolve these fractions selectively and thereby yield evidence for the dominant chemical locations of the trace elements.

An early method which passed into general use was that of Bascomb (1968) in relation to iron and organic carbon. His classification of iron compounds is a good example of one based on previous field and laboratory experience. He postulated that inorganic iron compounds could be classified as (1) silicates, (2) well crystallised oxides, (3) amorphous but 'aged' hydrous oxides and (4) amorphous 'gel' hydrous oxides. His fractionation method was based, in part, on an assumption that a potassium pyrophosphate solution (0.1 M) at pH 10.0 would extract only category (4) whereas the same reagent at pH 7.0 would also extract category (3). Other reagents were used for the organic fractions and the remaining iron categories.

McLaren and Crawford (1973) studied the fractionation of soil copper and proposed that it could occur in soils in several forms, namely:

(a) in the soil solution as ions or complexes
(b) on normal cation exchange sites
(c) on specific sorption sites, ie, adsorbed copper which cannot be removed by the reagents normally used for determining the exchangeable ions
(d) occluded in soil oxide material
(e) in living organisms or in organic residues
(f) in the lattice structure of soil minerals

This classification of fractions was also based on a generally agreed understanding of the composition of soils derived from many years of field and laboratory experience. The reagents proposed were also a selection of those in common use in soils laboratories. The soil solution and exchangeable copper were extracted by 0.05 M $CaCl_2$; specifically sorbed copper was extracted with 2.5% (v:v) acetic (ethanoic) acid; 1.0 M potassium pyrophosphate extracted organically bound copper; an ammonium oxalate/oxalic acid mixed solution at pH 3.25 extracted copper occluded by free iron oxides while the residual lattice copper was determined using hydrofluoric acid. There are now many schemes in use (eg, Clevenger, 1990) and that of Tessier *et al.* (1979) is often followed or adapted.

Some of the more commonly used reagents are listed in Table 1-6 together with an indication of which chemical fraction they are thought to extract. The order in the table is one of increasing vigour of attack on the soil so that a solution which extracts, say, metals from the humus in soil will also extract from the soil solution and from the exchangeable pool. All are empirical and, while the fractions proposed are based on much research, the fractions as measured are usually operationally defined in the sense that an individual fraction is presumed to attack a certain chemical reservoir. Shuman (1991) should be consulted for a recent critical review of the chemical forms of micronutrients in soils. Fig. 1-3 uses data from Shuman (1985) to illustrate the results from a typical fractionation study.

Table 1-6. Some common diagnostic soil extractants and the chemical forms they are thought to extract.

Soil fraction	Common extractant
Soil solution	H_2O; 0.01M $CaCl_2$
Readily exchangeable	0.5M CH_3COONH_4; 0.2M $MgSO_4$; 0.1M NH_4Cl; 1M NH_4NO_3
Specifically sorbed	0.5M CH_3COOH; 0.1M HCl; 0.1M HNO_3
Organically bound	0.05M EDTA[a]; 0.05M EDDAH[b]; 0.005M DTPA + 0.01M TEA + 0.05M $CaCl_2$[b]
Hydrous oxide bound	1M CH_3COONH_4 + 0.002M $C_6H_5(OH)_2$ 0.2M $(COO)_2(NH_4)_2$ + 0.15M $(COOH)_2$ @pH 3.3
Residual	HF; mixtures of mineral acids, concentrated hot, namely: HNO_3 + $HClO_4$; HNO_3; HNO_3 + HCl; also fusion mixtures

[a] ethylene diamine tetraacetic acid, disodium or diammonium salt
[b] ethylene diamine di-*o*-hydroxyphenyl acetic acid
[c] diethylene triaminepentaacetic acid and triethanolamine with calcium chloride at pH 7.3

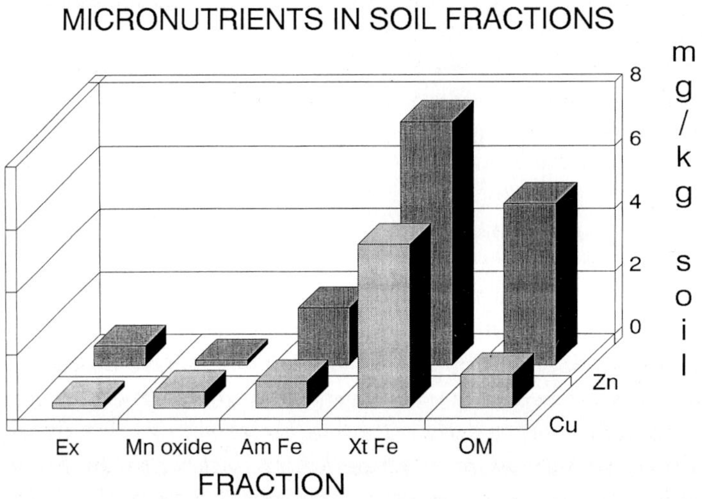

Fig. 1-3. Copper and zinc concentrations in soil chemical fractions (data from Shuman, 1985).

Bioavailability

General

The total content of the major elements in soil is of little practical significance since only a tiny, soluble fraction is available for absorption by plant roots (West, 1981; Tinker, 1986). To some extent this is less true for trace elements and micronutrients where for example, analyses of total copper or zinc can be used to assess the likelihood of plant deficiencies or toxicities. Nonetheless, for an element to be bioavailable it has to be relatively soluble.

Rather than considering soil as a mass of altered rocky material it is sometimes helpful to consider it as a three-dimensional network of pipes and voids which is characterised by abrupt changes in diameter: larger cavities open through narrow necks into a capillary system. Collectively, the non-solid part is referred to as the pore space and, typically, is 30–40% of the bulk volume of the soil with pore diameters from as small as a few picometres to cracks or earthworm burrows measured in centimetres. The pore space is occupied by water or air in constantly varying proportions. The chemical processes in soil which are of interest both to soil scientists and plant physiologists or agronomists are those which occur at the interfaces between the soil matrix and the soil water (soil solution) or at the air-water meniscus or in the contact zone between organisms and the soil solution, eg, the rhizosphere.

The Soil Solution

The immediate source of nutrient elements for plants is the soil solution but there is a paucity of information concerning its composition especially for micronutrients. A number of practical problems arise in studying the soil solution and the first is actually getting a sample of it. The soil solution can be removed for analysis by centrifugation or direct suction. Some workers prefer a 'saturation extract', when the soil is allowed to equilibrate with water which is then extracted by mild suction. The solutions extracted by these techniques are very dilute and consequently not stable chemically. They present considerable analytical problems although these have eased in recent years now that very sensitive flameless atomic absorption methods have become widely available.

The soil solution is dominated by the major cations calcium, magnesium, potassium and sodium whose solution concentrations are typically up to 2 to 3 mmol L^{-1}. In contrast, the concentrations of microntrients tend to be at the micromolar level. For example, in saturation extracts of Californian soils the following concentrations were measured: Co = 1.0, Cu = 0.6, Fe = 0.9, Mn = 3.1, Mo = 7.6 and Zn = 1.1 μmol L^{-1}, respectively (Bradford *et al.,* 1971).

Ions rarely exist in the soil solution as simple 'naked' forms since they are liable to complex formation with both inorganic and organic ligands. Moreover, not all solutes are completely dissociated in solution and are present as ion pairs. For example, soil solution calcium may be present as Ca^{2+}, $CaSO_4^0$ or $CaHPO_4^0$. Adams (1971) has suggested the following general principles for ion pairing common soil solution anions and cations:

(1) there is no ion pairing with Cl^-
(2) ion pairing of cations with NO_3^- is small enough to be neglected
(3) ion pairing with SO_4^{2-} is general; it is slight with univalent cations but extensive with multivalent cations
(4) ion pairing with $H_2PO_4^-$ or HPO_4^{2-} can be ignored for univalent cations but not with multivalent cations
(5) for HCO_3^-, ion pairing with univalent cations is insignificant whereas for multivalent cations it becomes significant at higher pH values.

For further information on this subject the reader is referred to Sposito (1989) who has also discussed the use of computer programs to predict the species likely to occur in soil solutions under various conditions such as pH differences (Sposito, 1981). Besides the formation of inorganic complexes there are many opportunities for organic complexation. Copper is particular prone to organic complexation. In displaced solution from 20 calcareous soils in Colorado 98–99% of the dissolved copper was in the form of an organic complex (Hodgson *et al.,* 1966).

Equilibria in the Soil Solution

The composition of the soil solution changes constantly as elements dissolve, desorb or decomplex from the solid phases, as plant roots and lower plants withdraw nutrients or release organic substances, as rainwater brings in solutes from outside and as the soil water passes down the profile into the drainage water. Figure 1-4 summarises these processes. It should be noted that equilibria are assumed between the solution and the matrix components but there is no certainty that rapidly fluctuating moisture contents will allow true equilibrium to be attained.

Much has been written in recent years about atmospheric inputs to the soil since the 'acid rain' problem has been recognised. Soils receive both wet deposition, ie, precipitation as rain or snow and dry deposition in which aerosols are deposited on the surface. Although precipitation is very dilute it can be an important source for some elements, cyclic salts which are derived from the oceans and transported to the land surface in the normal hydrological cycling of elements. Fuge and Johnson (1986) concluded that most iodine in soils is derived from the atmosphere and, hence, the oceans. Normally, however, rain makes only a very small contribution to the composi-

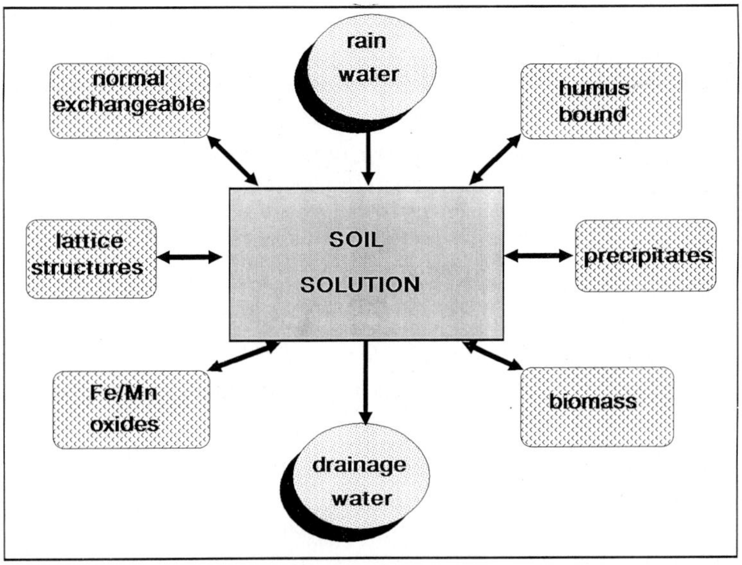

Fig. 1-4. Inter-relationships between the composition of the soil solution, soil components and the passage of water through the soil profile.

tion of soil. The pH of natural rain water is assumed to be controlled by the dissociation of dissolved CO_2 and has a value of 5.6 but variations in the natural sulphur cycle may depress it to as low as 4.5 (Charlson and Rodhe, 1982). However in many countries the pH of the precipitation is now regularly 4.5 or lower (Seip and Tollan, 1978; Fowler *et al.,* 1982) because of vehicle and industrial emissions. Soils formed on soft limestone or easily weathered parent materials and having pH values near neutrality are well buffered and are not affected by the excess of hydrogen ions in low pH rain, but other soils are impoverished and, in particular, aluminium ions are lost to the drainage water and hence to water bodies when aquatic organisms are seriously affected (Kaplan *et al.* 1981; Bache, 1983; Breemann *et al.,* 1984; Mason, 1984).

Isomorphous substitution of silicon and aluminium in the lattices of clay minerals leads to the development of a negative charge on their surfaces. Hydroxyls at the surfaces of the clay minerals dissociate above pH 2 and especially above pH 4 ($-SiOH-SiO^-$) and add to the negative charge. Carboxyl and phenolic groups in soil organic matter also progressively ionise as pH is raised (Talibudeen, 1981). Soils are seen to have both a fixed and a variable charge which is neutralised by cations and protons through adsorption onto the clay–humus complex. An electrical double layer results comprising the negative charges of the clay–humus complex and a diffuse cloud of cations and anions adjacent to the surface and the net positive charges neutralise the net negative charges of the soil surface. The concentration of cations is greatest close to the charged surface and decreases exponentially to the bulk solution. In contrast, the anion concentration increases exponentially from the charged surface (again, see Talibudeen, 1981). The sorbed cations *exchange* for cations in the soil solution and thereby from a pool of plant available nutrients. The *cation exchange capacity* of the clay–humus complex is measured by displacing all adsorbed cations with one other cation, in turn displacing that cation and measuring its concentration in the equilibrium solution (eg, Hendershot and Duquette, 1986). A high cation exchange capacity favours the retention of nutrients against leaching losses. Soil organic matter has a higher cation exchange capacity than the clays and is one of the reasons that declining organic contents is synonymous with declining fertility. There is a voluminous literature on soil cation exchange and the reader is referred to Talibudeen (1981) or Sposito (1989) for more information.

Besides its role in cation exchange, the soil organic matter is a significant reservoir of nutrients and other elements through complexation. The relatively stable soil *humus* arises from the degradation of plant and other tissues. Its composition is complex, heterogeneous and, in part, lignin-like. It is impossible to separate it from the mineral matrix of the soil without using vigorous chemical procedures, usually dilute aqueous sodium hydroxide (Schnitzer and Schuppli, 1989) and the extraction procedures probably produce misleading artifacts (Farmer and Pisaniello, 1985).

Some fractions of soil organic matter mineralise (oxidise) readily whereas others are relatively stable. Balesdent *et al.* (1988) have shown, for prairie soils in Missouri, USA, that there is an easily mineralisable component that was exhausted in 30–40 years whereas at least 50% of the total organic matter persisted for over 100 years. As the organic matter mineralises metals are released to the soil solution. Decomplexation is also important however, and these reactions have been discussed by Berrow and Mitchell (1980) and Livens (1991) and the reader is also referred to Sposito (1989).

Figure 1-4 shows an exchange between the soil solution and 'precipitates'. Both soluble iron and manganese will precipitate out from the soil solution as hydrous oxides, as pH or redox potential change, for iron, the changes can be represented very simply as:

$$Fe^{2+} = Fe^{3+} + e^-$$
$$Fe^{3+} + 3OH^- = Fe(OH)_3$$

The second equation clearly demonstrates the control of acidity (remembering that $H_2O = H^+ + OH^-$) and the first equation is redox controlled. Traditionally, redox potential is represented as *Eh* and measured in millivolts. However, Lindsay (1979) has argued for its replacement by pe (where *Eh* (millivolts) = 59.2 pe) and hence the combination of pe and pH to provide single equations to describe the chemistry of multivalent ions. The classical text dealing with this very large subject is Garrels and Christ (1965) and experimental approaches are exemplified by Gotoh and Patrick (1972) and Schwab and Lindsay (1983). There are many studies that have shown that the hydrous oxides, in turn, can control the solubilities of trace metals (Santiallan-Medrano and Jurinak, 1975; Hem, 1977; Loganathan *et al.,* 1977; Okazaki *et al.,*

1986). Cobalt is well known as having its bioavailability controlled by manganese oxides (Jarvis, 1984).

The plethora of many experimental chemistry papers dealing with hydrous oxides and their control of other metal solubilities in relation to redox and pH should not distract attention from the equally important role of microorganisms and soluble organic materials. It has long been known that soil microorganisms can significantly alter the pH for the Mn(II)/Mn(IV) equilibrium (Bromfield, 1956). Similarly, reduction of iron compounds in the formation of gley soils is very dependent on a supply of readily oxidisable organic matter. Couto *et al.* (1985) studied oxidation–reduction processes in a Brazil oxisol with a seasonal water table. Although the water table stood at 0.4–2 m of the surface for more than 90 days each year there was little pedological evidence of gley formation and bright platinum electrodes demonstrated that *Eh* values stayed within the range 500–700 mV. Fe(II) formed in samples from below 40 cm only when incubated with sucrose. The authors interpreted these data as showing that in natural conditions iron is reduced only when there is an energy source for microbial activity.

Plant Roots and the Rhizosphere

Plant roots derive their micronutrients from the soil solution but as they grow they alter their immediate solution environment, the rhizosphere. Van Noordwijk and de Willigen (1986) have discussed the relation between root density, nutrient mobility and actual uptake.

In water logged soils radial oxygen loss from the root raises the redox potential in the rhizosphere as a consequence of which iron oxide plaques are seen to develop on root surfaces. Bacha and Hossner (1977) removed the coatings on rice roots growing under anaerobic conditions. The coatings were composed primarily of the iron oxide mineral lepidocrocite (γ-FeOOH) as the only crystalline component. St-Cyr and Crowder (1990) studied the iron oxide plaque on roots of *Phragmites* and detected both Fe and Mn. The Fe:Mn ratio of the plaque resembled the ratio of Fe:Mn in substrate carbonates. The plaque material also contained Cu.

The form of fertiliser nitrogen can affect the pH of the rhizosphere. Ammonium nitrogen has been shown to decrease the pH in the vicinity of roots of wheat whereas fertilisation with nitrate nitrogen increased the rhizosphere pH compared with the pH of the bulk soil. Differences of up to 1.2 units occurred under field conditions (Smiley, 1974). Hauter and Mengel (1988) used Sb electrodes to measure soil pH at the root surface of red clover. Lowest pH values were found at the root tips and in a sandy soil the pH at the surface of root laterals was about one unit lower than in the bulk soil.

The rhizosphere is complex and heterogeneous. It contains exudates, lysates, mucilage, secretions, dead cell material and carbon dioxide all of which affect substrate flow to the root (Lynch and Whipps, 1991). Moreover, certain soil microorganisms colonise roots of higher plants and form symbiotic relationships. In this way the availability of nutrients, especially P, can be greatly increased (Bolan, 1991). Jenny and Grossenbacher (1963) were the first to draw attention to the possibility of an intimate contact between roots and minerals when they published electron micrographs which showed that the root can be surrounded by a mucigel which fills the space between the cell wall and a soil particle. Morel *et al.* (1986) have discussed the importance of this mucilagenous layer in retaining heavy metals in the rhizosphere. The flux of Cu and Pb to the root can be reduced whereas Cd is more easily transported.

Measurement of the Bioavailable Fraction

Only a tiny fraction of the total content of the macronutrients in soil is available for absorption by plant roots and the soil to soil variability in available macronutrients is far greater than the variability of total concentrations. To some extent this is less true for the micronutrients where for example, analyses of total copper or zinc can be used to assess the likelihood of plant deficiencies or toxicities occurring. However, the assessment of plant-available levels using soil extractants is more useful in practice. These are generally dilute (< 1 M) solution of mineral or organic acids, simple salts of organic and inorganic complexing agents (Jones and Benton, 1990; Peck, 1990).

Extractants are used in two ways. Historically the earliest, and still the most common use is to identify soils where a yield response may be expected if a micronutrient fertiliser is applied or to confirm a micronutrient deficiency following field observations of growth abnormalities. Thus, Junus and Cox (1987) began their paper dealing with a zinc soil test by saying ‘A soil test is desirable to determine whether or not fertilisation with Zn is required. Calibration of a Zn soil test would best be based on yield response to applied Zn’. The second use is to attempt to predict crop uptake of trace elements and this is especially important where plants are growing on soils contaminated by elements like lead or arsenic which are toxic to animals. A third use, already described, is in the chemical fractionation of soil in order to identify and evaluate the major chemical pool of micronutrients and toxic metals (see Table 1-6). The same extractants are used for all three purposes and none is based on any firm theoretical foundation. Yet, despite the empiricism involved in their use they have proved of great practical assistance both in advisory work and in laboratory studies. There are many published reports on the use of extractants to diagnose micronutrient responsive soils. The reader is advised to consult Shuman (1991) and Sims and Johnson (1991) for recent reviews and to obtain general details of methods. For specific information alluded to in the following paragraphs the reader should refer either to the papers quoted or to Needham (1983).

Soil extractants have proved least successful for those elements whose oxidation state is easily altered in the soil environment. There are no reliable soil tests for iron and a determination of soil pH combined with a knowledge of the susceptibility of a particular crop to either iron deficiency or toxicity is generally more helpful. In the case of manganese, deficiencies are possible where soil pH is greater than 6.0 and where humus levels are moderate to high. However, a determination of 'active' or easily reducible manganese has proved valuable on some occasions. This is done by adding a mild reducing agent to an extractant which accesses the readily exchangeable soil fraction. Hydroquinone (0.2%) in neutral molar ammonium acetate is widely used when <25 mg Mn per kg is a critical level. But the widely differing susceptibilities of crops to manganese deficiencies make the method at best only a rough guide. Where micronutrients are less liable to changes in their oxidation state or exist in only one valency extractants have proved more helpful.

The strong association of copper and zinc with soil humus has stimulated a wide interest in solutions of chelating agents (eg, Viro, 1955; Tills and Alloway, 1983). These are supposed to act by mimicking the activities of the plant roots in decomplexing metals from the organic matter. However, the quantities of trace metals which they liberate are far higher than those mobilised by the roots. In England and Wales the National Agricultural Advisory Service presently uses disodium EDTA (ethylene-diamine-tetraacetic acid) for copper and soil concentrations in the range 0–1.6 mg Cu L^{-1} indicate the likelihood of deficiencies while concentrations between 1.7 and 2.4 mg L^{-1} suggest the possibility of crop deficiencies. In the USA a mixed reagent containing (0.005 M) DTPA (diethylenetriaminepentaacetic acid) with (0.1 M) TEA (triethanolamine) and 0.01 M calcium chloride ($CaCl_2$) at pH 7.3 has been widely adopted (Lindsay and Norvell, 1978). Copper deficiencies are associated with soil levels <0.2 mg kg^{-1} and zinc deficiencies with concentration <0.5 mg kg^{-1} while 0.5–1 mg Zn kg^{-1} are regarded as marginal. The same reagent is used to test for manganese when the critical level is 1 mg kg^{-1}. Trierweiler and Lindsay (1969) have proposed the use of a mixture of EDTA with ammonium carbonate as a test for zinc and they established 0.4 mg Zn kg^{-1} as the critical soil level.

Although cobalt is involved in nitrogen fixation in legumes there is no evidence that low soil cobalt levels are anywhere limiting crop yields. However, the widespread occurrence of cobalt deficiencies in cattle and sheep (pining) has stimulated a search for a soil extractant (Mitchell, 1974). Both in Britain and the USA it is accepted that herbage cobalt contents less than 0.07–0.08 mg kg^{-1} dry matter indicate the likelihood of pining. The most useful extractant appears to be 2.5% acetic acid when soil concentrations <0.25 mg Co kg^{-1} imply the probability of low cobalt herbage and 0.25–0.35 mg Co kg^{-1} are marginal. Molybdenum(IV) has a charge/radius ratio of 57 which indicates its tendency to accumulate in weathering residues together with iron and manganese and it is interesting that the most widely used extractants involve low pH oxalate solutions which dissolve this fraction. The results need interpreting carefully, particularly with regard to soil pH and, as a rough guide, for soils of pH 5 an oxalate extractable value of 0.2 mg Co kg^{-1} indicates the possibility of plant deficiencies and at soil pH 6.5 the critical value falls to 0.05 mg Co kg^{-1} since the availability of molybdenum is pH dependent. Total molybdenum concentrations can be used to identify soils where cattle may suffer from molybdenosis. Most normal soils tend to have 1 – 3 mg Mo kg^{-1} and a value >20 mg kg^{-1} indicates the need for veterinary investigations.

Finally, the most popular extractant for boron is boiling water (eg, McGeehan *et al.,* 1989) and it is agreed in many countries that concentrations below 1 mg B kg^{-1} soil imply the likelihood of deficiencies in susceptible crops. In England and Wales, the National Agricultural Advisory Service classifies boron levels as follows: 0 – 0.5 mg L^{-1} indicates severe deficiency, 0.6 – 1.0 mg L^{-1} implies a likelihood of deficiency and values >4.1 mg L^{-1} indicate the possibility of boron toxicity.

References

Adams, F. (1971). *Soil Science Society of America Proceedings, 35,* 420 – 426.

Ahrens, L. H. (1954). *Geochimica et Cosmochimica Acta, 5,* 49 – 73.

Ahrens, L. H. (1966). *Geochimica et Cosmochimica Acta, 30,* 109 – 122.

Altschuler, Z. S., Schnepfe, M. M., Silber, C. C., Simon, F. O. (1983). *Science, 221,* 221 – 227.

Bacha, R. E., Hossner, L. R. (1977). *Soil Science Society of America Journal, 41,* 931 – 936.

Bache, B. C. (1983). *Water Science Technology, 15,* 33 – 45.

Balesdent, J., Wagner, G. H., Mariotti, A. (1988). *Soil Science Society of America Journal, 52,* 118 – 124.

Bascomb, C. L. (1968). *Journal of Soil Science, 19,* 251 – 268.

Berggren, D., Bergkvist, B., Falkengren-Grerup, U., Folkeson, L., Tyler, G. (1990). *Society of the Total Environment, 96,* 103 – 114.

Berrow, M. L., Mitchell, R. L. (1980). *Transactions Royal Society Edinburgh: Earth Sciences, 71,* 103 – 121.

Berrow, M. L., Reaves, G. A. (1986). *Geoderma, 37,* 15 – 27

Bolan, N. S. (1991). *Plant and Soil, 134,* 189 – 207

Bolviken, B., Låg, J. (1977). *Transactions Institute Mining and Metallurgy, 86,* B173 – B180.

Bradford, G. R., Bair, F. L., Hunsaker, V. (1971). *Soil Science, 112,* 225 – 230.

Breeman, N., van, Discoll, C. T., Mulder, J. (1984). *Nature (London), 307,* 599 – 604.

Bromfield, S. M. (1956). *Australian Journal of Biological Science, 9,* 238 – 252.

Burns, R. G. (1983). *Chemistry in Britain, 19,* 1004 – 1007.

Burridge, J. C., Ahn, P. M. (1965). *Journal of Soil Science, 16,* 296 – 309.

Calmano, W., Forstner, U., Kerstein, M. (1986). *Toxicological and Environmental Chemistry, 12,* 313 – 321.

Charlson, R. J., Rodhe, H. (1982). *Nature (London), 295,* 683 – 685.

Choppin, G. R., Stout, B. E. (1991). *Chemistry in Britain, 27,* 1126 – 1129.

Clevenger, T. E. (1990). *Water, Air and Soil Pollution, 50,* 241 – 254.

Couto, W., Sanzonowiecs, C., Barcellos, A. O. (1985). *Soil Science of America Journal, 49,* 1245 – 1248.

Davies, B. E, (ed.) (1980). *Applied Soil Trace Elements,* Chichester: John Wiley and Sons; pp. 287 – 253.

Davies, B. E. (1983). *Geoderma, 29,* 67 – 75.

Degens, E. T. (1989). *Perspectives on Biogeochemistry,* Berlin: Springer-Verlag.

De Laune, R. D., Reddey, C. N., Patrick, W. H. (1981). *Journal of Environmental Quality, 10,* 276 – 279.

Dercourt, J., Paquet, J. (1985). *Geology: Principles and Methods,* London: Graham and Trotman Ltd.

El Shazly, E. M., Webb, J. S., Williams, D. (1956). *Transactions Institute Mining and Metallurgy, 66,* 241–271.

Emsley, J. (1989). *The Elements,* Oxford: Clarendon Press.

Farmer, V. C., Pisaniello, D. L. (1985). *Nature (London), 313,* 474–475.

Finch, T. F., Fleming, G. A., Ryan, P. (1964). *Proceedings 8th International Congress Soil Science, Bucharest, Romania,* 1 Vol. 4ff., 425–434.

Fowler, D., Cape, J. N., Leith, D., Paterson, I. S., Kinnaird, J. W. Nicholson, I. A. (1982). *Nature (London), 297,* 383–386.

Fuge, R., Johnson, C. C. (1986). *Environmental Geochemistry and Health, 8,* 31 –54.

Garrels, R. M., Christ, C. L. (1965). *Solutions, Minerals and Equilibria,* New York: Harper and Row.

Gotoh, S., Patrick, W. H. (1972). *Soil Science Society of America Proceedings, 36,* 738–741.

Hauter, R., Mengel, K. (1988). *Biology Fertility of Soils, 5,* 295–298.

Hem, J. D. (1977). *Geochimica et Cosmochimica Acta, 41,* 527–538.

Hendershot, W. H., Duquette, M. (1986). *Soil Science Society of America Journal, 50,* 605–608.

Hewitt, C. N., Kelly, M. (1990). *Environmental Technology, 11,* 387–382.

Hodgsen, J. F., Lindsay, W. L., Trierweiler, J. F. (1966). *Soil Science Society of America Proceedings, 30,* 723–726.

Jarvis, S. C. (1984). *Journal of Soil Science, 35,* 431–438.

Jenne, E. A. (1968). In: *Trace Inorganics in Water:* Baker, R. A., (ed.) Washington DC: American Chemical Society.

Jenny, H., Grossenbacher, K. (1963). *Soil Science Society America Proceedings, 27,* 273–277.

Jones, J., Benton, (1990). *Communications in Soil Science and Plant Analysis, 21,* 1091–1101.

Jones, K. C., Davies, B. E., Peterson, P. J. (1986). *Geoderma, 37,* 157–174.

Junus, M. A., Cox, F. R. (1987). *Soil Science Society of America Journal, 51,* 678–683.

Kaplan, E., Thode, H. C., Protas, A. (1981). *Environmental Science and Technology, 15,* 539–544.

Khan, M. A., Nortcliff, S. (1982). *Journal of Soil Science, 33,* 763–770.

Le Riche, H. H. (1959). *Journal of Soil Science, 10,* 133.

Lindsay, W. L. (1979). *Chemical Equilibria in Soils,* New York: John Wiley and Sons.

Lindsay, W. L., Norvell, W. A. (1978). *Soil Science Society of America Journal, 42,* 421–428.

Livens, F. (1991). *Environmental Pollution, 70,* 183–208.

Loganathan, P., Burau, R. G., Fuerstenau, D. W. (1977). *Soil Science Society of America Journal, 41,* 57–62.

Lynch, J. M., Whipps, J. M. (1991). In: *The Rhizosphere and Plant Growth:* Keister, L., Cregan, P. B. (eds.) Amsterdam: Kluwer Academic Publishers.

Mason, B. J. (1984). *The Current Status of Research on Acidification of Surface Waters,* London: The Royal Society.

McGeehan, S. L., Topper, K., Naylor, D. V. (1989). *Communications in Soil Science and Plant Analysis, 20,* 1777–1786.

McLaren, R. G., Crawford, D. (1973). *Journal of Soil Science, 24,* 172–181.

Morel, J., Mench, M., Guckert, A. (1986). *Biology Fertility of Soils, 2,* 29–34.

Mitchell, R. L. (1974). *Netherlands Journal of Agricultural Science, 22,* 296–304.

Needham, P. (1983). In: *Diagnosis of Mineral Discorder in Plants.* Volume 1. *Principles:* Bould, C., Hewitt, E. J., Needham P. (eds.) London: HMSO.

Oertel, A. C. (1961). *Journal of Soil Science, 12,* 119–128.

Okazaki, M., Takamidoh, K., Yamane, I. (1986). *Soil Science and Plant Nutrition, 32,* 523–533.

Pearson, R. G. (1967). *Chemistry in Britain, 3,* 103–107.
Peck, T. R. (1990). *Communications in Soil Science and Plant Analysis, 21,* 1165–1186.
Pierce, F. J., Dowdy, R. H., Grigal, D. F. (1982). *Journal of Environmental Quality, 11,* 416–422.
Purves, F., Ragg, J. M. (1962). *The Journal of Soil Science, 13,* 241–246.
Rai, D., Serne, R. J., Moore, D. A. (1980). *Soil Science Society of America Journal, 44,* 490–495.
Rose, A. W., Hawkes, H. E., Webb, J. S. (1979). *Geochemistry in Mineral Exploration,* 2nd edn., London: Academic Press.
Santillan-Medrano, J., Jurinak, J. J. (1975). *Soil Science Society of America Proceedings, 39,* 851–856.
Schnitzer, M., Schuppli, P. (1989). *Soil Science Society of America Journal, 53,* 1418–1424.
Schwab, A. P., Lindsay, W. L. (1983). *Soil Science Society of America Journal, 47,* 210–205.
Seip, H. M., Tollan, A. (1978). *Science of the Total Environment, 10,* 253–270.
Shuman, L. M. (1985). *Soil Science, 140,* 11–22.
Shuman, L. M. (1988). *Soil Science, 146,* 192–198.
Shuman, L. M. (1991). In: *Micronutrients in Agriculture,* 2nd edn.: Mortveldt, J. J., Cox, F. R., Shuman, L. M., Welch, R. M. (eds.) Madison, Wisconsin: Soil Science Society of America.
Sims, J. T., Johnson, G. V. (1991). In: *Micronutrients in Agriculture,* 2nd edn.: Mortvedt, J. J., Cox, F. R., Shuman, L. M., Welch, R. M. (eds.) Madison, Wisconsin: Soil Science Society of America.
Sinclair, A. J. (1974). *Journal of Geochemical Exploration, 3,* 129–149.
Smiley, R. W. (1974). *Soil Science Society of America Proceedings, 38,* 797–799.
Sposito, G. (1981). *Environmental Science and Technology, 15,* 396–403
Sposito, G. (1989). *The Chemistry of Soils.* New York: Oxford University Press.
St-Cyr, L., Crowder, A. (1990). *Soil Science, 149,* 191–198.
Swaine, D. J., Mitchell, R. L. (1960). *Journal of Soil Science, 11,* 347–368.
Talibudeen, O. (1981). In: *The Chemistry of Soil Processes:* Greenland, D. J., Hayes, M. H. B. (eds.) Chichester: John Wiley and Sons.
Tennant, C. B., White, M. L. (1959). *Economic Geology, 54,* 1281–1290.
Tessier, A. P., Campbell, G. C., Bisson, M. (1979). *Analytical Chemistry, 51,* 844–851.
Thomson, I., Thornton, I., Webb, J. S. (1972). *Journal of the Science of Food and Agriculture, 23,* 879–891.
Tills, A. R., Alloway, B. J. (1983). *Journal of the Science of Food and Agriculture, 34,* 1190–1196.
Tinker, P. B. (1986). *Journal of Soil Science, 37,* 587–601.
Trierweiler, J. F., Lindsay, W. L. (1969). *Soil Science Society of America Proceedings, 33,* 49–54.
Ure, A. M., Berrow, M. L. (1982). In: *Environmental Chemistry* Volume 2: Bowen, H. J. M. (ed.) London: The Royal Society of Chemistry.
Van Noordewijk, M., de Willigen, P. (1986). *Netherlands Journal of Agricultural Science, 34,* 273–281.
Viro, P. J. (1955). *Soil Science, 79,* 459–465.
West, T. S. (1981). *Philosophical Transactions Royal Society London, B, 294,* 19–39.

2 Metal Ions and Plant Nutrition

Aradhana Mehra and Margaret E. Farago

Environmental Geochemistry Research
Royal School of Mines
Imperial College of Science, Technology and Medicine
Prince Consort Road, London SW7 2BP, UK

Contents

Introduction

Plants are composed largely of water and complex organic molecules, which are synthesised from water, nutrients from the soil, CO_2 from air, and energy from sunlight. Six nutrients, the major- or macronutrients, are used by plants in relatively large amounts, N, P, K, S, Ca and Mg. In addition all plants need a variety of metal ions for the optimum integration of growth and metabolic activity (Hewitt and Smith, 1975; Hewitt, 1983 a, b; Marschner, 1986). Elements can thus be grouped as essential macronutrients, essential micronutrients and elements having beneficial or restricted essentiality. Metals and metalloids are found associated with a wide variety of cellular activities in plants, either as structural components of key molecules, (eg, chlorophyll, cytochrome) or they are involved in a wide range of metabolic processes such as enzymes (eg, peroxidases, polyphenol oxidases, nitrate reductase). Hewitt (1983 b), Robb und Pierpoint (1983), Kabata-Pendias and Pendias (1984) and Fitzpatrick (1986) have discussed the role of elements in plant life. The effects of metal deficiencies and excesses on plant growth and composition have been discussed by several authors (Hewitt and Smith, 1975; Lepp 1981; Bould *et al.*, 1983; Merian, 1991).

Essentiality

Not all elements detected in plant tissues are deemed to be essential. Criteria of essentiality have been proposed by Epstein (1972): an element is essential (a) if, without it, the plant cannot complete its life cycle, and (b) if it is part of an essential plant constitutent or metabolite. Mineral elements which either stimulate growth without being essential or which are essential only for certain plant species or families are usually defined as beneficial elements.

Besides major metal ions, there are only 10 trace metals among the 24 elements which are currently thought to be essential to plants (Hay, 1986; Farago and Cole, 1988; Kieffer, 1991). They are Fe, Cu, Mn, Zn, Co, Mo, Ni, V, Na and Rb. Other elements such as Al, Sn, Cr and Sr are regarded as elements beneficial to plants. Hay (1986) and Kieffer (1991) suggested that As, Ag, Cd and Pb may be required at very low levels for plants, although they are recognised to be toxic in higher concentrations and there is no evidence to show that they serve any essential biological functions. Rose *et al.* (1979) have even postulated that plants require certain small amounts of every element in the periodic table, but perhaps in quantities so small that the need would be extremely difficult to demonstrate in greenhouse experiments.

Concentrations of Elements in Plants

Plant micronutrients are those required in plant tissue at concentrations which are equal to or less than 100 mg kg^{-1} dry weight. Macronutrients or major elements are those which are needed in concentrations of 1000 mg kg^{-1} or more in dry weight of the plant. Typical concentrations of elements in plants are given by Bowen (1966), Kabata-Pendias and Pendias (1984) and Markert (1992); Angelone and Bini (1992) have detailed elemental concentrations for plants and soils of western Europe.

It has been pointed out by a number of workers (for example Jarrell and Beverley, 1981) that the terms used to designate the quantities of an element in plant tissues can be ambiguous. 'Content' has been used both for concentration and for 'total mass', ie. the concentration $\times$ the biomass. These authors further point out that decreasing concentrations could result from increased biomass, with the total taken up remaining constant, or even increasing: the 'dilution effect'.

Deficiency

If the supply of an element is inadequate, the growth of the plant is abnormal or stunted and its further development, especially its metabolic cycles, are disordered. Although deficiency symptoms are difficult to generalise, they may be quite characteristic for a particular element. Kabata-Pendias and Pendias (1984) and

Hewitt and Smith (1975) have listed descriptions of deficiency symptoms, the most frequent being chlorosis. Visible symptoms are important in diagnosis of deficiencies; however, disturbance of metabolic processes and consequent losses in production of biomass may occur before the deficiency symptoms are recognised.

Toxicity

Excessive concentrations of some metals in soils may produce toxic symptoms in plants. Levitt (1980) suggested that metals in the plant environment operate as stress factors in that they cause physiological reaction change (strain) and in so doing can reduce vigour, or in the extreme, totally inhibit plant growth. Sensitivity describes the effects of a stress which result in injury or death of the plant.

The physiology of metal toxicity in plants has been reviewed by Foy *et al.* (1978). The most widely described effects of metal toxicity in plants are inhibited root growth, depressed shoot and leaf growth, and general chlorosis of the younger leaves (Bradshaw and McNeilly, 1981; Baker and Walker, 1989). Literature on the relative toxicities of different metals to different species is limited to a few, mostly species-specific studies.

Critical concentrations of elements in plants have been defined by Beckett and Davis (1977) and Davis and Beckett (1978) as those concentrations which cause toxic reactions and reduce yield (biomass). They presented yield curves of plants grown in the presence of toxic metals:

1. There is a yield plateau Y_0 over which the element is neither toxic nor deficient, and the biomass is independent of the tissue concentration of the element.
2. There is an upper critical tissue concentration T_c, and above this tissue concentration the metal is toxic and the yield is reduced.
3. With plants grown in culture media there is a similar critical solution concentration S_c.

Values of T_c were found to be remarkably free of growing conditions and characteristic of each element (Beckett and Davis, 1977, 1978; Davis and Beckett, 1978).

The ED_{50} method (effective dose of metal that produces 50% of normal growth) has been used to test toxicities of a number of metals, for example Wong and Bradshaw (1982) found the order of increasing toxicities (from root growth) toward ryegrass (*Lolium perenne*) were

$$Cu > Ni > Mn > Pb > Cd > Zn > Al > Hg > Cr > Fe$$

Tolerance

Various mechanisms of tolerance have been proposed to explain how some plants cope with toxic conditions and how some species have developed tolerant ecotypes (Baker, 1989; Cumming and Tomsett, 1992; McNair and Baker in Chapter 3)

Methods of Study of Plant Responses

There are two general approaches to the study of plant responses. One approach is the carefully designed greenhouse, field or pot experiments, where the observation of plant responses to soil factors are made. The alternative approach is the study of soils and plants in their natural environment. There are a number of drawbacks to the greenhouse experiments, for instance, reduction of time-scale in experiments creates uncertainty as to field relevance although reliable qualitative indications may be obtained (Folkeson *et al.*, 1990). The responses of plants under artificial environmental conditions are usually not representative of the natural conditions (Irgolic and Martell, 1985).

Sources of Nutrients for Plants

The main supplies of nutrients for plants are: release from soil reserves, decomposing biological residues (eg, roots, straw, manures), biological nitrogen fixation, aerial deposition, and in the case of crop plants, fertilisers. Davies (1992) has pointed out that many soil factors, including acidity, organic content, cation exchange capacity, redox status, temperature and synergistic and antagonistic effects of other elements, must be taken into account when considering the availability of elements to plants. These points are further discussed in Chapter 1.

The main sources of elements in soils are the parent materials from which they are derived (Thornton, 1983). Nutrients may be naturally deficient in soils, and elements may be removed in crop and livestock products. Geochemical maps are helpful in the location of areas with deficiencies or excesses of particular elements (Webb *et al.*, 1978; Potter, 1992). Atmospheric metals, whether present as precipitation, or in an enriched form of throughfall or stem flow, represent a more dilute source, available on a less regular basis.

The biovailability of an element in the soil is the result of a competition between surface complexation at the plant root system, at various soil solid phases, and that remaining in solution. Organic and humic materials in soils act as ligands and the resulting complexes can be important in the movement of metal ions (Lindsay, 1974). At present there is little information on the relationship between concentration of the complexed species and their uptake rate by plants though some general evidence suggests that the metal complexes in soil must split before uptake occurs, indicating that it is the free ion which is absorbed (Bowie and Thornton, 1985).

Metals have been released into the environment from a wide variety of sources. Natural geochemical sources include the weathering of metal-rich parent materials such as marine black shales and metal mineral deposites. The elevated concentrations of elements in such soils can have effects on crops and vegetation. Anthropogenic sources include metal mine spoils; surface and percolating drainage water and flood water from metal and other mines; atmospheric emission from industrial activities, such as particulates from spoil tips and metal smelting; manufacturing processes using metals such as electroplating, battery, alloy and pigment production; fungicides and pesticides; power station ash, and sewage sludge (Gemmell, 1977; Peterson and Girling, 1981; Mattigod and Page, 1983).

The Rhizosphere

Conditions in the rhizosphere, the cylinder of soil that surrounds the plant root at a distance of up to 2 – 5 mm (Curl and Truelove, 1986) can be very different from those in the bulk soil. This is the local environment from which the root takes up nutrients, excretes inorganic and organic species, and in which there is shedding and decomposition of parts of the root surface. The pH and the microbial population can both deviate from those in the bulk soil, the latter showing a population density 2 – 20 times higher. The roots of many plant families are associated with particular fungi, mycorrhizae, which are very important for the mineral nutrition of plants (Tinker and Gilden, 1983; Marschner *et al.*, 1986, Streit and Stumm, 1993).

Fertilisers

Soil provides the principal natural source of nutrients, and fertilisers supply the additional input for increased crop yields (Olsen, 1987; Jenkinson 1982). However, agricultural exploitation of soils with marginal ‘available’ essential metal pools has led to the development of deficiency problems in certain types of crops which have either a high demand for particular elements, or which possess relatively inefficient uptake mechanisms. Such problems become particularly pronounced in perennial crops (such as fruit or timber trees) and require routine correction by supplementing natural metal sources. Artificial fertilisers, especially those containing N, P, K and Ca, are extensively used (Sutcliffe and Baker, 1981; Bockman *et al.*, 1990) and have been shown to stimulate growth (Trueby and Raba, 1991). Plants take up the macronutrients through the roots, however, deficiencies of micronutrients are usually corrected by spraying on the leaves, as some trace elements can be absorbed in this way.

Phosphate fertilisers also contain a number of other elements found in the parent phosphate rock (Bowen, 1979; Bockman *et al.*, 1990; Jackson and Alloway, 1992). Cadmium originating from sedimentary rock is particularly undesirable, and processes for the removal of Cd from such fertilisers are being developed (Bockman *et al.*, 1990). Fertilised soils have shown increases in Cd content after a number of years, but there appears to be little evidence for long-term Cd-increase in crop plants, except possibly for wheat (Jones and Johnston, 1989). Mortvedt (1984) determined the uptake of Cd and Zn by several vegetable crops heavily fertilised with triple superphosphate over a ten year period. Cd levels were found to be similar in fertilised and unfertilised snap bean seed, beet blades and roots, and in sweet corn leaves and grain. However, Zn concentrations were found to decrease with P application in all tissues except cabbage heads and cores. Claims that fertilisers promote the uptake of Al by plants have been refuted (Akerstrand *et al.*, 1988).

A number of waste materials have been considered for use as fertilisers, for example municipal sewage sludge can be a valuable source of N and P and some trace elements such as Cu, Fe, Mn and Zn. However, some sludges contain high metal concentrations that can be toxic to plants, cause long-term soil contamination and introduce toxic metals into the food chain (Jackson and Alloway, 1992). The long-term application of sewage sludge and consequent metal uptake by plants has been reviewed by Juste and Mench (1992). These authors concluded that sludge-borne metals do not cause toxicity to the majority of crops and that Zn is the most available of the sludge-borne metals.

Recently Chaney and Ryan (1992) and Henry and Harrison (1992) have reviewed the effects of heavy metals in sewage sludges and MSW-composts (municipal solid waste) used as soil amendments. Chaney and Ryan found that in contrast to sewage sludge, MSW-composts contain phytotoxic levels of boron. Application can also raise the pH of the soil-compost mixture, which can result in compost-induced Mn-deficiency. Chaney and Ryan conclude that uncontaminated sludges and MSW-composts comprise no risks in relation to Cd uptake by crops and vegetables, thus there is no food chain Cd risk to humans consuming western diets. This largely comes about because Zn is a natural limiting factor in the following way: either soil pH is kept at a reasonable level for crop production limiting Cd uptake; or if the soil pH drops enough to allow Cd uptake, then Zn-phytotoxicity reduces the yield and the Cd risk.

The potential of ammonia-based flue gas desulphurisation waste solution as a nitrogen fertiliser has been assessed by Gissel-Neilson and Bertelsen (1989) by field trials using barley and rye grass. The solution had the same fertiliser value as calcium-ammonium-nitrate. The toxic effects of sulphite were reduced by avoiding direct contact of the solution with the plants.

Elemental Interactions and Comparisons

In soil and soil solutions, plants are exposed to many elements simultaneously and to other factors, which could interact in different ways. Thus, the integrated effects may be quite different from those obtained when any of the elements or factors is acting alone.

Tinker (1981) compared uptake rates of trace metals with those of major elements and suggested that the absorbing power for trace metals is relatively low. Bowie and Thornton (1985) have summarised the effect of elemental interactions on their uptake. The uptake of trace metal ions is inhibited by the major cations, especially by Ca. Interaction of Cu and Zn is such that both ions reduce the uptake of the other. Zn is also known to affect Fe levels in plants. Interactions between major elements and trace elements have been listed by Kabata-Pendias and Pendias (1984) where Ca and Mg are seen to be mainly antagonistic to the absorption and metabolism of several trace elements. A list of some of the interactions for the heavy metals has been given by Fergusson (1990). Some elements can be both antagonistic and synergistic to another element, presumably in different biochemical processes. Fergusson (1990) has also listed the relative uptake of some heavy elements by selected plants.

Interactions both positive and negative may be species-, and conditions specific, for example, in a series of pot experiments to investigate the effects and interactions of Zn, Cu and Ni, present at toxic or near toxic levels, Davies (1980) showed that the crops had differing sensitivities to the three heavy metals. At the levels tested the toxic effect on each metal was increased by the others on clover, lettuce and red beet. Positive interactions between Cu and Zn were shown by barley and between Ni and Zn by French beans.

Davies (1992) has studied the uptake of Cd, Cu, Pb and Zn by radish plants grown in contaminated soils and has pointed out that many soil factors, including synergistic and antagonistic effects of other elements must be taken into account. Davies has also stressed the need for chemical extraction procedures which can be used as a reliable indication of the potential uptake by crops. Davies suggests that ideally it should be possible to relate quantitatively the concentration of a given metal in a plant to that in the soil, taking into account a number of soil factors, including P supply, and ion exchange capacity. It was found that acetic acid predicted Cd uptake, whereas that of Pb was best predicted by 'total' Pb (concentrated nitric acid extraction). Zn uptake was best described by exchangeable Zn: an improvement was obtained by inclusion of pH as a positive term, indicating that raising the soil pH would increase Zn uptake. The uptake of Cu was not satisfactorily predicted. The regressions were not improved by the inclusion of acid soluble P and exchangeable K and Mg.

Hassett *et al.* (1976) reported that the elongation of soil-grown maize seedlings was depressed more by Pb and Cd in combination than by Pb alone. Webber (1980) found that Cd reduced the uptake of Zn in lettuce and there was a tendency for the reverse effect also.

An investigation into the partitioning of Cu, Zn, Pb and Cd amongst various above-ground organs of plants by Kubota *et al.* (1992), demonstrated that Zn and Cu are selectively transferred into seed or grain. However, Pb and Cd showed restricted movement into these organs. The authors suggest that grain and seed crops thus provide natural barriers to the movement of the toxic elements Cd and Pb into the food chain, whilst conserving Zn and Cu levels in the edible portions of these crops.

Role of Some Elements in Plant Life

Macronutrients

Nitrogen

Nitrogen is derived from the atmosphere or dead tissues and in both cases it is transformed by bacteria in the soil into ammonia and nitrate which are taken up by the plant roots. It occurs in high quantities in young plants particularly in the leaves. Nitrogen forms a part of every living cell, occurring in chlorophyll and all proteins with many of the latter serving as enzymes. Abundance of nitrogen leads to green succulent growth while nitrogen deficiency causes a loss of colour, reduction in protein production and a gradual yellowing and stunted growth. The roots of leguminous plants are associated with the symbiotic bacterium *Rhizobium. Rhizobium* bacteria, which are located inside root nodules are able to convert elemental nitrogen to ammonia, nitrogen fixation. Ammonium is further converted to nitrate by the soil bacteria. Nitrate reduction to form other nitrogen compounds such as amino acids takes place in the roots. The nitrogen cycle has been discussed by Sprent (1988) and Wilson (1988), and nitrogen metabolism in plants has been recently reviewed (Mengel and Pilbeam, 1992).

Plants take up N both as nitrate (usually the most abundant form of soluble N in the soil), and as ammonium. Although the balance varies with species, nitrate is the major source. Numerous studies have shown that ammoniacal-N as a sole source of nitrogen is deleterious to the growth of many higher plants (Pill and Lambeth, 1977). The addition of nitrate however, may alleviate the inhibitory effects of NH_4^+ on growth (Ota and Yamamoto, 1989). In several crops, combination of NH_4^+ and NO_3^- usually results in greater vegetative growth than when either N form is used alone (Hartman *et al.*, 1986). However, some species of the genus *Vaccinium* (blueberry and cranberry) grow better when NH_4^+ is the N source rather than NO_3^- (Rosen *et al.*, 1990).

The influence of nitrogen concentration and ammonium/nitrate ratio on N-uptake of citrus was studied by Serna *et al.* (1992). In short-term water culture experiments with ^{15}N labelled NH_4^+ or NO_3^- at various concentrations, citrus seedlings absorbed

NH_4^+ at a higher rate than NO_3^-. $^{15}NH_4^+$ accumulated in roots and to a lesser degree in both leaves and stems. However, $^{15}NO_3^-$ was mostly partitioned between leaves and roots. In a long-term experiment, the effects of concentration and source of added N (NO_3^- or NH_4^+) on nutrient concentrations in leaves from plants grown in sand were evaluated. Leaf concentrations of N, P, Mg, Fe and Cu were increased by NH_4^+ versus NO_3^- nutrition, whereas the reverse was true for Ca, K, Zn and Mn. Nitrogen concentrations in leaves were highest when plants were provided with either NO_3^- or NH_4^+ as a sole source of nitrogen. Lowest nitrogen concentration in leaves was found with a 75:25 NO_3^--N/NH_4^+-N ratio. With increasing proportions of NH_4^+ in the nitrogen supply, leaf nutrients such as P, Mg, Fe and Cu increased, whereas Ca, K, Mn and Zn decreased. Yield in number of fruits per tree was increased significantly by supplying all nitrogen as NH_4^+, although fruit weight was reduced.

Studies on the effect of increased nitrogen supply to plants have shown varying results. In a study (Vos and Biemond, 1992) to determine the effects of nitrogen on the development and growth of the potato plant, (*Solarum tuberosum* L.), the rate of leaf appearance was negligibly affected by nitrogen supply, whereas the rate of leaf expansion was related to leaf number and nitrogen supply. Nitrogen promoted apical branching and hence the total number of leaves that appeared on a plant. Nams *et al.* (1993) studied the effects of nitrogen fertilisation on several woody and nonwoody boreal forest species in southwestern Yukon. Over the two years of fertilisation there were significant increases in growth over control levels for perennial grasses (*Festuca altaica* and *Calamagrostis lapponica*), two herbs (*Epilobium angustifolium* and *Achillea millefolium*), and two deciduous shrubs (*Salix glauca* and *Betula glandulosa*). However, the growth of white spruce trees (*Picea glauca*) increased only slightly in response to increasing nitrogen levels and the evergreen dwarf shrub *Arctostaphylos uva-ursi* showed no response.

Phosphorus

Phosphorus is a constituent of every living cell and occurs in the protoplasm, with its greatest concentration in seeds. Its main role is in the energy cycle. Phosphorus deficiency causes a purplish colouration at the seedling stage with later yellowing, stunted growth and delayed maturity. P deficiency often causes the accumulation of sugars and sometimes starch. Al may immobilise P in soils, and some crops (eg, barley) show deficiency symptoms when Al is present at low concentrations.

Phosphorus in the form of phosphate is the principal source of P for plants. Phosphorus is commonly applied in fertilisers although its uptake by plants is much less than that of N and K and its availability is one of the most common nutritional limitations to crop yield (Bould *et al.*, 1983). Some of the P in soils and soil solutions may be present in organic compounds. However, it is unlikely to be available to plants in its organic form and decomposition of organic matter releases P in its available form. The inorganic P in soils is present both in a labile form and a non-labile phosphate in soil minerals. The concentration of P needed for optimum growth depends on species and stage of growth. Uptake of P depends on the size of the root system, and this explains why P deficiency most often occurs in seedlings and young plants where the root system is small.

Mn nutrition of wheat was found to be dependent on the P status of soil as well as on its pH and Mn status (Neilson *et al.*, 1992). Concentrations of Mn were depressed in leaf tissue of plants from lime soils and also in high P soils. The depressed values for lime treatments were explained in terms of depressed soils solution Mn concentrations resulting from elevated pH. It was suggested that high soil P resulted in elevated plant P which interfered in the uptake and/or translocation of Mn.

Sulphur

Sulphate is the form in which S is normally absorbed by plant roots and the reduction of sulphate for the synthesis of sulphur-containing compounds is universal for plants. Sulphur occurs in some amino acids, proteins and also in the oils of some plants such as cabbages and turnips. The S—H group is very important for the action of some enzymes and coenzymes. S is also a constituent of ferredoxins. A deficiency leads to stunting and yellowing but can be cured. A classical case of S deficiency known as 'tea yellows' was identified in 1933 by Storey and Leach (reported in Hewitt and Smith, 1975). Sulphur deficiency leads to accumulation of amine and ammonia nitrogen, with little change in the total.

Sulphite is known to cause growth inhibition and injuries to plants (Garsed, 1981). The toxicity to *Zea mays* of root applied sulphite has been investigated by Bertelsen and Gissel-Nielsen (1987). Sulphite toxicity produced greyness of the leaves with flabbiness and desiccation. The symptoms were more pronounced at low pH and in non-aerated solutions at higher pH values. The toxicity was found to decrease in the order: $SO_2(aq) > HSO_3^- > SO_3^{2-}$.

There has been much interest in the injuries to plants from environmental SO_2 emissions. These injuries may be caused by aqueous SO_2 or by the entrance of SO_2 through the stomata (Taekemoto and Noble, 1986). The general symptom of SO_2 toxicity is water-soaked areas on the leaves which develop into well-defined dry white necrotic spots (Mansfield and Lucas, 1990). Although acute SO_2 damage is quite common, the occurrences are usually very localised near the source, and related to weather conditions favouring slow dispersion of the gas.

Potassium

Potassium is essential in all cell metabolic processes. K has three major roles in plant nutrition (Leonard, 1983): (i) as an enzyme activator for a number of processes including respiration, protein synthesis and photosynthesis; (ii) it is the most important ion involved in osmotic processes, including turgor driven cell growth and plant responses that are turgor generated, such as stomatal and leaf movements; (iii) it is involved in charge neutralisation, important for membrane processes and the synthesis and translocation of organic anions. K^+ has a role in the maintenance of cytoplasmic pH (Cumming and Tomsett, 1992). Under K-deficient conditions, plants accumulate basic amines, which can substitute for K^+ in charge neutralisation. Potassium deficiency can cause lodging in cereals and yellowing of leaf tips and margins. Early experiments on the growth of barley demonstrated that rubidium can partially replace K but addition of Na, Li or Cs was not effective (Hewitt and Smith, 1975).

Calcium and Magnesium

Calcium and magnesium are very abundant in soils, and soils deficient in Ca are rare, calcium status is maintained when lime is added to correct acidity. Plant cells contain relatively large concentrations of Ca, but most of it is bound in the cell-wall as the pectate (about 60%) or sequestered in different organelles (Clarkson, 1984). Ca affects the permeability of the cytoplasmic membrane and its deficiency leads to malformation of the growing parts of the plant. Hewitt and Smith (1975) have discussed the early experiments on the morphological effects of Ca deficiency. Calcium is often found in combination with organic acids, for example oxalic acid, a soluble, toxic metabolic by-product is converted to insoluble calcium oxalate.

It has long been known that the nature and extent of expression of environmental stresses, particularly ion stresses, on a whole plant depend on Ca. Ca ions alleviate toxic effects of Al (Foy, 1988; Brady *et al.*, 1993), Na (Cramer *et al.*, 1986) and various heavy metals (Baker and Proctor, 1990). It has been suggested that cytoplasmic levels of Ca^{2+} and Ca-related processes may represent a signal for metal toxicity and tolerance (Leonard and Hepler, 1990). Ca, or Ca bound to Ca^{2+} binding proteins (including calmodulin) is involved in the activity of a number of key enzymes, thus toxic metals may have an impact on normal Ca^{2+} fluxes and cytoplasmic concentrations, and thus alter cell metabolism (Siegel and Haug, 1983; Cheung, 1984). Calcium cycling amongst the various pools within the plant have been discussed by Cumming and Tomsett (1992).

Magnesium is an activator of more enzymes than any other element. It is an essential constituent of chlorophyll where about 2.7% of the weight of the chlorophyll molecule is magnesium. It is also involved in ion transport and cation balance in plants. Magnesium deficiency causes discolouration and chlorosis, and sometimes, premature defoliation of the plants.

Other Elements

The Halogens

Halogens in plants have been reviewed by Fuge (1988). Of the halogens only chlorine is considered to be essential for higher plants (Bowen, 1966; Gough *et al.*, 1979). Deficiency symptoms include wilting, chlorosis, necrosis bronzing of the tissues, and clubbing of roots. Susceptibility to Cl-deficiency varies among crop species. Tomato, lettuce and cabbage are susceptible, whereas some cereals and beans are resistant. Br can be partially substituted for Cl. Iodine has been found to be essential for seaweeds (Hewitt and Smith, 1975).

Chloride, although essential for photosynthesis, can be toxic to plants, and chlorosis can result from the use of saline irrigation water or high Cl-containing fertilisers (Gough *et al.*, 1979). A few organic chlorine compounds have been isolated from higher plants, although these are more abundant in fungi (Hewitt and Smith, 1975).

Fluorine can be acccumulated and metabolised by a number of plant species, in particular the African *Dichapetalum* species and the Australian *Acacia georginae*. These plants all contain fluoroacetate which may be up to 50 mg kg^{-1} dry weight. High concentrations of up to 1% F in dry plant tissues have been found in plants growing on fluorite-rich spoil (Cooke *et al.*, 1976) as compared with normal values of around 30 mg kg^{-1} (Kabata-Pendias and Pendias, 1984).

While marine plants are often rich in Br and I the concentrations of these elements are generally low in terrestrial plants (Kabata-Pendias and Pendias, 1984; Fuge and Johnson, 1986).

Boron

Boron is an essential plant nutrient and H_3BO_3 is the principal form in which it is taken up (IAEA, 1984). Boron appears to play a role in calcium utilisation and the development of the actively growing parts of the plant. It is also essential for the fixation of nitrogen by bacteria in the nodules of legumes. Parr and Loughman (1983) have pointed out that one of the main difficulties in determining the primary role of boron in plants is the minute amount involved (1 – 100 mg kg^{-1} dry weight). These authors have reviewed studies on the role of B and suggest that there is one underlying aspect; that of impaired membrane function with B deficiency, this affects transport of all chemical species needed for normal growth and development.

The basic characteristics of B deficiency show that the youngest leaves are the first affected, stems are short, severely affected plants liable to have a shrunken appearance, growing points die, root growth is impaired, fruit formation is irregular and the fruits are deformed, and necrotic and watery patches develop in storage tissue. Although plants require B for normal development, additions must be small or boron toxicity will result. Plants greatly vary in their tolerance to boron.

The symptoms of B toxicity are similar in most plants. It is somewhat curious that while B toxicity symptoms can be readily induced by excessive B application, significant reductions in yield are seldom associated even with slight to moderate symptoms in the field. This may be due to the fact that the only damage caused is to the leaf tip or margin and that this has little or no effect on growth. However, Francois (1992) found reduced fruit yield in summer and winter squash cultivars grown in large, outdoor sand cultures treated with B by irrigation. Reduced yields of all cultivars were attributed to a reduction in fruit number and not fruit size. B concentrations in leaves and fruit were directly correlated to B in the soil solution.

Halophytic and non-halophytic species may show variable depression of growth when treated with B in their growth medium. Rozema *et al.* (1992) found that the growth of halophytes in nutrient solutions with increasing B concentrations was less reduced than that of some non-halophytes. In view of the high concentration of B in sea water, the resistance of the halophytic species to high B concentrations relative to non-halophytes might be of adaptive value.

Uptake of B is species dependent. Greenhouse pot experiments were conducted to study B uptake by different plant species, *Brassica juncea* (L.) *Czern* (wild brown mustard), *Festuca arundinacea* Schreb. L. (tall fescue), and *Brassica napus* (canola) (Banuelos *et al.*, 1993). Results showed that the greatest total amount of above ground biomass and below surface biomass was produced by tall fescue and the highest concentrations of tissue B were from roots of tall fescue. A correlation for B and Ca has been reported by Dixon *et al.* (1973), where B sprays have been shown to increase Ca transport into apples (*Malus* spp.).

Selenium

Soluble selenium, normally the selenate (Carlsen *et al.*, 1991), is readily taken up by plants and its assimilation is analogous to that of sulphate (Dawson and Anderson, 1989). The question of selenium essentiality in plants has been discussed by Anderson and Scarf (1983). A number or organoselenium compounds have been identified in plants (Adriano, 1986; Peterson *et al.*, 1981; Anderson and Scarf, 1983; Harborne, 1988).

The selenium content of a plant is dependent on the plant species and on whether it is grown in seleniferous or non-seleniferous soil. Generally, Se content in plants increases with the level of Se available and decreases with the sulphur supply. Se concentrations in plant tissues were found to be similar in three different plant species: *Brassica juncea* (L.) Czern (wild brown mustard), *Festuca arundinacea* Schreb L. (tall fescue) and *Brassica napus* (canola) when these were grown in soil containing naturally occurring concentrations of 1.77 mg total Se kg^{-1} in greenhouse pot experiments (Banuelos *et al.*, 1993).

Some plants can hyperaccumulate Se, for example some *Astragalus* species of the Fabaceae family accumulate up to 0.5% growing on Se-rich soils in North America. The mechanism for selenium adaptation has been discussed by Harborne (1988) and Streit and Stumm (1993). In non-tolerant plants grown in the presence of Se, SO_4^{2-} and SeO_4^{2-} are not distinguished by the plant, leading to the partial synthesis of the Se-analogues of the S-containing amino acids, selenomethionine and selenocysteine. When these are incorporated into enzyme proteins, the enzymes are non-active and toxicity symptoms and reduced biomass have been recorded. Se-tolerant plants are able to distinguish between S as SO_4^{2-} and Se as SeO_4^{2-} or SeO_3^{2-} and direct the latter into the synthesis of non-protein amino acids, which can be stored in leaf vacuoles away from metabolism. Such plants can cause Se toxicity in grazing animals.

Much agricultural interest in selenium has thus been in toxicity when the element is present in excess (Mikkelson *et al.*, 1986). Dhillon and Dhillon (1991) have surveyed the accumulation of Se in sugarcane plants in seleniferous areas of the Punjab, and found that leaves accumulate Se to levels which may be injurious to animals. The application of gypsum to the soils was found to decrease the uptake in both cane and leaves. Recently there has been concern over 'selenium responsive diseases' that result from selenium deficiencies and the resultant agricultural uses of selenium supplementation. The risks and benefits of the agricultural uses of selenium have been reviewed by Oldfield (1992). Both the pH of soils and the presence of other elements at high concentrations can have an effect on Se uptake by plants, in addition the chemical species of Se in the soil can have a marked effect. Selenate is more available than selenite (Carlsen *et al.*, 1991). The use of Se-rich plants as amendments to improve the selenium status of soils has been suggested by Banuelos *et al.* (1991). Recent studies by Hossner *et al.* (1992) showed that shoot weights for five of the ten plant species grown in soils with medium and high Se status were reduced.

Copper

Copper is an essential element. Copper plays a significant role in several physiological processes – photosynthesis, respiration, carbohydrate distribution, nitrogen reduction and fixation, protein metabolism, and cell wall metabolism. Many plant metalloenzymes contain copper. It also influences water permeability of xylem vessels and thus controls water relationships. It is mainly complexed with organic compounds of low molecular weight and with proteins (Henze and Umland, 1987). Kabata-Pendias and Pendias (1984) have compiled data on the Cu concentrations in plants. About 5 to 20 mg kg^{-1} in tomato tissue is reported to ensure normal growth (Roads *et al.*, 1989). Concentrations below this range are considered to be deficient while those above were considered to be toxic. Thus the most important practical implications of the role of copper in plants are related to its deficiency and toxicity. Copper is involved in the mechanisms of disease resistance. Resistance of plants to fungal diseases can be related to an adequate copper supply. There is also evidence that plants with enriched copper concentrations are susceptible to some diseases.

Several field disorders of crops are caused by copper deficiency (Hewitt and Smith, 1975); symptoms are species specific. Copper deficiency is common on peat leading to growth abnormalities such as rapid wilting and weak stalks, spiralling of leaves and no grain formation. Deficiency levels of copper in plants show large differences; however, some generalisations suggest that copper levels less than 2 μg g^{-1} are likely to be inadequate for most plants (Kabata-Pendias and Pendias, 1984).

The utilisation of Cu by the plant depends predominantly on its movement among various parts of the plant. Lepp (1981) has reviewed Cu uptake and distribution in plants. Root tissues have been observed to hold on to Cu strongly against the transport to shoots under conditions of both Cu deficiency and excess (Kabata-Pendias and Pendias, 1984). Jarvis and Jones (1979) have pointed out that plants growing in high copper environments often translocate surprisingly little copper to their leaves but generally accumulate large amounts in their roots. Such a situation was reported by Farago *et al.* (1980): roots of copper-tolerant *Armeria maritima* growing in a copper-impregnated bog accumulated very high levels of Cu a small amount of which is also transported to the leaves. Farago and Mullen (1981) found that the roots of the Cu-tolerant *Armeria maritima* contain very high levels of the amino acid proline.

In spite of the general copper tolerance of some plant species and genotypes, it is also considered to be toxic, Cu-induced chlorosis and root malformation being the most common characteristic symptoms. The reduced root development may result in lowered water and nutrient uptake, leading to disturbances in the metabolism and to growth retardation. In a study by Heale and Ormond (1982) four woody plant species were grown in culture solutions. Visible symptoms of toxicity were observed in the leaves and shoots. At 4 mg L^{-1} of copper, two-year old seedlings of *Pinus resinosa* were wilted, and the biomass was reduced by 30% compared with controls. In a later study by Burton *et al.* (1986) similar concentrations of copper similarly affected seedlings of sitka spruce (*Picea sitchensis*). Both shoot and root growth were significantly decreased by 5 mg L^{-1} of copper in the growth medium.

Many complex interactions of Cu with other elements are observed within plant tissues and also in the external root media, particularly in the uptake. These interactions have been reviewed by Kabata-Pendias and Pendias (1984). The interactions may be both antagonistic and synergistic. For example, high levels of Cu in the plant are known to decrease the Fe content in chloroplasts (Reilly and Reilly, 1973). However, a synergistic effect of Cu and Fe absorption by rice seedlings was reported by Kitagishi and Ymane (1981).

Zinc

Zinc is an essential micronutrient and plays an important part as a specific activvator of certain enzymes such as peptidases, dehydrogenases, proteinases and phosphohydrolases. It is taken up readily as the hydrated Zn^{2+} ion. Zn concentration in plants is species specific and is influenced by the age and vegetation state of the plant (Siegert *et al.*, 1986). Usually, the highest Zn content is found in young plants, concentrations decreasing during aging as a result of dilution (Mugwira and Knezek, 1971). In general, the plant can take up large quantities of Zn without evidence of exclusion when Zn concentrations in soil are less than 500 mg kg^{-1} (Brooks, 1972). Zn is rather soluble in soil solutions and very mobile in acid soils, uptake being promoted by low soil pH, it is easily translocated to different parts of the plant.

Its deficiency symptoms vary with species. Kiekens (1990) reported that Zn deficiency produces interveinal chlorosis, stunted growth, malformation of stems and leaves. Several field symptoms of Zn-deficiency are well known, such as 'little leaf' of citrus, and 'white bud of maize'. The toxicity limit for Zn depends on the plant species and genotypes, as well as on the growth stage.

Excess concentrations of Zn can affect both shoot and root growth and the shoots may become stunted and chlorotic. Paivoke (1983) has reported that the epidermis of roots and the cells in the epidermis may become lignified in excess Zn. In a study with *Picea abies* seedlings grown for five weeks, shoot and root dry weights decreased at a Zn concentration of 1.1 mg L^{-1} (Godbold *et al.*, 1985) as did root elongation of *Deschampsia caespitosa* (Godbold and Hutterman, 1983). Toxicity of Zn may be reduced in the presence of competing Ca ions (Wainwright and Woolhouse, 1977; Wong and Bradshaw, 1982; Brown and Wilkins, 1985). Chaudhry and Wallace (1976) found that Fe competitively inhibited Zn absorption by rice (*Oryza sativa* L. cv Earlirose) grown in solution culture. The effect was more marked for shoots since Fe also had a competitive effect on Zn translocation from roots to shoots. Elevated concentrations of Zn in crops and vegetables growing near industrial installations have been reported by a number of workers, for example corn near electrical pylons (Jones *et al.*, 1988) and mushrooms grown in the vicinity of a lead smelter (Kalac *et al.*, 1991).

Molybdenum

Molybdenum is necessary for the reduction of nitrate in the plant, otherwise nitrate will accumulate and interfere with protein synthesis. Nitrogen fixation by legumes is also dependent upon molybdenum. Small amounts of Mo appear to be essential for maximum growth, the high levels found in forages as a result of industrial contamination do not appear to depress feed production (Gardner and Hill-Patch, 1962).

Molybdenum deficiency is of two types. The first is associated with Mo deficiency combined with nitrate as the source of N, and is a result of accumulation of nitrate in the leaves associated with the lack of the Mo-containing enzyme nitrate reductase. The other type, which is almost limited to brassicas, results in plant abnormalities such as the formation of narrow leaves curled around the midrib in cauliflowers (Hewitt, 1983a). Its deficiency in legumes usually shows as effects of nitrogen deficiency because of the role of Mo in nitrogen fixation.

The content of Mo in plants may vary widely. Warren *et al.* (1971) reported that some vegetables may have Mo contents 500 times more than in others. Also, individual samples of the same vegetables from different areas contain different amounts of Mo. Moreover, plants growing next to each other on the same soil may contain different levels of Mo as found by Ylaranta and Sillanpaa (1984).

It is believed that plants take up Mo mainly as molybdate (MO_4^{2-}) ions (Kabata-Pendias and Pendias, 1984) and it is easily transported to different parts of the plant. The authors have also suggested that sulphur depresses Mo uptake by plants in areas with high Mo. They also report that Mo is more available both in acid and alkaline soils, thus liming to increase Mo availability is a common agricultural practice. Farago *et al.*(1992) have reported that bioavailability of Mo in some Portuguese soils increases at high pH where liming has taken place.

Iron

Iron has roles in enzyme systems and is necessary for the synthesis of chlorophyll (Hipkins, 1983). Chlorosis, the lack of chlorophyll is the first and obvious result of iron deficiency. Chloroplasts are reduced in size and without chlorophyll. Iron deficiency also has an effect on respiration and on cell division and thus growth. An excess of phosphate, bicarbonate, Cu, Zn, Co, Cd, Mn or Ni in the growth medium may cause Fe deficiency (Hewitt, 1948, 1953; Chaney and Giordano, 1977).

Fe and Mn oxides play an important role in the soil in immobilising trace elements such as Co, Cu, Zn and Ni as well as pollutants like Pb (Norrish, 1975) and rendering them unavailable to plants (Peterson and Girling, 1981). Uptake of Fe is promoted by low soil pH, waterlogged soils, mobile organic complexes, and chelates; it is restrained by high pH and well-aerated soils.

A good deal is known about Fe uptake in dicotyledons. In aerobic soils Fe is usually in the form of highly insoluble Fe(III) compounds, it must be solubilised before it can be taken up, a process that involves acidifying the medium surrounding the root by proton extrusion (Zocchi and Coccucci, 1990), reduction to Fe(II), or is accomplished by phenolic compounds which are exuded by certain plants (Chaney and Bell, 1987). Reduction takes place at the root surface (Brüggeman *et al.*, 1990). The Fe is reoxidised and is transported within the plant as the citrate complex (Valenti *et al.*, 1991). Excessive amounts of free Fe^{2+} ions in the plant cells are undesirable, and yet plants have ongoing need for Fe. The system that regulates uptake and availability of Fe involves the iron storage protein ferritin (Bienfait and Van der Mark, 1983).

Manganese

Mn is an essential element and is taken up in the reduced Mn^{2+} form. Often the first symptom of Mn-deficiency is Fe-deficiency, but the appearance of a number of necrotic areas is a distinguishing feature. Mn-deficiency is widespread in areas of high organic carbon, particularly where these have been limed. Under these conditions Mn^{2+} is oxidised by soil microorganisms to Mn^{3+}, followed by irreversible dismutation to Mn^{2+} and Mn(IV) as MnO_2. The MnO_2 is not available under alkaline conditions and Mn deficiency results.

To determine the effects of the deprivation of specific micronutrients on the water hyacinth (*Eichhornia crassipes*), Colley *et al.* (1979) studied the rate of uptake of iron and manganese in comparison with phosphorus. Results indicated that all three elements were actively absorbed by the root systems, but the rates of absorption differed markedly. The rate of absorption of manganese by roots was 13 and 21 times that for radio-iron and -phosphorus, and iron was taken up by the roots at nearly twice the rate of phosphorus. Manganese translocation appeared to be faster than phosphorus translocation by an order of magnitude and 65 times faster than iron translocation.

Foy and Oakes (1984) tested limpograss, *Lemarthris altissima,* for its tolerance to excess Mn. Cuttings from the same plant were grown in Mn-toxic soil with pH 5.1, in soil with 1250 mg kg^{-1} calcium carbonate and pH 6.3, and in nutrient soil with varying amounts of Mn and pH 4.0. Although there were differences among top and root dry weights, limpograss was found to be tolerant to Mn in all situations.

Sodium

Sodium is an essential element and additions of sodium chloride to soils can provide increased yields of some plants. There is some degree of overlap in the roles of sodium and potassium in plant nutrition. Both Na and Rb are beneficial in K-deficiency.

Sodium is toxic to some plants at high concentrations. It has been shown by Cramer *et al.* (1985) that excess NaCl leads to the loss of K^+ due to membrane depolarisation and that there is a loss of Ca^{2+} ions from the cell plasmalemmae and internal pool due to displacement by Na^+ ions (Cramer *et al.*, 1987, 1991; Lauchli, 1990).

Salinity also interferes with N acquisition and utilisation (Lewis, 1986). Adaptation to salinity and models of salt tolerance have been discussed by Stewart and Ahmad (1983). Hawkins and Lewis (1993) found that in both NO_3^- and NH_4^+ supplied *Triticum aestivum* L., the shoot and root morphology of the salinities plants was visibly different from the controls. The salinised shoots were smaller and the stems were less fibrous compared with control plants. Moreover, with increasing NaCl concentrations, there was a concomitant decrease in the shoot : root ratio and reduction in the moisture content of both shoots and roots.

Cobalt

Cobalt is specifically required for nitrogen fixation and for plant growth when dependent on symbiotic atmospheric nitrogen fixation. Its deficiency produces chlorosis of leaves and reduced growth of plant (Wilson and Nicholas, 1967). Translocation of Co from the roots to other plant parts is through the transpiration flow in the xylem (Jarosik *et al.*, 1988). Bollard (1983) found that Co is present as its cation in the xylem exudate, however, in phloem exudate it is largely present as negatively charged complex.

Co uptake by plants is species dependent. It is hardly detectable in green beans; however it is present in low concentrations in radishes, although it has been found in leaves of the plant (Kloke, 1980). Co is acccumulated in considerably lower concentrations in the plant storage organs or seeds as compared with the vegetative parts.

Vanadium

Vanadium has been demonstrated to be a biologically active metal. It has the ability to influence soil enzyme systems which affect mineralisation processes and release of plant nutrients, thus indirectly affecting primary production (Bengtsson and Tyler, 1976). High levels of V may cause colour deepening in the shoots followed by chlorosis. Bowen (1979) has reported that concentrations of 10–40 mg L^{-1} of V in nutrient solutions is toxic for plant seeds.

Astragalus species are reported to tolerate very high concentrations of V which they utilise in nitrogen fixation (Cannon, 1963). Tolerance of high amounts of V in plants is due to immobilisation of the element as an insoluble product in the root, partial exclusion mechanisms also preventing uptake of the element from enriched soils (Peterson and Girling, 1981).

Most plants accumulate V in very small amounts from the soil in relation to the total soil content. In an experiment where maize, *Zea mays* var. Fronica, was treated with both vanadate and vanadyl ions at different concentrations (2, 20, 100 and 200 μmol L^{-1}), no progressive reduction in root and shoot biomass was observed as the applied V concentration was increased (Morrell *et al.*, 1983). There was a general stunting of growth although no specific toxicity symptoms could be attributed to the V treatment. The study, however, revealed a clear difference in uptake and subsequent transport between the vanadate and vanadyl ions in the maize plants. The authors found that more V was accumulated when supplied as the vanadyl cation than when supplied as the vanadate anion from solutions of the same concentrations with respect to vanadium. Subsequent transport of the ions revealed the vanadyl cation to be more labile than the vanadate anion. Vanadium uptake by higher plants has been further discussed by Morrell *et al.* (1986). Barley roots grown in the presence of V(V) supplied as NH_4VO_3 showed the characteristic ESR spectrum of the vanadyl V(IV) species (see Chapter 9) demonstrating the bioreduction of V(V) to V(IV) during uptake by plants.

Calcium seems to have a significant effect on the accumulation and transport of the vanadate anion. Morrell *et al.* (1983) found that low levels of Ca in the nutrient solution resulted in less vanadate accumulation in the root and proportionally more of it being transported to the shoot. These results agree with the hypothesis of Cannon (1963) who suggested that V was immobilised in the root as the sparingly soluble calcium vanadate.

Vanadium accumulates in some species of *Amanita* toadstools (Frausto da Silva, 1989). Amavadine, the vanadium containing substance, is discussed in Chapter 9.

Chromium and Nickel

It has not been possible so far to establish that Cr is an essential element required by plants, however, addition of Cr to soils deficient in the element has been shown to increase growth rates and yields of potatoes, maize, rye, wheat or oats (Scharrer and Schropp, 1935; Huffman and Allaway, 1973; Bertrand and De Wolf, 1986). Nickel appears to be an essential element for plants (Farago and Cole, 1988). Zerner and coworkers (Dixon *et al.*, 1975) demonstrated that urease isolated from jack bean (*Canavalia ensiformis*) was a nickel enzyme. Eskew *et al.* (1983) have shown that Ni is an essential micronutrient for legumes. Most plants contain nickel in the range $1-6$ mg kg^{-1} (Vanselow, 1966; Hutchinson, 1981). The uptake of Ni is enhanced by low pH values, and available nickel increases at pH less than 6.5 as a consequence of the breakdown of Ni complexes in the soil with Fe and Mn oxides. Uptake of nickel by plants and questions of toxicity and tolerance have been reviewed by Farago and Cole (1988). Nickel toxicity toward plants has been reviewed by Vanselow (1966) and Hutchinson (1981).

Serpentine soils throughout the world contain elevated levels of Mg, Cr, Ni and Co. They are generally low in Ca and other micronutrients and generally have distinctive vegetation (Cole, 1992; Streit *et al.*, 1993) and in general they lack fertility. Generally, the level of available Cr is very low. Hence plants growing in high Cr concentrations may not appear to be different from those on soils with relatively low Cr values (Wild, 1974, 1978). Cr and Ni levels in plants have been found to be species specific (Shewry and Peterson, 1976) for a range of species from various serpentine sites. Cr concentrations in roots were well below those in soils. The Cr was non-available, and strongly bound in the soil. Nickel concentrations in plants, however reflect those in the soil. Many plants growing on serpentine soils accumulate nickel (Farago and Cole, 1988; Brooks, 1983, 1993; Brooks and Mailaise, 1985; see also Chapter 4). It appears likely that organic acids are involved in Ni tolerance and accumulation (Farago and Cole, 1988).

The oxidation state of Cr strongly influences the rate of Cr-uptake (Bianchi and Levis, 1984). It is usually assumed that Cr(VI) is much more toxic to plants than

Cr(III), however, Cr(VI) may be rapidly reduced to Cr(III) (Fenke, 1977), thus reducing the problems of Cr toxicity. It has been demonstrated that reduction of Cr(VI) to Cr(V) and Cr(III) can be brought about in the cortex of garlic (*Allium sativum*) roots (Micera and Dessi, 1988). Cr concentrations may differ largely between different parts of the plant (Sykes *et al.*, 1981). Transport of Cr up the root is very slow (Skeffington *et al.*, 1976), hence, low levels of Cr are found in the tops of plants. Shewry and Peterson (1974) reported less than 1% of the accumulated Cr to be transported to the shoots. In iron-deficient conditions, increased transport of Cr to plants has been observed (Cary *et al.*, 1977).

Strontium

The most important factor influencing the uptake of strontium by plants is the Ca status of the soil, other factors include the pH and organic matter content. Lembrechts *et al.* (1990) have investigated the relation between soil solution composition and ^{85}Sr uptake by young lettuce plants. A close relationship was shown between the Sr concentrations in the plants and the Sr/Ca ratio in solution either in nutrient medium, or in soil solution. It was suggested that plants primarily withdraw their nutrients from the soil liquid phase.

Aluminium

The physiological function of aluminium in plants is not clear, although there is some evidence that low levels of aluminium can have a beneficial effect on plant growth, especially in Al-tolerant plant species (Clark, 1977). The Al toxicity syndrome comprises short-term effects (measurable within minutes or even seconds following an exposure to Al) and long-term responses that are measured in hours or even days after addition of Al. Long-term responses are not necessarily caused by Al directly but may rather be a consequence of Al-related impairment of numerous other biochemical and physiological processes. Al has become recognised as the most limiting factor in many acid soils throughout the world (Haug and Caldwell, 1985; Foy, 1988). The most prominent effect of Al disorder is reduction in root growth. Besides rhizotoxicity, many other Al-related physiological disorders in plants have been summarised and various mechanisms of Al tolerance have been suggested in a number of recent reviews (Foy, 1988; Taylor, 1988, 1991; Haug and Shi, 1991; Runge and Rode, 1991).

Solubilisation of Al-containing minerals is enhanced in acidic environments (Russel, 1961), and plants growing in acid conditions are generally Al-tolerant (Clarkson, 1966).

There are reports claiming decreased or increased K concentration in plants grown in the presence of Al (Roy *et al.*, 1988). A large number of reports show decreased concentrations of Ca and Mg in roots and shoots of plants grown under Al stress for extended periods of time (Kruger and Sucoff, 1989; Godbold, 1991; Huang and Bachelard, 1993). Al reduces accumulation of Ca in plant tissues to such an extent that symptoms of Al toxicity are often similar to those of Ca deficiency. Root growth of *Glycine max* was similarly inhibited by either addition of Al or deprivation of Ca^{2+} (Cakmak and Horst, 1991).

Huang and Bachelard (1993) examined the effects of Al on the growth of *Pinus radiata* (D. Don) and *Eucalyptus mannifera* subsp. *mannifera* (Mudie) seedlings in an attempt to explain the failure of *Radiata* pine trees on some acid, low fertility soils in Australia on which the native *Eucalyptus* flourish. Al in culture solution increased the growth of roots and shoots of seedlings of both species, but while growth of the *Eucalyptus* continued to increase with increases in Al to 2222 μmol L^{-1}, growth of the pine was largest at 370 μmol L^{-1} Al. In addition to total root length, specific root length (length per unit dry weight), a measure of fitness of the root, increased in the *Eucalyptus* seedlings as the substrate Al increased. Growth of the shoots and roots of the pine in the absence of any added Al was extremely poor suggesting that Al, in low concentrations, may be an essential element or ameliorate some other factors in solution culture at low pH.

Arsenic

Arsenic is a constituent of most plants. As can be phytotoxic and the toxicity of arsenite is greater than that of arsenate (Peterson *et al.*, 1981). It is suggested that As uptake is passive (Streit and Stumm, 1993), and that it is translocated to most parts of the plant, most being found in roots and old leaves. Natural As levels in plants seldom exceed 1 mg kg^{-1}, but the leaf content may be higher if arsenic pesticides have been used. As accumulates in *Pseudotsuga mensiesii*, up to 2000–5000 mg kg^{-1} have been reported in the plant ash (Hewitt and Smith, 1975).

The uptake of As into crop plants has been of concern, since this provides an entry of the toxic element into the food chain. Total concentrations of As in six garden crops (lettuce, carrot, onion, beetroot, peas and beans) in Cornwall, England (an area rich in As) were not high (Xu and Thronton, 1985), and with the exception of lettuce, did not exceed 1 mg kg^{-1} dry matter. The relationships between As contents of beetroot, lettuce, onion and pea and both total and extractable As in soil were significant, showing As levels in the edible tissues to increase with increasing soil content. As in carrots and beans was not significantly related to soil content. Rice grown in polluted soil had a mean concentration of 0.19 mg kg^{-1} but the roots contained a range of 936–1182 mg kg^{-1} (Kitagishi and Yamane, 1981). Yields of rice, barley and alfalfa in cotton fields previously treated with As may be reduced (Peterson *et al.*, 1981). The authors have also suggested that low levels of phosphates displace As from soil particles which increases its uptake and phytotoxicity, whereas, larger amounts of phosphates compete with As at root surfaces to decrease uptake and phytotoxicity.

As concentrations in plants may also reveal contamination. High concentrations of up to 16 mg kg^{-1} in dandelion (*Taraxacum officinale*) growing around smelters in Bulgaria have been recorded by Djingova and Kuleff (1993). Otte *et al.* (1988) studied the uptake of As by grasses, rush, reed and nettle from the Rhine estuary and from waters near mines. Some grasses showed higher As concentrations in dead leaves than in roots. Growth of reed was more inhibited than growth of nettle. Fe plays a key factor in determining the mobility of As in salt marsh soils and in the translocation processes in the plants (Otte *et al.*, 1991).

Cadmium

Cadmium is not an essential element for plants, which take up only small amounts under normal conditions (Adriano, 1986). Background concentrations in a number of crops and plants have been compiled by Kabata-Pendias and Pendias (1984). Cadmium is likely to be concentrated in the protein fractions of plants, where it is bound to low molecular weight proteins called phytochelatins (see below). There are no known enzymes that depend on cadmium for their normal activity. It is considered to be a toxic element to plants, and the main cause of its toxicity lies in that it disturbs enzyme activities. Soil pH is a major factor in determining the bioavailability of Cd (Eriksson, 1989; Alloway, 1990), Cd being most bioavailable below pH 6. When the *Eh* of soil drops to about -0.13 V, and sulphate is reduced to sulphide, the proportion of available Cd drops significantly (Imura, 1981) since insoluble CdS is produced. This has consequences for rice growing, the uptake of Cd being much lower on submerged soil than on soil that has been drained after tillering.

Since cadmium is toxic to both plants and animals, there is concern about the transfer of Cd along the soil-plant-human pathway, and this transfer has been discussed in detail by Jackson and Alloway (1992). These authors list Cd concentrations in a number of food crops including those grown on fertiliser amended soils. Moir and Thornton (1989) reported cadmium levels in crops from urban areas in England to be generally similar to 'background levels'. The crops studied were carrot, iceberg lettuce, spinach, broccoli, cabbage and parsnip. Highest Cd levels were found in spinach and lettuce, which are known Cd accumulators (Davis and Carlton-Smith, 1980). Among vegetables, lettuce and spinach leaves contain the highest amounts of Cd (Alloway *et al.*, 1990; Jackson and Alloway, 1991).

The effects of Cd and Zn interactions on the accumulation and tissue distribution of Zn and Cd in lettuce and spinach were studied by McKenna *et al.* (1993). Compared with Cd-only pollution, Zn-Cd combined pollution may not decrease Cd concentrations in lettuce and spinach edible tissues, but because it increases their Zn it lowers plant Cd bioavailability.

Cd is especially absorbed by plants in the vital rooting zones. From laboratory experiments it is known that Cd concentrations, on average, decrease significantly in plants from roots to shoots. Commonly, the root contains ten times as much cadmium as is contained in the shoot, the root acting as a barrier (Jarvis *et al.*, 1976). However, the extent is strongly species dependent. Many species show significantly lower Cd levels in fruits than in the shoot. In various plants, eg, spinach, soy beans, and wheat, these barriers are only poorly effective. In rice grains, there can even be an enrichment of Cd compared with the leaves (Bingham *et al.*, 1975).

High applied cadmium can induce growth retardation and root damage, chlorosis of leaves, and red-brown colouration of leaf margins or veins. Russo and Brennan (1979) reported that the leaves of two-year old *Quercus rubra* seedlings were chlorotic when grown for seven weeks in a nutrient solution supplied with 100 $\mu g\ L^{-1}$ of cadmium. The volume of roots was also significantly affected. In a similar experiment, in which the growth and development of the seedlings were followed for two years, a biomass increase was obtained during the first year of cadmium treatment. In the second year, cadmium caused no effect on tree growth, but the percentage of foliar toxicity symptoms was increased (Kazimir and Brennan, 1986).

A number of Cd-binding proteins have been found in vascular plants that are similar to metallothioneins, the differences have been discussed by Jackson and Alloway (1992) and by Rauser (1987). Studies have been made on the Cd-binding proteins of wheat (Wagner *et al.*, 1984); cabbage (Wagner, 1984); soybean (Casterline and Barnett, 1982); rice; (Kaneta *et al.*, 1983); tomato (Bartoff *et al.*, 1980); lettuce (Henze and Umland, 1987); and *Agrostis gigantea* (Rauser, 1984). Methods involved in the study of Cd-binding proteins are discussed by Grill (1987).

Lead

Lead is considered to be a non-essential metal to plants, and only a small proportion of the lead in soils is biovailable to plants (Alloway, 1990). Visible symptoms of toxici ty, though unspecific to Pb, are smaller leaves and a stunted growth. Leaves may become chlorotic and reddish with necrosis and the roots may turn black. Several plant species, ecotypes and bacterial strains have been known to develop Pb tolerance. The phytotoxicity of Pb is low as it has very limited availability and uptake from soil and soil solutions. However, plant roots are usually able to take up and accumulate large quantities of Pb^{2+} in soil and culture solutions but translocation to aerial shoots is generally limited due to binding at root surfaces and cell walls (Lagerwerff, 1971; Jones *et al.*, 1973; Lane and Martin, 1977).

Pb can be found on the surface of leaves, trunk and stem where it is deposited from the atmosphere by wet or dry deposition. Lead and other elements in tree bark have been reviewed by Walkenhorst *et al.* (1993). Translocation of Pb from the outer surface to the inner parts of leaves, tubers, fruits or seeds is very small. Peterson (1978) has reviewed the mechanisms of Pb uptake and transport in plants and of the Pb concentrations found in different plant species.

The toxic effects of lead may vary considerably and are quite often conflicting. Pb concentrations greater than those normally found in the field, except near smelters or mines, have been found to interfere with the photosynthesis and growth of plants (Koeppe, 1981). In nutrient solutions containing 104 $\mu g\ L^{-1}$ of Pb as the chloride, the root elongation of *Picea abies* seedlings decreased by more than 30% compared to the control. At four times higher Pb levels, the root elongation was half of that of the control. Moreover, growth retardations were greater in an artificial soil solution than in full strength nutrient solution (Godbold and Hutterman, 1985). Stiborova *et al.* (1986a, b) reported a decrease in biomass and a lower root/shoot ratio of *Hordeum vulgare* and *Zea mays* when treated with Pb at 207 $\mu g\ L^{-1}$.

Elemental Uptake Processes

General mechanisms of uptake, assimilation, distribution and transport have been suggested by a number of authors (Bowling, 1976; Baker, 1983; Marschner, 1983; Azpiazu *et al.*, 1986; Davies and Jeffcoat, 1990; Fergusson, 1990; Streit and Stumm, 1993).

Root Uptake

Roots modify their environment quite extensively in many ways. The most important of these are pH change, exudation and microbiological activity in the rhizosphere. Root exudates contain compounds such as hydroxycarboxylic acids and amino acids and these are capable of complexing trace metals. Bowling (1976), Farago (1986) and Streit and Stumm (1993) have discussed the theories of mineral uptake by plant roots; the first suggests that there are four links in the uptake chain: movement of ions or complexes in the soil to the roots; uptake into the root; transport across the root to the vascular system; and movement to the shoot.

The epidermis of the root has extensions, the root hairs, with a pectic coating which allows them to adhere to soil particles. The hairs also greatly enhance the area of contact with the soil. Figure 2-1 shows a transverse section of a typical root. A large part of the root consists of relatively large and loosely arranged parenchyma cells, with air spaces, collectively known as the cortex. The central portion of the root, the stele, contains the vascular system which is responsible for the transport of food, water and minerals throughout the plant. These tissues contain the xylem, which conducts water and nutrients up to the aerial parts, and the phloem, which functions as a conductor of organic material from the leaves to the roots. The stelar portion of the root is surrounded by a distinct layer of cells, the endodermis, which separates the stele from the cortex. The chief feature of the endodermis is the Casparian strip or band which surrounds the walls of each cell. The band is composed of suberised material which

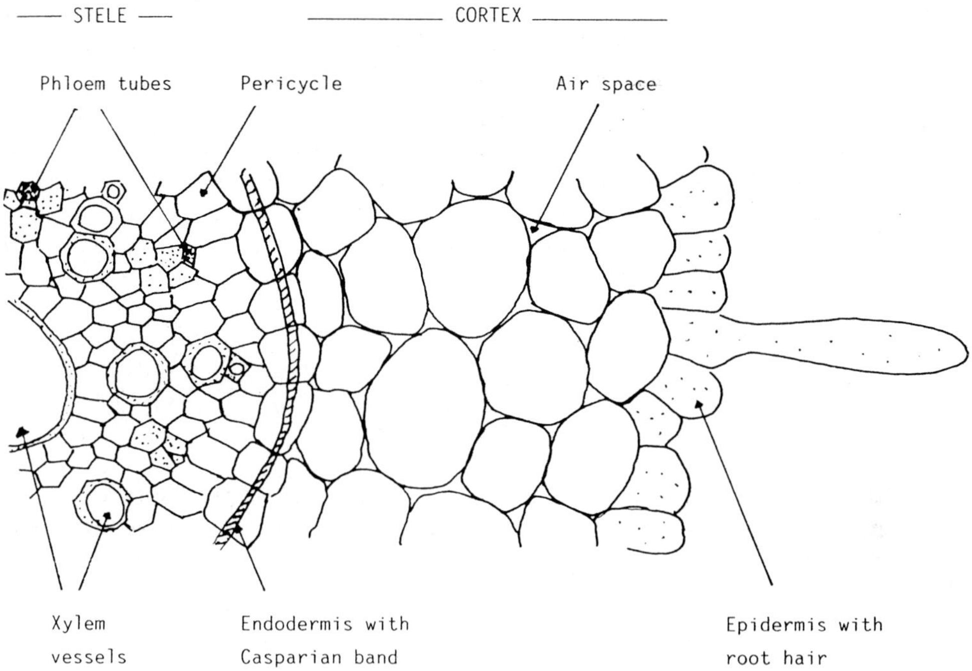

Fig. 2-1. Transverse section of a typical root (From Farago 1986).

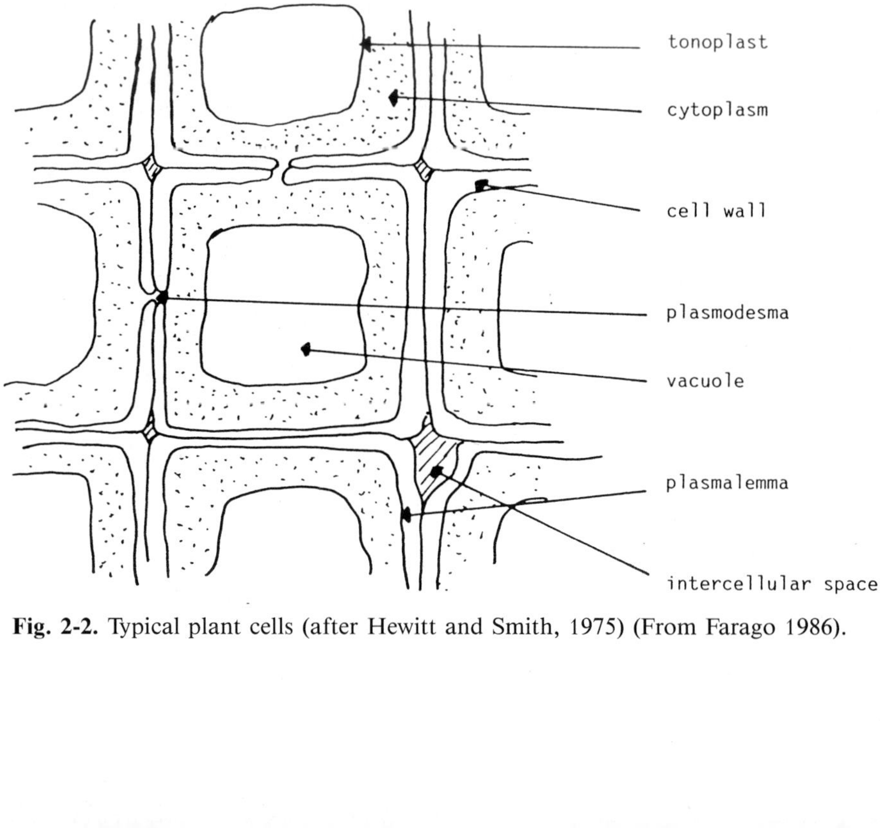

Fig. 2-2. Typical plant cells (after Hewitt and Smith, 1975) (From Farago 1986).

forms an impermeable barrier to water and ionic solutes. In order to reach the xylem, water and dissolved salts must pass through the living portions of the cells by passing through a membrane, the plasmalemma (Fig. 2-2).

The absorption of elements by roots can be both active (metabolic) and passive (nonmetabolic). It was known from early experiments that a number of metal ions are able to diffuse passively in and out of the root up to the Casparian band although uptake in some cases, such as K^+ and Na^+, by epidermal and cortex cells seems to be by active transport. There is evidence that ions cross the root by two general pathways:

1. cell wall pathway: ions cross the cortex by means of the cells walls;
2. symplasm pathway: ions cross the cortex by cytoplasmic drift through the continuum of cytoplasm that extends from cell to cell through the plasmodesmata, and is known as the symplasm.

Transport into the xylem can be passive or active and is a problem still to be resolved (Streit and Stumm, 1993).

Transport of Ions to the Aerial Parts of the Plant

There is a close relationship between the metabolism of the shoot and the root. It is generally accepted that the xylem forms the main path for upward movement of water and ions from the roots to the leaves. Most of the essential major elements are transported in the xylem as inorganic ions. Nitrogen may be transported along the xylem as NO_3^- if it is present in the external solution as nitrate. However, the plant sap may also contain organic nitrogen compounds such as amino acids. In the xylem, heavy metals will usually only be transported if special chelates are formed, eg, by citrate (Streit and Stumm, 1993). Iron is taken up and transported more readily when supplied as a chelated complex, such as ferric ethylenediamine tetraacetate (FeEDTA) or as ferric diethylenetriamine pentaacetate (FeDTPA) (Wallace and North, 1953). Calcium may also be transported in a chelated form (Jacoby, 1966).

The phloem tubes consist of adapted cells that transport the products of photosynthesis, and other compounds such as growth regulators, to the growing parts of the plant including the roots. Since phosphate concentration in the phloem is high, those elements with insoluble phosphates are not usually transported in the phloem. Na^+, K^+, Rb^+, Cs^+ and Mg^{2+} are easily transported, whereas, Pb^{2+}, Ca^{2+}, Sr^{2+} and Ba^{2+} are not.

The mechanisms of metal uptake may be modified in those plants where the roots are associated with mycorrhizal fungi. Streit and Stumm (1993) suggest that metal ions might form complexes with the chitin of the fungal cell wall. However is it not clear whether this would facilitate the uptake of metals by the plant host. Some selection seems to be involved since not all the elements taken up by the fungus are released to the host. These authors suggest that in part metals may simply be released to the host by the local decay of the fungal hyphae.

Uptake by Aerial Parts of the Plant

Many fertilisers for the treatment of deficiencies of trace elements are added to the leaves. In addition, uptake of trace elements from aerial sources through the leaves may have a significant impact on plant contamination, especially of elements such as Fe, Mn, Zn and Cu. Anthropogenic aerosols have been reported to bring about foliar deposition of Pb to a great extent (Streit and Stumm, 1993). Zieve and Peterson (1984) have reported the assimilation of diemethyl selenide through leaves of terrestrial plants. Roberts *et al.* (1975) found a significant absorption of foliar-applied Zn, Fe, Cd and Mg.

Foliar uptake of elements is believed to consist of two phases: non-metabolic cuticular absorption which is generally considered to be the major route of entry; and metabolic mechanisms which account for element accumulation against a concentration gradient. Passsage through the cuticle is possible only by diffusion (Ziegler, 1988; Reiderer, 1991). The second process is responsible for transporting ions across the plasma membrane and into the cell protoplast.

References

Adriano, D. C. (1986). *Trace Elements in the Terrestrial Environment*, New York: Springer Verlag.

Angelone, M., Bini, C. (1992), in: *Biogeochemistry of Trace Metals:* Adriano, D. C. (ed.) Boca Raton: Lewis Publishers; pp. 109–158.

Alloway, B. J. (ed.) (1990). *Heavy Metals in Soils*, London: Blackie.

Alloway, B. J., Jackson, A. P., Morgan, H. (1990). *Sci Tot Environ 91*, 223–236.

Akerstrand, K., Kardell, L., Moller, T. (1988). *Var Foda 40*, 259–270 (In Swedish, cited in Bockman, 1990.)

Anderson, J. W., Scarf, A. R. (1983), in: *Metals and Micronutrients, Uptake and Utilization by Plant:* Robb, D. A., Pierpoint, W. S. (eds.) London: Academic Press; pp. 242–273.

Azpiazu, M. N., Romero, F., Diaz, J. M. (1986). *Water Air Soil Pollut 28*, 1–26.

Baker, D. A. (1983), in: *Metals and Micronutrients: Uptake and Utilization by Plants:* Robb, D. A. Pierpoint, W. S. (eds.) London: Academic Press.

Baker, A. J. M., Proctor, J. (1990). *Plant Systematics and Evolution 173*, 91–108.

Baker, A. J. M., Walker, P. L. (1989). *Chemical Speciation and Biovailability 1*, 7–17.

Baker, A. J. M., Walker, P. L. (1990), in: *Heavy Metal Tolerance in Plants: Evolutionary Aspects:* Shaw, A. J. (ed.) Boca Raton: CRC Press Inc.; pp. 155–177.

Banuelos, G. B., Mead, R., Akhouse, S. (1991). *J. Plant Nutr 14*, 46–50.

Banuelos, G. S., Cardon, G. E., Phene, C. J., Wu, L., Akhouse, S., Zambrzuski, S. (1993). *Plant and Soil 148*, 253–263.

Bartolf, M., Brennan, E., Price, C. A. (1980). *Plant Physiol 66*, 438–331.

Beckett, P. H. T., Davis, R. D. (1977). *New Phytol 79*, 95–106.

Beckett, P. H. T., Davis, R.D. (1978). *New Phytol. 81,* 153–157.
Bengtsson, S., Tyler, G. (1976). *Vanadium in the Environment – A Technical Report.* Monitoring and Assessment Research Centre, University of London.
Bertelsen, F., Gissel-Nielsen, G. (1987). *Environ Geochem Health 9,* 12–16.
Bertrand, D., de Wolf, A. (1986). *C R Acad Sci Ser D Paris 266,* 1494–1495.
Bianchi, V., Levis, A. G. (1984). *Toxicol Environ Chem 9,* 1–25.
Bienfait, H. F., Van der Mark, F. (1983), in: *Metals and Micronutrients, Uptake and Utilization by Plants:* Robb, D. A., Pierpoint, W. S. (eds.) London: Academic Press; pp. 112–123.
Bingham, F. T., Page, A. L. (1975), in: *International Conference on Heavy Metals in the Environment:* Edinburgh: CEP.
Bockman, O. C., Kaarstad, O., Lie, O. H., Richards, I. (1990). *Agriculture and Fertilizers.* Oslo: Norsk Hydro.
Bollard, E. G. (1983), in: *Inorganic Plant Nutrition, Part B. Encyclopedia of Plant Physiology, 15 B:* Lauchli, A., Bieleski, R. L. (eds.) Berlin: Springer Verlag; pp. 695–744.
Bould, C., Hewitt, E. J., Needham, P. (1983), in: *Plants, Vol. 1. Principles:* Robinson, J. B. D. (ed.) London: Ministry of Agriculture, Fisheries and Food/Agricultural Research Council; pp. 54–110.
Bowen, H. J. M. (1966). *Trace Elements in Biochemistry,* London: Academic Press.
Bowen, H. J. M. (1979). *Environmental Chemistry of the Elements,* New York: Academic Press.
Bowie, S. H. U., Thornton, I. (1985). *Environmental Geochemistry and Health,* Amsterdam: D. Reidel Publishing Company.
Bowling, D. J. F. (1976). *Uptake of Ions by Plant Roots,* London: Chapman and Hall.
Bradshaw, A. D., McNeilly, T. (1981). *Evolution and Pollution. Institute of Biology Studies in Biology No. 130,* London: Arnold.
Brady, D. J., Edwards, D. G., Asher, C. J., Blamey, F. P. C. (1993). *The New Phytologist 123,* 531–538.
Brooks, R. R. (1972). *Geobotany and Biogeochemistry in Mineral Exploration,* New York: Harper and Row.
Brooks, R. R. (1983). *Biological Methods of Prospecting for Minerals,* New York: Wiley.
Brooks, R. R. (1993), in: *Plants as Biomonitors. Indicators for Heavy Metals in the Terrestrial Environment:* Markert, B. (ed.) Weinheim: VCH; pp. 127–153.
Brooks, R. R., Malaise, F. (1985). *The Heavy Metal Tolerant Flora of South Africa,* Rotterdam: Balkema.
Brown, M. T., Wilkins, D. A. (1985). *New Phytol 99,* 91–100.
Brüggeman, W., Moog, P. R., Nakagawa, H., Janiesch, P., Kuiper, P. J. C. (1990). *Physiol Plant 79,* 339–346.
Burton, K. W., Morgan, E., Roig, A. (1986). *New Phytol 103,* 549–557.
Cakmak, I., Horst, W. J. (1991). *Physiologia Plantarum 83,* 463–468.
Cannon, H. L. (1963). *Soil Sci 96,* 196–204.
Carlsen, C. L., Adriano, D. C., Dixon, P. M. (1991). *J. Environ Qual 20* 363–368.
Cary, E. E., Allaway, W. H., Olson, O. E. (1977). *J Agric Fd Chem 25,* 300–304.
Casterline, J. L., Barnett, N. M. (1982). *Plant Physiol 69,* 1004–1007.
Chaney, R. L., Bell, P. F. (1987). *J Plant Nutr 10,* 963–994.
Chaney, R. L., Giordano, P. M. (1977), in: *Soils for Management of Organic Wastes and Waste Waters:* Elliot, L. F., Stevenson, F. J., (eds.) Madison: Soil Science Society of America; pp. 235–279.
Chaney, R. L., Ryan, J. A. (1992), in: *Proceedings of the International Composting Symposium:* Hoitink *et al.* (eds.), in press.
Chaudhry, F. M., Wallace, A. (1976). *Plant Soil 45,* 697–700.

Cheung, W. Y. (1984). *Fed Proc 43*, 2005–2009.

Clark, R. B. (1977). *Plant Soil 47*, 653.

Clarkson, D. T. (1966). *J Ecol 54*, 167–178.

Clarkson, D. T. (1984). *Plant Cell Environment 7*, 440–456.

Cole, M. M. (1992), in: *The Ecology of Areas with Serpentanized Rocks. A World View:* Roberts, B. A., Proctor, J. (eds.), Dordrecht: Kluwer Academic Publications; pp. 343–373.

Colley, T. N., Gonzales, M. H., Martin, D. F. (1979). *Economic Botany 32 (4),* 371–8.

Cooke, J. A., Johnson, M. S., Davison, A. W., Bradshaw, A. D. (1976). *Environ Pollut 11*, 9–23.

Cramer, G. R., Lauchli, A., Polito, V. S. (1985). *Plant Physiology 79*, 207–211.

Cramer, G. R., Lauchli, A., Epstein, E. (1986). *Plant Physiology 81*, 792–797.

Cramer, G. R., Lynch, J., Lauchli, A., Epstein, E. (1987). *Plant Physiology 83,* 510–516.

Cramer, G. R., Epstein, E., Lauchli, A. (1991). *Physiologia Plantarum 80* 83–88.

Cumming, A. R., Tomsett, A. B. (1992), in: *Biogeochemistry of Trace Metals:* Adriano, D. C. (ed.) Boca Raton: Lewis Publishers; pp. 109–158.

Curl, E. A., Truelove, B. (1986). *The Rhizosphere,* Berlin: Springer-Verlag.

Davies, B. E. (1992). *Water Air Soil Pollution 63*, 331–342.

Davies, G. R. (1980), in: *MAFF Reference Book 326, Inorganic Pollution and Argriculture:* London: HMSO; pp. 191–204.

Davies, W. J., Jeffcoat, B. (eds.) (1990). *Importance of Root to Shoot Communication in the Responses to Environmental Stress. British Society for Plant Growth Regulation, Monograph No. 21,* Proceedings of a meeting held at the University of Lancaster, Sept, 1989.

Davis, R. D., Beckett, P. H. T. (1978). *New Phytol 80*, 23–32.

Davis, R. D., Carlton-Smith, C. (1980). *Crops as Indicators of the Significance of Contamination of Soil by Heavy Metals. WRC Technical Report 140.* Water Research Council, Medmenham, UK.

Dawson, J. C., Anderson, J. W. (1989). *Phytochem 28*, 51–55.

Dhillon, K. D., Dhillon, S. K. (1991). *Environ Geochem Health 13*, 165–170.

Dixon, B., Sagar, G. R., Shorrocks, V. W. (1973). *Hortic Sci 48*, 403–411.

Dixon, N. E., Gazzola, C., Blakeley, R. L., Zerner, B. (1975). *J Am Chem Soc 97*, 4131–4133.

Djingova, R., Kuelff, I. (1993), in: *Plants as Biomonitors. Indicators for Heavy Metals in the Terrestrial Environment:* Markert, B. (ed.) Weinheim: VCH; pp. 436–460.

Epstein, A. (1972). *Mineral Nutrition in Plants*, London: John Wiley.

Eriksson, J. E. (1989). *Water Air Soil Pollution 48*, 371–335.

Eskew, D. L., Welch, R. M., Carey, E. E. (1983). *Science 222*, 621–623.

Farago, M. E. (1986), in: *Frontiers in Bioinorganic Chemistry:* Xavier, A. V. (ed.) Weinheim: VCH.

Farago, M. E., Cole, M. M. (1988). *Metal Ions in Biological Systems. Vol. 23. Nickel and its Role in Biology:* Sigel, H., Sigel, M. (eds.) New York: Marcel Dekker; pp. 47–90.

Farago, M. E., Mullen, W. A. (1981). *Inorg Nucl Chem Lett 17,* 9–12.

Farago, M. E., Mullen, W. A., Cole, M. M., Smith, R. F. (1980). *Environ Pollut (Series A) 21*, 225–244.

Farago, M. E., Cole, M. M., Xiao, X., Vaz, M. C. (1992). *Chem Speciat Bioavail 4*, 19–27.

Fenke, K. (1977). *Dissertation*, University of Bonn, Germany.

Fergusson, J. E. (1990). *Heavy Elements in Plants. Chemistry, Environmental Impact and Health Effects,* London: Pergamon Press.

Fitzpatrick, E. A. (1986). *An Introduction to Soil Science, 2nd edn.,* London: Longman Scientific and Technical.

Folkeson, L., Nyholm, N. E. L., Tyler, G. (1990). *Sci Total Environ 96*, 211–233.

Foy, C. D. (1988). *Communications in Soil Science and Plant Analysis 19,* 959–987.

Foy, C. D., Oakes, A. J. (1984). *J Plant Nutrition 7,* 95–98.

Foy, C. D., Chaney, R. L., White, M. C. (1978). *Ann Rev Plant Physiol 29,* 511.

Francois, L. E. (1992). *Plant Soil 147,* 163–170.

Frausto da Silva, J. J. R. (1989). *Chem Speciation Bioavail 1,* 139–150.

Fuge, R. (1988). *Environ Geochem Health 10,* 51–61.

Fuge, R., Johnson, C. C. (1986). *Environ Geochem Health 8,* 31–54.

Gardner, A. W., Hill-Patch, P. K. (1962). *Vet Rec 74,* 113–115.

Garsed, S. G. (1981), *Environ Pollut (Series A) 24,* 303–311.

Gauglhofer, J. (1991), in: *Metals and their Compounds in the Environment: Occurrrence, Analysis, and Biological Relevance:* Merian, E., (ed.) Weinheim: VCH Publishers; pp. 853–878.

Gemmel, R. P. (1977). *Colonization of Industrial Wasteland. Institute of Biology Studies in Biology No. 80.*

Gissel-Neilson, G., Bertelsen, F. (1989). *Environ Geochem Health 11,* 54–56.

Godbold, D. L. (1991), in: *Plant-Soil Interactions at Low pH:* Wright, R. J., Baligar, V. C., Murrmann, R. P. (eds.) Dordrecht: Kluwer Academic Publishers; pp. 747–753.

Godbold, D. L., Hutterman, A. (1985). *Environ Poll (Series A) 38,* 375–381.

Godbold, D. L., Horst, W. J., Marschner, H., Collins, J. C., Thurman, D. A. (1983). *Z Pflanzenphysiol 112,* 315–324.

Gough, K. P., Shacklette, H. T., Case, R. A. (1979). *Element Concentrations Toxic to Plants, Animals and Man. United States Geological Survey Bulletin 1466.*

Grill, E. (1987) in: *Metallothionein II: Prodeedings of the Second International Conference on Metallothionein an Other Low Molecular Weight Metal-Binding Proteins:* Kägi, J. H. R., Kojima, Y. (eds.) Basel: Birkhäuser Verlag; pp. 317–322.

Harborne, J. B. (1988). *Introduction to Ecological Biochemistry 3rd edn,* London: Academic Press.

Hartman, P. L., Mills, H. A., Jones, J. B., Jr (1986). *J Am Soc Hortic Sci 111,* 487–490.

Hassett, J. J., Miller, J. E., Koeppe, D. E. (1976). *Environ Pollut 11,* 297–301.

Haug, A., Caldwell, C. R. (1985), in: *Frontiers of Membrane Research in Agriculture:* St John, J. B., Berlin, E., Jackson, P. C. (eds.) Ottawa: Rowman and Allanheld; pp. 359–381.

Haug, A., Shi, B. (1991), in: *Plant-Soil Interactions at Low pH:* Wright, R. J., Baligar, V. C., Murrmann, R. P. (eds). Kluwer Academic Publishers: Dordrecht; pp. 839–850.

Hawkins, H. J., Lewis, O. A. M. (1993). *The New Physiologist 124,* 161–170.

Hay, R. W. (1986). *Bio-inorganic Chemistry,* London: Ellis Horwood.

Heale, E. L., Ormond, D. P. (1982), *Can J Bot 60,* 2674–2681.

Henry, C. L., Harrison, R. B. (1992), in: *Biogeochemistry of Trace Metals:* Adriano, D. C. (ed.) Boca Raton: Lewis Publishers; pp. 195–216.

Henze, W., Umland, F. (1987). *Med Biol 4,* 501–507.

Hewitt, E. J. (1948). *Nature (Lond) 161,* 489–490.

Hewitt, E. J. (1953). *J Exp Bot 4*59–64.

Hewitt, E. J. (1983 a), in: *MAFF/ARC Diagnosis of Mineral Disorders in Plants, Vol. 1, Principles:* Bould, C., Hewitt, E. J., Needham, P. (eds.) London: HMSO; pp. 54–110.

Hewitt, E. J. (1983 b), in: *Metals and Micronutrients, Uptake and Utilization by Plants:* Robb, D. A., Pierpoit, W. S. (eds.) London: Academic Press; pp. 277–323.

Hewitt, E. J., Smith, T. A. (1975). *Plant Mineral Nutrition,* London: English Universities Press.

Hipkins, M. F. (1983), in: *Metals and Micronutrients, Uptake and Utilization by Plants:* Robb, D. A., Pierpoint, W. S. (eds.) London: Academic Press; pp. 147–168.

Hossner, L. R., Woodard, H. J., Bush, J. (1992). *J Plant Nutrition 15,* 2743–2761.

Huang, J., Bachelard, E. P. (1993). *Plant Soil 149,* 121–127.

Huffman, E. W. D., Allaway, W. H. (1973). *Plant Physiology 52,* 72–75.
Hutchinson, T. C. (1981), in: *Effect of Heavy Metal Pollution on Plants. Vol 1: Effect of Trace Metals on Plant Function:* Lepp, N. W. (ed.) London: Applied Science; pp. 171–211.
IAEA (1984). *Soil and Fertilizer Nitrogen. International Atomic Energy Authority Technical Report No. 244.* Vienna.
Imura, K. (1981), in: *Heavy Metal Pollution in Soils in Japan:* Kitagishi, K., Yamane, I. (eds.) Tokyo: Japan Scientific Society Press; pp. 37–63.
Irgolic, K. J., Martell, A. E. (1985). *Environmental Inorganic Chemistry,* Deerfield Beach, Florida: VCH Publishers.
Jacoby, E. (1966). *Nature (London) 211,* 212.
Jackson, A. P., Alloway, B. J. (1991). *Plant Soil 132,* 179–186.
Jackson, A. P., Alloway, B. J. (1992), in: *Biogeochemistry of Trace Metals:* Adriano, D. C. (ed.) Boca Raton: Lewis Publishers; pp. 109–158.
Jarosik, J., Zvara, P., Konecny, J., Obdrzalek, M. (1988). *Sci Total Environ 71,* 225–229.
Jarrell, W. M., Beverley, R. B. (1981). *Advances in Agronomy 34,* 197–221.
Jarvis, S. G., Jones, L. H. P. (1979). *J Sci Food Agric 30,* 742.
Jarvis, S. C., Jones, L. H. P., Hopper, M. H. (1976). *Plant Soil 44,* 179–191.
Jenkinson, D. S. (1982). *Proc Roy Soc Lond B296,* 9–31.
Jones, K. E., Johnston, A. E. (1989). *Environ Pollut 57* 199–216.
Jones, L. H. P., Clement, C. R., Hopper, M. J. (1973). *Plant Soil 38,* 403.
Jones, R., Prohaska, K. A., Burgess, M.S.E. (1988). *Water Air Soil Pollut 37,* 355–363.
Juste, C., Mench, M. (1992), in: *Biogeochemistry of Trace Metals:* Adriano, D. C. (ed.) Boca Raton: Lewis Publishers; pp. 159–193.
Kabata-Pendias, A., Pendias, H. (1984). *Trace Elements in Soils and Plants,* Boca Raton: CRC Press.
Kalac, P., Burda, J., Staskova, S. (1991). *Sci Total Environ 105,* 109–119.
Kaneta, M., Hikischi, H., Endo, S., Sugiyama, M. (1983). *Agric Biol Chem 47,* 417–418.
Kazimir, J., Brennan, E. (1986). *Forest Sci 32,* 1061–1066.
Kieffer, F. (1991), in: *Metals and Their Compounds in the Environment:* Merian, E. (ed.) Weinheim: VCH.
Kienkens, L. (1990), in: *Heavy Metals in Soils:* Alloway, B. J. (ed.) London: Blackie.
Kitagishi, K., Yamane, I. (1981). *Heavy Metals Pollution in Soils of Japan,* Tokyo: Japan Science Society Press; p. 302.
Kloke, A. (1980). Lecture at GDI Institute, Zurich. Cited in *Chem Rundsch 35,* (10) 3 (March 3, 1982).
Koeppe, D. E. (1981), in: *Effects of Heavy Metals in Plants, Vol. 1:* Lepp, N. W. (ed.) London: Applied Science Publishers; pp. 55–76.
Kruger, E., Sucoff, E. (1989). *J Experim Bot 40,* 653–658.
Kubota, J., Welch, R. M., Van Campen, D. R. (1992). *Environ Geochem Health 14,* 91–100.
Lagerwerff, J. V. (1971). *Soil Sci 3,* 129.
Lane, S. D., Martin, E. S. (1977). *New Phytol 79,* 281–286.
Lauchli, A. (1990), in: *Current Topics in Plant Physiology, Vol. 4, Calcium in Plant Growth and Development:* Leonard, R. T., Hepler, P. K. (eds.) Rockville, MD: American Society of Plant Physiologists; pp. 26–35.
Lembrechts, J. F., Van Ginkel, J. H., Dermet, D. M. (1990). *Plant Soil 125,* 63–69.
Leonard, R. T. (1983), in: *Metals and Micronutrients, Uptake and Utilization by Plants:* Robb, D. A., Pierpoint, W. S. (eds.) London: Academic Press; pp. 71–86.
Leonard, R. T., Hepler, P. K. (eds.) (1990). *Current Topics in Plant Physiology, Vol. 4, Calcium in Plant Growth and Development,* Rockville, MD: American Society of Plant Physiologists.

Lepp, N. W. (ed.) (1981). *Effect of Heavy Metal Pollution on Plants. Vol. 1. Effects of Trace Metals on Plant Function,* London: Applied Science Publishers.
Levitt, J. (1980). *Responses of Plants to Environmental Stress, 2nd edn. Vol. 2,* New York: Academic Press.
Lewis, O. A. M. (1986). *Plants and Nitrogen:* London: Edward Arnold; pp. 21–41.
Lindsay, W. L. (1974), in: *The Plant Root and its Environment:* Carson, E. W. (ed.) Charlotsville: Charlotsville University Press; pp. 509–511.
Mansfield, T. A., Lucas, P. W. (1990), in: *Pollution: Causes, Effects and Control:* Harrison, R. M. (ed.) Cambridge: Royal Society of Chemistry; pp. 237–259.
Markert, B. (1992), in: *Biogeochemistry of Trace Metals,* Adriano, D. C. (ed.) Boca Raton: Lewis Publishers; pp. 401–428.
Marschner, H. (1983), in : *Encyclopedia of Plant Physiology New Series Vol 15A, Inorganic Plant Nutrition,* Lauchli, A., Breleski (eds.) Berlin: Springer-Verlag; pp. 5–49.
Marschner, H. (1986). *Mineral Nutrition of Higher Plants,* London: Academic Press.
Marschner, H., Romheld, V., Horst, W. H., Martin, P., (1986). *Z Pflanzenernähr Bodenk 149,* 441–456.
Mattigod, S. V., Page A. L. (1983), in: *Applied Environmental Geochemistry:* Thornton, I. (ed.) London: Academic Press.
McKenna, I. M., Chaney, R. L., Williams, F. M. (1993). *Environ Pollution 79,* 113–120.
Mengel, K., Pilbeam, D. J. (eds.) (1992). *Nitrogen Metabolism in Plants. Proc. Phytochem. Soc. Oxford. Vol 33,* Oxford: Clarendon Press.
Merian, E. (1991). *Metals and Their Compounds in the Environment: Occurrence, Analysis, and Biological Relevance,* Weinheim: VCH Publishers.
Micera, G., Dessi, A. (1988). *J Inorg Biochem 35,* 157–166.
Mikkelson, R. L., Page, A. L., Bingham, F. T. (1986), in: *Trace Substances in Environmental Health-XX:* Hemphill, D. D. (ed.) Columbia: University of Missouri.
Moir, A M., Thornton, I. (1989). *Environ Geochem Health 11,* 113–119.
Morrell, B. G., Lepp, N. W., Phipps, D. A. (1983). *Minerals and the Environment 5,* 79–81.
Morrell, B. G., Lepp, N. W., Phipps, D. A. (1986). *Environ Geochem Health 8,* 14–18.
Mortvedt, J. J. (1984), in: *Trace Substances in Environmental Health-XVIII:* Hemphill, D. D. (ed.) Columbia: University of Missouri.
Mugwira, L. M., Knezek, B. D. (1971). *Commun Soil Sci Plant Anal 2,* 337–343. Cited in Ohnesorge, F. K., Wilhelm, M. (1991), in: *Metals and Their Compounds in the Environment: Occurrence, Analysis and Biological Relevance:* Merian, E. (ed.) Weinheim: VCH Publishers; pp. 1309–1342.
Nams, V. O., Folkard, F. G., Smith, J. N. M (1993). *Can J Bot 71,* 93–97.
Neilsen, D., Neilsen, G. H., Sinclair, A. H., Linehan, D. J. (1992). *Plant Soil 145,* 45–50.
Norrish, K. (1975), in: *Trace Elements in Soil-Plant-Animal Systems:* Nicholas, D. J. D., Egan, A. R. (eds.) New York: Academic Press; pp. 55–82.
Oldfield, J. E. (1992). *Environ Geochem Health 14,* 81–86.
Olsen, R. A. (1987), in: *Land Transformations in Agriculture:* Woolman, M. G., Fournier, F. G. A. (eds.) Chichester: John Wiley; pp. 203–226.
Ota, K., Yamamoto, Y. (1989). *Plant Cell Physiol 30,* 365–371.
Otte, M. L., Dekkers, M. J., Rozema, J., Broekman, R. A. (1991). *Can J Bot 69,* 2670–2677.
Otte, M. L., Rozema, J., Beek, M. A., Ernst, W. H. O., Broekman, R. A. (1988), in: *Proceedings of the 3rd Conference on Environmental Contamination, Venice:* Edinburgh: CEP Consultants Ltd.
Paivoke, A. (1983). *Ann Bot Fennici 20,* 197–203.
Parr, A. J., Loughman, B. C. (1983), in: *Metals and Micronutrients, Uptake and Utilization by Plants:* Robb, D. A., Pierpoint, W. S. (eds.) London: Academic Press; pp. 87–107.

Peterson, P. J. (1978), in: *The Biogeochemistry of Lead in the Environment, Part B:* Nriagu, J. O. (ed.) Amsterdam: Elsevier; pp. 335–384.

Peterson, P. J., Girling, C. A. (1981), in: *Effect of Heavy Metal Pollution on Plants, Vol. 1:* Lepp, N. W. (ed.) London: Applied Science Publishers; pp. 210–277.

Peterson, P. J., Girling, C. A., Benson, L. M., Zieve, R. (1981), in: *Effect of Heavy Metal Pollution on Plants, Vol 1:* Lepp, N. W. (ed.) London: Applied Science Publishers; pp. 279–342.

Pill, W. G., Lambeth, V. N. (1977). *J Am Soc Hortic Sci 102,* 78–81.

Potter, J. F. (1992). *Environ Professional 14,* 143–150.

Rauser, W. E. (1984). *Plant Physiol 74,* 1025–1029.

Rauser, W. E. (1987), in: *Metallothionein II: Proceedings of the Second International Conference on Metallothionein and Other Low Molecular Weight Metal-Binding Proteins:* Kägi, J. H. R., Kojima, Y. (eds.) Basel: Birkhäuser Verlag; pp. 301–308.

Reilly, A., Reilly, C. (1973). *Plant Soil 38,* 671.

Rhoads, F. M., Olson, S. M., Manning, A. (1989). *J Environ Qual 18,* 195–197.

Riederer, M. (1991). *Naturwissenschaften 78,* 201–208.

Robb, D. A., Pierpoint, W. S. (1983). *Metals and Micronutrients: Uptake and Utilization by Plants,* London: Academic Press.

Roberts, T. M., Gizyn, W., Hutchinson, T. C. (1975), in: *Trace Substances in Environmental Health-VIII:* Hemphill, D. D. (ed.) Columbia: University of Missouri.

Rose, A. W., Hawkes, H. E., Johns, W. (1979). *Geochemistry in Mineral Exploration, 2nd edn.*, London: Academic Press.

Rosen, C. J., Allan, D. L., Luby, J. J. (1990). *J Am Soc Hortic Sci 115,* 83–89.

Roy, A. K., Sharma, A., Talukder, G. (1988). *Botanical Review 54,* 145–178.

Rozema, J., Bruin, J. De, Broekman, R. A. (1992). *The New Phytologist 121,* 249–256.

Runge, M., Rode, M. W. (1991), in: *Soil Acidity:* Ulrich, B., Summer, M. E. (eds.) Berlin: Springer-Verlag; pp. 183–202.

Russel, E. W. (1961). *Soil Conditions and Plant Growth,* London: Longmans.

Russo, F., Brennan, E. (1979). *For Sci 25,* 328–332.

Scharrer, K., Schropp, W. (1935). *Z Pflanzenernähr Düng Bodenkde 37,* 137–149.

Serna, M. D., Borras, R., Legaz, F., Primo-Millo, E. (1992). *Plant Soil 147,* 13–23.

Shewry, P. R., Peterson, P. J. (1974). *J Exp Bot 25,* 785–797.

Shewry, P. R., Peterson, P. J. (1976). *J Ecol 64,* 195–212.

Siegel, N., Haug, A. (1983). *Ann Rev Plant Physiol 59,* 285–291.

Siegert, E., Anke, M., Szentmihalyi, S., Regius, A., Lokyay, D., Pavel, J., Grun, M., Hora, K. (1986), in: *5. Spurenelement-Symposium:* Anke, M., Baumann, W., Braunlich, H., Bruckner, C., Groppel, B. (eds.) Leipzig: Karl-Marx-Universität; pp. 487–493.

Skeffington, R. A., Shewry, P. R., Peterson, P. J. (1976). *Planta (Berl) 132,* 209–214.

Sprent, J. I. (1988). *The Ecology of the Nitrogen Cycle. Cambridge Studies in Ecology,* Cambridge: Cambridge University Press.

Stewart, G. R., Ahmad, I. (1983), in: *Metals and Micronutrients, Uptake and Utilization by Plants:* Robb, D. A., Pierpoint, W. S. (eds.) London: Academic Press; pp. 33–50.

Stiborova, M., Dubravova, M., Brezinova, A., Friedrich, A. (1986a). *Photosynthetica 20,* 418–425.

Stiborova, M., Hromadkova, R., Leblova, S. (1986b). *Biologia (Bratislava) 41,* 1221–1228.

Streit, B., Stumm, W. (1993), in: *Plants as Biomonitors. Indicators for Heavy Metals in the Terrestrial Environment:* Markert, B. (ed.) Weinheim: VCH; pp. 31–62.

Streit, B., Hobbs, R. J., Streit, S. (1993), in: *Plants as Biomonitors. Indicators for Heavy Metals in the Terrestrial Environment:* Markert, B. (ed.) Weinheim: VCH; pp. 167–178.

Sutcliffe, J. F., Baker, D. A. (1981). *Plants and Mineral Salts, 2nd edn.,* London: Edward Arnold.

Sykes, R. L., Corning, D. R., Earl, N. J. (1981). *J Am Leather Chem Assoc (JALCA) 76,* 102–125.
Taekemoto, B. K., Noble, R. D. (1986). *New Phytol, 103,* 525–539.
Taylor, G. J. (1988): in: *Metal Ions in Biological Systems, Vol. 24, Aluminium and its role in Biology,* Sigel, H., Sigel, A., (eds.) Marcel Dekker: New York; pp. 123–163.
Taylor, G. J. (1991). *Current Topics in Plant Biochemistry and Physiology 10,* 57–93.
Thornton, I. (ed.) (1983). *Applied Environmental Geochemistry,* London: Academic Press.
Tinker, P. B. (1981), in: *Metabolic and Physiological Consequences of Trace Element Deficiency in Animals and Man:* Fowden, L., Gaston, G. A., Mills, C. F. (eds.) *Phil Trans R Soc Lond B294,* 41.
Tinker, P. B., Gildon, A. (1983), in: *Metals and Micronutrients, Uptake and Utilization by Plants:* Robb, D. A., Pierpoint, W. S. (eds.) London: Academic Press; pp. 21–32.
Trueby, P., Raba, A. (1991). *Angewandte Botanik 65 (3–4),* 253–264.
Valenti, V., Scalorbi, M., Gurrini, F. (1991). *Plant Physiol Biochem 29,* 249–255.
Vaselow, A. P. (1966), in: *Diagnostic Criteria for Plants and Soils:* Chapman, E. D. (ed.) University of California Riverside; pp. 302–309.
Vos, J., Biemond, H. (1992). *Annal Bot 70,* 27–35.
Wagner, G. J. (1984). *Plant Physiol 76,* 797–805.
Wagner, G. J., Nulty, E., LeFevre, M. (1984). *J Toxicol Environ Health 13,* 979–989.
Wainwright, S. J., Woolhouse, H. W. (1977). *J Exp Bot 28,* 1029–1036.
Walkenhorst, A., Hagemeyer, J., Breckle, W. (1993), in: *Plants as Biomonitors. Indicators for Heavy Metals in the Terrestrial Environment:* Markert, B. (ed.) Weinheim: VCH; pp. 524–540.
Wallace, A., North, C. P. (1953). *California Agriculture 7,* 10–12.
Warren, H. V., Delavault, R. E., Fletcher, K., Wilks, E. (1971), in: *Trace Substances in Environmental Health, Vol. IV:* Hemphill, D. D. (ed.) Columbia: University of Missouri; pp. 94–103.
Webb, J. S., Thornton, I., Howarth, R. J., Thompson, M., Lowenstein, P. L. (1978). *The Wolfson Geochemical Atlas of England and Wales,* Oxford: Clarendon Press.
Webber, J. (1980), in: *MAFF Reference Book 326, Inorganic Pollution and Agriculture:* London: HMSO; pp. 205–210.
Wild, H. (1974). *Kirkia 9,* 233–241.
Wild, H. (1978), in: *Biogeography and Ecology of Southern Africa:* Weger, M. J. A. (ed.) The Hague: Dr W. Junk Publishers; pp. 1301–32.
Wilson, J. R. (1988). *Advances in Nitrogen Cycling in Agricultural Ecosystems,* Wallingford: CAB International.
Wilson, S. B., Nicholas, D. J. D. (1967). *Phytochemistry 6,* 1057–1066.
Wong, M. H., Bradshaw, A. D. (1982). *New Phytol 91,* 255–261
Xu, J., Thornton, I. (1985). *Environ Geochem Health 7,* 131–133.
Ylaranta, T., Sillanpaa, M. (1984). *Ann Agric Fenn 23,* 158–170.
Zocchi, G., Cocucci, S. (1990). *Plant Physiol 92,* 908–911.
Ziegler, H. (1988), in: *Schadwirkungen auf Pflanzen,* 2nd edn., Hock, B., Elstener, E. (eds.) Mannheim: BI Wissenschaftsverlag.
Zieve, R., Peterson, J. (1984). *Sci Tot Environ 32,* 197–202.

3 Metal-Tolerant Plants: An Evolutionary Perspective

Mark R. Macnair[1] *and Alan J. M. Baker*[2]

[1]Department of Biological Sciences, University of Exeter, Hatherly Laboratories
Prince of Wales Road, Exeter, EX4 4PS, UK
[2]Department of Animal and Plant Sciences, University of Sheffield
PO Box 601, Sheffield, S10 2UQ, UK

Contents

Introduction

Most metals are phytotoxic when present in soils to excess, even those that are essential plant micronutrients. There are various places in the world where, either naturally or through anthropogenic pollution, the soils contain elevated concentrations of metals that will kill most normal plants. These places are rarely devoid of plant life, however, even though the range of species present on the contaminated site will frequently be much less than in uncontaminated but otherwise similar sites in the immediate vicinity. When those species are investigated it is normally found that the populations inhabiting these sites have evolved tolerance to the metals present in the soils, and that populations of the same species from uncontaminated areas show normal levels of susceptibility to the metal toxicity. Since many of the sites where tolerant plants have been found have been contaminated in the very recent past (sometimes as little as 30 years before), this indicates very rapid evolution of an adaptation. This phenomenon has proved to be one of the clearest examples of 'microevolution', and has excited much interest and research on a number of fronts. There have been several reviews on this topic in recent years (Baker, 1987; Macnair, 1987; Shaw, 1990). In this chapter we will concentrate on these topics in which research is most active, and where we believe that further research effort is most needed.

The Phenomenon of Tolerance

Detection and Measurement of Tolerance

Metal tolerance can be defined as the ability of a plant to survive on soils that are toxic or inimical to most other plants because the soils contain elevated levels of one or more metals. This definition of tolerance requires us to be able to show (a) that the soil is toxic and (b) that a particular genotype or species can grow better on this soil than others. The toxicity of the soil can be inferred from the reduction in species diversity, or from direct tests in which plants are grown on the soil and their survivorship assessed; however the causes of the toxicity (metals or other possible causes) cannot so easily be determined. Again, tolerance of the soil may be suggested by the survivorship of plants on the soils in which other plants perish (Bradshaw, 1952; Watkins and Macnair, 1991); whether this is tolerance of the metals present in the soils or other factors cannot be assumed without more detailed experiments.

It is for these reasons that much of the experimental work on metal tolerance has relied on laboratory tests for tolerance that seek to isolate the toxic effects of the metal from other possible factors. The most widely used technique is the root elongation test, in which the toxic effects of metals on roots are used to assess tolerance. Cuttings or tillers (for grasses) are grown in solutions containing varying concentrations of the metal, and tolerance measured by the relative root growth in the metal solution and a control containing none of the metal. There are various modifications of the test in which the control is missing (Macnair, 1983); multiple concentrations are used (Craig, 1977; Nicholls and McNeilly, 1979); or the same individual is used sequentially (Wilkins, 1978). The advantages and problems of this test and its variants have been widely discussed (Wilkins, 1978; Baker, 1987; Baker and Walker, 1989; Macnair, 1990). The test works well in making within-spieces comparisons where the two (or more) individuals or populations are from very different environments and thus can be expected to possess large phenotypic differences in tolerance (Fig. 3-1); it is less easy to interpret when comparing individuals or populations which differ rather little in phenotype, or in making inter-species comparisons. It is also easy to use for grasses and herbs; trees and woody species are much more difficult to experiment on in this way (but see, eg, Brown and Wilkins, 1985), and therefore the literature suffers from a serious bias in the species that have been shown to be tolerant.

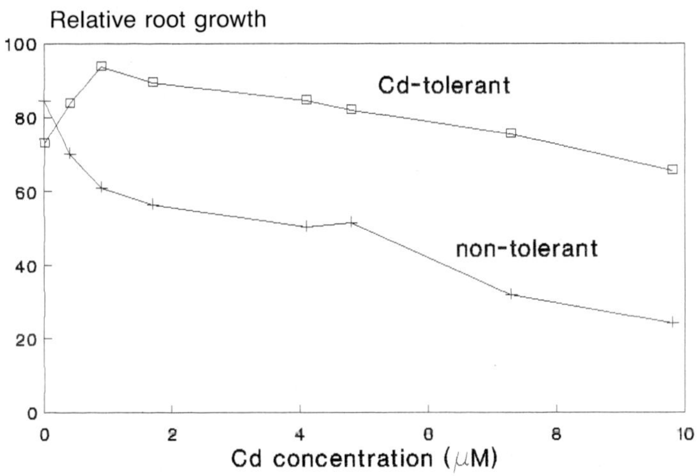

Fig. 3-1. The effect of increasing cadmium concentrations on sequential indices of cadmium tolerance for a non-tolerant (□) and a cadmium-tolerant (+) clone of *Holcus lanatus*, (from Baker and Walker, 1989).

The Genetics of Tolerance

Fundamental to a proper understanding of the evolution and mechanisms of tolerance is a knowledge of the genetic bases for tolerance. Unfortunately there has still been too little detailed work in this area. In particular, two questions are of great interest. (1) How many genes are involved in tolerance? Is there only a single gene for tolerance, or are there many (ie, polygenic inheritance)? As we shall see, the distinction between these two genetic scenarios is important for both physiological and evolutionary models of tolerance. (2) Is tolerance recessive? Again, the difference between recessive and dominant genes is similarly important. The early literature asserted that tolerance was polygenic, largely on the basis that all degrees of metal tolerance were detectable, rather than it being an all-or-nothing property of tolerant plants. Unfortunately, this conclusion could not follow from the experiments conducted (Macnair, 1990). In order to distinguish clearly between a polygenic and a major gene model requires extensive breeding and progeny testing. There have been only a limited number of studies on the scale required. Macnair (1983) was the first to demonstrate a single gene for tolerance in a higher plant, though single gene inheritance had been found in microorganisms previously (Seno, 1962; Antoine, 1965). Watkins and Macnair (1991) have found evidence of a major gene for arsenic tolerance in *Agrostis capillaris*, and Macnair *et al.* (1992) have shown that there is probably only one major gene responsible for arsenic tolerance in *Holcus lanatus*. Schat and ten Bookum (1992 a) have found a major gene for copper tolerance in *Silene vulgaris*. In all cases, evidence for additional modifying genes has been found, illustrating the problems of identifying a single gene in species where inadequate progeny testing has been conducted.

Tolerance has normally been demonstrated to be dominant, either completely or partially. This is the situation expected from evolutionary theory (see below). There has been some confusion in the literature about whether tolerance is always dominant, or whether dominance can be variable. The first point to make is that dominance is not a property of genes, but is highly environment-dependent: the environment includes both the physical environment and the genomic environment. We would thus expect dominance to vary both with the testing environment of the plants, and with their genetic constitution. Figure 3-2 illustrates the possible effects of the testing

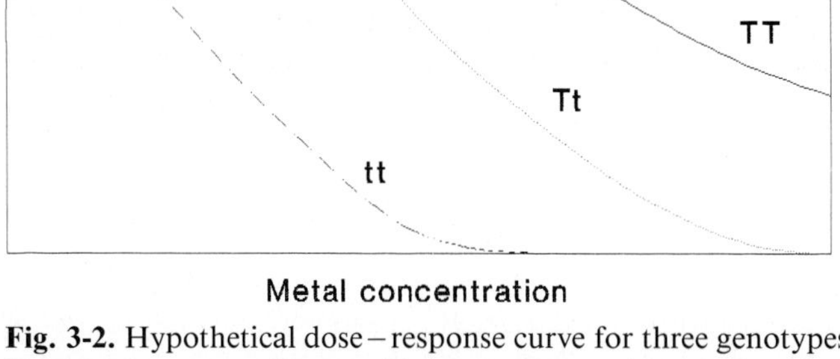

Fig. 3-2. Hypothetical dose – response curve for three genotypes: TT homozygous tolerants; tt homozygous non-tolerants; Tt heterozygotes. Depending on the testing concentration, the tolerance will appear to be either dominant (a), show no dominance (b), or be recessive (c).

environment. If the heterozygote shows a dose–response curve intermediate between the two homozygotes, then it is inevitable that tolerance will appear to be dominant at low concentrations, but recessive at higher levels (Allen and Sheppard, 1971). What is important, however, is the phenotype of the heterozygote at physiologically-important concentrations, and this is never known. In *Mimulus guttatus*, it has been possible to produce more-or-less isogenic lines which differ solely in the copper tolerance locus. When the dose–response curve of these lines is studied, no differences in the behaviour of homozygote and heterozygote tolerance are detected (Strange and Macnair, 1991). This implies that the differences in dominance observed in non-isogenic lines by Allen and Sheppard (1971) in this species may be due to the effects of genetic modifiers of tolerance, rather than tolerance genes *per se*.

The Specificity of Tolerance

Early work suggested that tolerance was very metal-specific, with, say, copper-tolerant plants not being tolerant to zinc (Gregory and Bradshaw, 1965; Antonovics *et al.*, 1971). Recently this paradigm has been challenged, with the suggestion that tolerance to one metal may frequently be associated with tolerance, albeit at a much lower level, to other metals (Baker, 1987; Baker and Walker, 1989). For instance, Schat and ten Bookum (1992b) studied the tolerances of a number of populations of *Silene vulgaris* to a variety of metals, and related these to the metals actually found at the sites (Fig. 3-3). They found a number of instances where a high level of tolerance to a particular metal could not be explained by elevated soil levels of that metal. Thus the Imsbach population has a very low level of zinc in the soil, but a very high zinc tolerance; all the mine populations appear to show elevated levels of cadmium tolerance, even though most of the mines do not contain detectable levels of this metal; all but the Marsberg population show an increase in cobalt and nickel tolerance. They suggest that selection for copper tolerance also increases tolerance to zinc, cadmium, cobalt and nickel, while selection for zinc, cadmium and lead tolerance also leads to tolerance to cobalt and nickel. However, it is important to draw a distinction here between cotolerance, or cross-tolerance, in which the gene(s) for tolerance to one metal also confer resistance to another, and multiple tolerance, in which the genes for different metals just happen to be found together, because both metals were present in the environment in which tolerance was selected. To a geneticist, this is the distinction between pleiotropy and linkage disequilibrium. It is not possible to distinguish between the two by a phenotypic analysis. Thus in the example discussed above of *S. vulgaris*, it is not possible from Fig. 3-3 to exclude common evolutionary history (ie, there may have been some migration between mines, perhaps mediated by human migration) or heterogeneity of the mine environments, as possible causes of the phenotypic correlations between metal tolerances. It is essential to conduct a proper breeding experiment in which the degree of genetic correla-

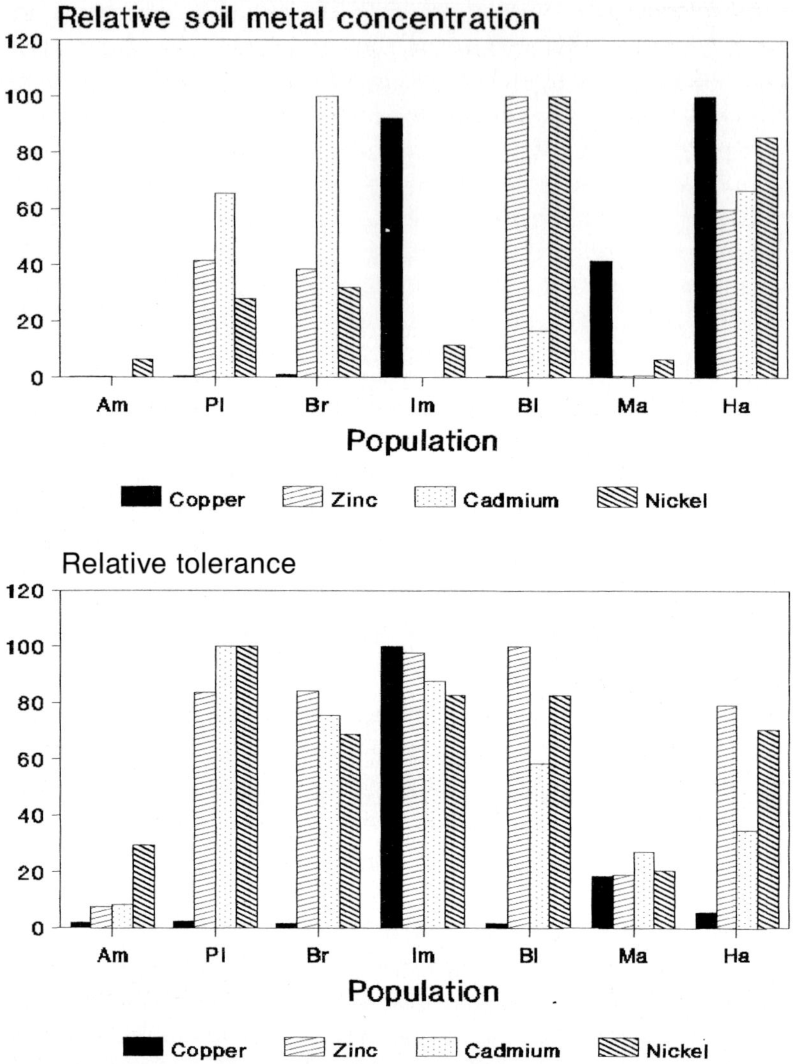

Fig. 3-3. Relative available metal ion concentrations (top) and relative metal tolerances to the ions (bottom) for six populations of *Silene vulgaris*. In both graphs, for each metal the population with the highest tolerance or soil metal concentration has been given the value of 100; all the other populations have been expressed relative to this. Key to populations: Am: Amsterdam (normal, control); Pl: Plombières; Br: Breinig: Im: Imsbach; Bl: Blankenrode; Ma: Marsberg; Ha: Harlingerode. Drawn from data in Schat and ten Bookum (1992b).

tion between the characters can be assessed. This is not easy or quick, and it is not surprising that it has rarely been attempted. In *Agrostis capillaris*, Humphreys and Nicholls (1984) found no evidence of a genetic correlation between copper and lead tolerance. In *M. guttatus*, it is clear that the copper tolerance gene does not automatically confer zinc tolerance, since in the isogenic lines copper-tolerant plants are no more zinc tolerant than non-tolerant plants (Strange, 1988; Macnair, unpublished work). However, in *S. vulgaris* preliminary genetic experiments suggest that there may be a genetic correlation between zinc tolerance and some of the genes responsible for copper tolerance (Schat *et al.*, quoted in Schat and ten Bookum, 1992b).

Inducible Tolerance

It has always been asssumed, if not demonstrated unequivocally, that metal tolerance is an entirely genetically-based phenomenon, unaffected by environmental influences (Baker and Walker, 1989). Carefully controlled pretreatment and cultivation experiments with clonal plant materials have shown that a certain degree of metal tolerance can be induced by metal pretreatment and subsequently 'lost' when metal influences are removed. Thus Baker *et al.* (1986) were able to demonstrate the tolerances of both cadmium-tolerant and non-tolerant populations of the grasses *Holcus lanatus, Agrostis capillaris, Festuca rubra* and *Deschampsia caespitosa* were reduced by an overall 13% when clonal materials were cultivated in an uncontaminated potting compost instead of native soils. Tolerance could also be induced in non-tolerant *H. lanatus* by transplantation of tillers into metalliferous soil. Subsequent experiments by Walker and Baker (1991) with seedlings of this grass have shown that zinc tolerance is similarly inducible, but like cadmium tolerance, much influenced by the nutrient status of the plant and growth medium. Outridge and Hutchinson (1991) have shown similar effects with the induction of cadmium tolerance in the clonal fern *Salvinia minima* where changes in tolerance of daughter ramets could be shown from prior acclimation to cadmium of parent plants. The evolutionary significance of inducible tolerances requires urgent investigation as the ubiquity of the phenomenon is not known. It could represent an important mechanism for survival of a pollution episode or establishment in marginally-contaminated habitats, such as those resulting from the deposition of airborne metal particulates (Baker *et al.*, 1990).

Mechanisms of Tolerance

Metals damage cells in a variety of ways, and there are intuitively many possible ways in which plant cells could protect themselves against this damage. Figure 3-4 shows some of these possibilities. The challenge facing us when we try to understand the physiology of tolerance is to work out which of these many possible mechanisms actually operate in any particular plant. It has become abundantly clear that there exists a 'syndrome of tolerance' (Baker, 1987), whereby several mechanisms may be developed to different extents in different species, so resulting in metal avoidance and tolerance (see, eg, Baker and Walker, 1990; Verkleij and Schat, 1990; Verkleij *et al.*,

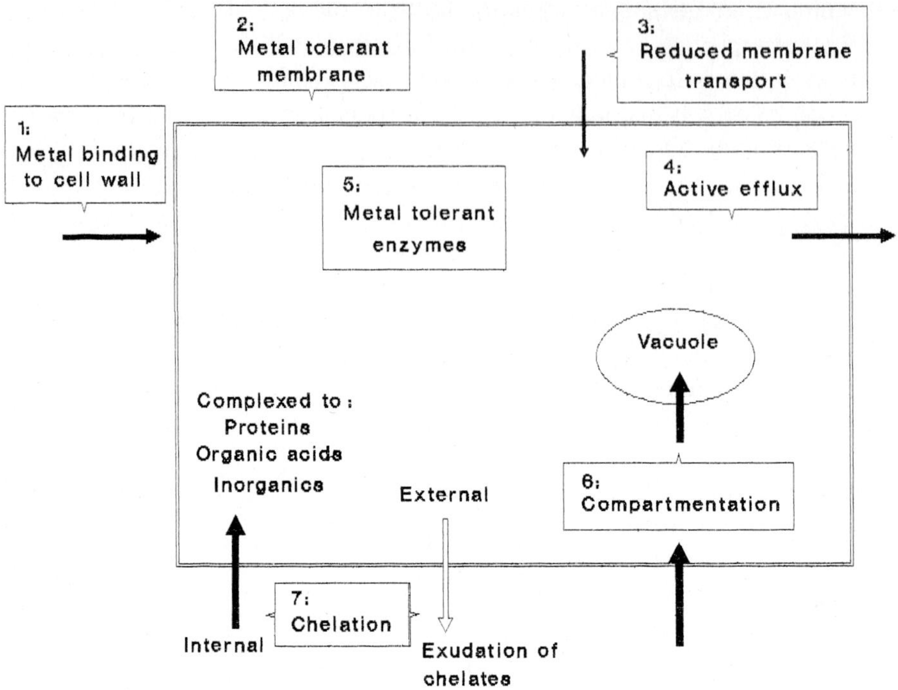

Fig. 3-4. A diagrammatic representation of seven possible mechanisms of metal tolerance in higher plants.

1991). However, it is important to distinguish between primary and secondary mechanisms of tolerance. Primary mechanisms are those mechanisms present in the tolerant plant, but not in the susceptible. These mechanisms allow the cell to survive the inimical concentrations of metal. There are probably other mechanisms that operate that any cell could produce if its metabolism was functioning properly. In tolerant plants these mechanisms operate, but in susceptible cells, because the cells are damaged, they do not. These mechanisms are of intrinsic interest to a physiologist trying to understand how a cell or plant copes with excess metal, but are not of themselves tolerance mechanisms. We call these secondary mechanisms. It is not easy to distinguish the two. The problem is that in investigations in which tolerant and non-tolerant plants are challenged with metal, and their physiology or biochemistry compared, in practice the investigator is frequently comparing live and dead cells. It is hardly surprising that many differences become apparent, but they are not necessarily primary tolerance mechanisms. The difference between primary and secondary mechanisms relates to the genetical difference between the tolerant and non-tolerants. If, for instance, there is only a single major gene differentiating the two, then there must be only a single primary mechanism, and the task facing the investigator is to identify the gene product and its function. On the other hand, if the difference between the two is polygenic, or governed by many major genes, then there must be many different physiological processes altered. The genetical evidence so far suggests that the former is more likely to be true (Macnair, 1990).

The dominance of the tolerance gene can also give some indication of the sorts of gene mutation that may be responsible. The majority of mutations lead to the loss of activity of the gene product. Since the concentration of a particular gene product is rarely the rate-limiting step in determining the phenotype, most of these loss-of-function mutations are recessive (Kacser and Burns, 1981). Dominant mutations are normally ones in which the gene product is altered, or is produced in much greater quantities. Mutations in the gene control mechanisms can also be dominant. The fact that metal tolerance is normally dominant therefore suggests that the mutations involved are not loss-of-function mutations, but rather we must look for differences in proteins, or differences in concentrations of proteins, between tolerants and non-tolerants.

To illustrate the integration of physiological and biochemical mechanisms of metal tolerance, we have selected tolerance to copper as an example. There has been a great deal of work on the mechanisms of copper tolerance in plants, but as yet we are still very far from resolving the problem. An early paper by Wainwright and Woolhouse (1977) pointed out that copper appeared to damage the cell membranes of non-tolerant plants, and produced evidence that cell-wall bound acid-phosphatases of tolerant and non-tolerant plants might differ in their sensitivity to copper. In 1980, Rauser and Curvetto (1980) published the first paper suggesting that copper-tolerant plants of *A. gigantea* might possess a metallothionein which was involved in tolerance. Due to the known importance of metallothioneins in producing metal tolerance in animals, this discovery opened up considerable possibilities in plants, and substantial research effort was devoted to the study of these interesting compounds. A decade later, we know a great deal about their biochemistry (see Rauser, 1990; Robinson, 1990; Steffens, 1990 for reviews), but their role in metal tolerance is increasingly uncertain. In summary, much of the copper within cells grown in high concentrations of copper is found bound to peptides, frequently called phytochelatins (Grill *et al.*, 1987). These peptides are synthesized by an enzyme, γ-glutamylcysteine dipeptidyl transpeptidase, which is produced constitutively in plant cells (Loeffler *et al.*, 1989). It is activated by the addition of heavy metals. However, these compounds are found ubiquitously in higher plants (Gekeler *et al.*, 1989), and no difference in the activity of this enzyme or in the rate of synthesis of these compounds between tolerant and non-tolerant plants has been discovered (Robinson and Thurman, 1986; Verkleij *et al.*, 1989). For these to be responsible for copper tolerance, some difference should have been detectable in genetically-defined clones of *M. guttatus*, known to differ in only a single gene for copper tolerance. No such difference has been detected (Lucas, 1981; Robinson and Thurman, 1986). It is possible that these compounds have a role in detoxifying copper in live cells: the primary tolerance mechanism enables the cells to survive, so that the synthesis of phytochelatins is possible; in non-tolerant cells, death occurs before significant synthesis has been possible. Recently, a peptide that appears to be a true metallothionein has been discovered in *M. guttatus* (de Miranda *et al.*, 1990); it remains to be seen whether the protein or its synthesis differs between tolerant and non-tolerant genotypes.

Recent work with both *Silene vulgaris* (de Vos *et al.*, 1989) and *M. guttatus* (Strange and Macnair, 1991) has focussed attention back onto the pioneering work of Wainwright and Woolhouse (1977). Both studies have found that the initial effect of copper is to damage the cell membrane, and that tolerant and non-tolerant genotypes differ markedly in the degree to which the membrane is damaged. In *M. guttatus*, the difference can be detected in a few minutes, and the leakiness of the membrane results in the non-tolerants showing a greater short-term uptake of copper by diffusion. This greater short-term uptake can be expected to lead to greater damage by the copper, and accelerated death of the cells. These results tend to suggest that the mechanism of tolerance resides in some way in the membrane, either by structural alteration, or by the production of some compound constitutively that can protect the membrane from copper. Further work is required to confirm and differentiate these hypotheses.

Evolution of Tolerance

Evolution *de novo*

Not all species have evolved races tolerant to metals. In species which have evolved one or more tolerant races, tolerance has often only been evolved to one or a few metals. What causes the differences between these species in their ability to evolve this adaptation?

There have been a number of attempts to relate the ability to evolve tolerance to taxonomy (Antonovics *et al.*, 1971). However, when the distribution of tolerance is considered on a world-wide basis, there appears to be no clear pattern that can be attributed to phylogeny (Baker, 1987). Certain plant families do emerge as being well represented in metallophyte floras: the Poaceae, Caryophyllaceae and Brassicaceae, include most of the temperate taxa which readily evolve tolerance. In the tropics a much greater range of families is represented including the Fabaceae and Lamiaceae, families which are rarely present on metal-contaminated soils in temperate regions. The reasons for this climatic contrast are not known. Similarly, the reasons why some common and large plant families, such as the Ranunculaceae, Rosaceae and Apiaceae have not apparently evolved tolerance presently remains obscure.

The ability to evolve tolerant races has been studied experimentally by Bradshaw and McNeilly and their students. Thus Walley *et al.* (1974) showed it was normally possible to select tolerant individuals of the grass *Agrostis capillaris* directly out of non-tolerant populations. This species, or other *Agrostis* species, is found ubiquitously on mines and contaminated sites in Europe. Gartside and McNeilly (1974) and Ingram (1988) extended this work, and studied a number of species, some of which were found on mines, and some of which were not. In general they found that it was possible to select tolerant individuals from non-tolerant races of species that were found on mines, but not from those that were not (Table 3-1). This work has led to the generalisation that for a species to be able to evolve tolerance, it must possess the appropriate genetic variance before the selection is imposed. Bradshaw (1984) has referred to the phenomenon by which species are constrained from evolving by the lack of genetic variability as *genostasis*.

Table 3-1. The percentage of copper-tolerant individuals found in populations of grass species collected from a range of non-cupriferous soils, in relation to the presence or absence of the species on copper-polluted wastes, and the copper tolerance of collected adult plants (data of Ingram, 1988, from Baker and Proctor, 1990).

Species	% Occurrence of tolerant individuals[a]	Species presence/absence		Tolerance of adult plants
		On mines	At margins	
Holcus lanatus	0.16	+	+	+
Agrostis capillaris	0.13	+	+	+
Festuca ovina	0.07	–	+	–
Dactylis glomerata	0.05	+	+	+
Deschampsia caespitosa	0.03	+	+	+
Anthoxanthum odoratum	0.02	–	+	–
Festuca rubra	0.01	+	+	+
Lolium perenne	0.005	–	+	–
Poa pratensis	0	–	+	–
Poa trivialis	0	–	+	–
Phleum pratense	0	–	+	–
Cynosurus cristatus	0	–	+	–
Alopecurus pratensis	0	–	+	–
Bromus spp.	0	–	+	–
Arrhenatherum elatius	0	–	+	–

[a] Seeds (20000) were screened by germination on solutions containing copper concentrations at least 0,1 mg/L greater than that estimated to prevent root growth in non-tolerant individuals. It thus varied from species to species tested, and was in the range 0.15–0.85 mg/L

There is another important determinant of which species do and can evolve tolerance and those that do not. Symeonidis *et al.* (1985) screened cultivars of *Agrostis capillaris* for metal tolerances and demonstrated differential sensitivities to metals, suggesting the existence of non-specific low level tolerances to metals other than those present at toxic levels in soils from which they were derived. Other screening experiments have implicated such constitutive properties in the ability to evolve full tolerances (Baker and Walker, 1989; Baker and Proctor, 1990; Baker *et al.*, 1990). Table 3-2 shows the results of a screening experiment in which non-tolerant populations of eleven common grasses were tested for their sensitivity to a sub-lethal cadmium dose. It emerged from the ranked tolerance indices that there were very great differences in response to cadmium at a species level, and that those species which possessed a strong intrinsic resistance to cadmium (constitutive tolerance), were those which frequently evolve full tolerance under appropriate selection. The species which proved very sensitive to cadmium are not known to evolve cadmium-tolerant populations. Constitutive tolerance, like inducible tolerance, can allow a species to survive in a polluted environment until selection can operate on the appropriate genetic variability, leading ultimately to the evolution of a fully tolerant population.

Table 3-2. Mean indices of cadmium tolerance (±SE) for seedlings of eleven grass species grown from seed collected from populations on soils with low cadmium concentrations, screened at 10 μM Cd in full-strength Rorison nutrient solution. The distribution of these species on cadmium-rich soils is also indicated. (Unpublished data of K. Ewart, from Baker and Proctor, 1990).

Species	Index of tolerance (%) ± SE	Species presence/absence on cadmium-rich soils
Deschampsia caespitosa	135 ± 40	+
Festuca rubra	101 ± 21	+
Agrostis capillaris	83 ± 20	+
Festuca ovina	82 ± 19	+
Arrhenatherum elatius	67 ± 25	+
Koeleria macrantha	65 ± 10	+
Holcus lanatus	65 ± 12	+
Brachypodium pinnatum	53 ± 17	−
Poa annua	47 ± 13	−
Bromus sterilis	26 ± 6	−
Lolium perenne	24 ± 6	−

Macnair (1991) has recently considered the effect of differences in genetic architecture on the ability to evolve tolerance over a relatively short period of time. He considered two contrasting genetic scenarios: polygenic inheritance, with many genes present in the initial population at intermediate frequencies, and major gene inheritance, with a major gene for tolerance present at low frequency in the base population. Both scenarios will produce populations that have moderate to high genetic variance for tolerance, so that selection will lead to an increase in mean tolerance of the population in the short term. However, if the variance is based on many genes at intermediate frequency, this genetic variance is rapidly exhausted, without the population having achieved a very significant shift in mean. With a major gene, a much larger shift in mean is possible. Macnair suggests that for many environmental challenges of anthropogenic origin, such as metal pollution, a large shift in mean is required, and thus argues that adaptation is only possible by major genes. This argument both explains the observed genetic bias towards major genes, and at the same time predicts considerable stochastic variability in the frequency of these tolerance genes in normal populations, so that only some species and some populations will have the major genes at low frequency from which a tolerant race could be selected.

Al-Hiyaly *et al.* (1989) have investigated this aspect of the evolution of metal tolerance. They studied the evolution of zinc tolerance in grass populations under electricity pylons on acid soils in North Wales. These pylons were galvanized with a zinc coating when originally erected, and over the years the zinc has washed off the pylons and killed most of the vegetation under them. This has created a situation where there are replicated islands of zinc-contaminated soil, which can presumably only be colonized from the normal populations surrounding each island. They found that in some of these islands zinc-tolerant populations of *A. capillaris* had evolved, but in others they had not. The most likely explanation of this is that tolerance only evolved where the surrounding population had genetic variance for tolerance, and for stochastic reasons only some of the local populations of *A. capillaris* had such variance.

Finally, it must be noted that the evolution of tolerance is a necessary condition for the evolution of a population able to colonize a mine, but it may not be sufficient. Mines differ from normal habitats in many ways apart from mere metal contamination: the soil structure is usually worse, and the organic content less, so that the soils dry out quickly; the soils are frequently very deficient in nitrogen and phosphorus and other essential elements; and wind and water erosion may mean that seedling establishment is very difficult (Baker and Proctor, 1990). The result is that plants have to be able to adapt to all these conditions as well as to the metal contamination: in practice, only those species which show at least some preadaptation to these harsh conditions are going to be able to evolve tolerant races. It may well be that it is this factor, rather than the evolution of tolerance *per se*, that is most important in determining which species are able to evolve tolerant races. Such arguments may be relevant when comparing the evolution of tolerance in mine environments and aerially-contaminated sites such as those around smelters, where soil conditions and selective forces are markedly different (Baker, 1987).

The Costs of Tolerance

There is considerable circumstantial evidence that the evolution of tolerance involves a cost, in the sense that plants adapted to high metal concentrations are less well adapted to living in an uncontaminated environment than their non-tolerant counterparts. The evidence comes from a variety of sources. Firstly, there is the observation (see above) that tolerance is normally maintained at a very low frequency in populations from uncontaminated areas. At the edge of a mine or contaminated site very steep clines for tolerance are frequently observed, with, in extreme cases, the mean phenotype of the population switching from full tolerance to complete non-tolerance over the distance of a few metres (Hickey and McNeilly, 1975). These observations suggest a disadvantage of tolerance in normal soils.

Secondly, experiments comparing plants from different environments frequently find that mine plants are inferior to plants of the same species from an uncontaminated grassland. Thus copper-tolerant plants of *Silene vulgaris* grow slower than plants from an uncontaminated site (Ernst *et al.*, 1990), while Wilson (1988) found differences in relative growth rate between a number of tolerant races of *A. capillaris* that were correlated with tolerance. Morishima and Oka (1977) demonstrated a negative correlation between growth rate and copper tolerance in barnyard grass (*Echinochloa* sp.). Hickey and McNeilly (1975) compared mine and pasture populations of four species of plant in competition with ryegrass (*Lolium perenne*): again they found that the mine plants were competitively inferior.

Thirdly, there is evidence that tolerant plants may have a greater requirement for the metal to which they are tolerant in order to grow normally. There are many examples where tolerant plants show a stimulation of root growth in low concentrations of metal (Fig. 3-1), and some reports that tolerant plants (particularly zinc-tolerant plants) need to be given exogenous metal in order for them to be grown normally in potting compost (Antonovics *et al.*, 1971). Schat and ten Bookum (1992b) report that copper-tolerant plants from their Imsbach site, which are very highly copper-tolerant (Fig. 3-3), show symptoms of copper deficiency when grown in soil from a lead/zinc mine. However, while this is very good evidence that mine plants are competitively inferior to 'normal' plants, it is unfortunately not clear that these differences are necessarily caused by metal tolerance. The problem is the same as that discussed above concerning multiple tolerance. A cost implies a pleiotropic effect of metal tolerance: how can one distinguish a pleiotropic effect from linkage disequilibrium? In practice, as discussed above, mine plants will show many other adaptations to the stressful mine environment, including drought tolerance, low-nutrient tolerance, and possibly adaptations increasing the probability of persistence in this environment. It is not unlikely that many of these adaptations may also reduce growth rate or competitive ability, and it is not possible to disentangle the effects of metal tolerance from these other possible adaptations by the types of experiments and observations discussed above. There have been only three experiments in which some attempt was made to examine tolerance by itself: Nicholls and McNeilly (1985) compared high, medium and low tolerant plants selected from a non-tolerant population of *A. capillaris*, and Macnair and Watkins (1983) compared plants of *Mimulus guttatus* that had been bred so that they differed essentially only at the copper tolerance locus. Walker (1990) selected cadmium-tolerant individuals from a non-tolerant population of *Holcus lanatus* and then performed a competition experiment between adult plants derived from these selections and non-tolerants from the same population. In no case was a cost of tolerance manifested, but the scale of the experiments was such that the cost would have had to have been enormous to have been demonstrated.

Speciation

One of the features of areas of the world that have had naturally occurring toxic soils for many thousands or millions of years is that these soils often support species that are found nowhere else: so-called edaphic endemics (Kruckeberg and Rabinowitz, 1985). Examples of such soils include the Great Dyke in Zimbabwe and serpentine soils in many areas of the world. It has been argued on many occasions (eg, Proctor and Woodell, 1975; Kruckeberg, 1984) that the phenomenon of metal tolerance provides a model of how these species may have started on the evolutionary route that finally led to full edaphic endemism.

The evolution of the tolerant ecotype is conceptually easy. As described above, first tolerance evolves, and then further adaptations to the other features of the mine environment can be expected to spread. Other genes that enhance the tolerance produced by the initial gene(s) (modifiers) increase the adaptation (Macnair, 1983; Schat and ten Bookum, 1992a). The problem is, how does an ecotype evolve into a full species? Kruckeberg (1984) points out that on the California serpentines there are about 250 endemic taxa, but a further 1000 species that are found both on and off the serpentine (bodenvag species), most of which can be expected to have evolved serpentine-tolerant taxa. What is the difference between a bodenvag species and a species that has evolved into an edaphic endemic?

One possibility is that having evolved the tolerant ecotype, the progenitor population becomes locally extinct, leaving the tolerant population allopatrically isolated from its congeners. The ecotype can now evolve into a new species by the classic allopatric model. This model would predict that the process would take a very long time, and that we would certainly not expect to see any evidence of speciation taking place on mines that have only been available for colonization during historical times. The model may well have validity, especially where edaphic species appear to have no close relatives off the contaminated area, such as with some of the species on the Great Dyke in Zimbabwe, but is more difficult to apply where derivative and progenitor species coexist. It is also reasonable to ask whether any other models are possible.

Two processes are involved in speciation. Firstly, the two populations must evolve sufficient ecological differences that they do not suffer from competitive exclusion. In the case of metal-tolerant ecotypes, this condition has been fulfilled. Secondly, they must evolve sufficient barriers to gene exchange that the two gene pools are effectively isolated, and will evolve independently. There are a number of ways in which this could occur, and the first stages of this have been observed in a number of studies of metal-tolerant plants.

Metal-tolerant populations often flower earlier than local non-tolerant populations (McNeilly and Antonovics, 1968; Macnair *et al.*, 1989 and unpublished work). Differences in flowering time can be a very potent isolating barrier in flowering plants, particularly those with a relatively short flowering season. A number of possible explanations for the proximate cause of the earlier flowering have been advanced. It could be as a result of selection for reduced outcrossing between the two ecotypes, in other words direct selection for isolation: this process, sometimes called the Wallace effect (Grant, 1966) is controversial (Butlin, 1987; Coyne and Orr, 1989), and the selective forces are weak. Alternative explanations are that early flowering has been selected as an adaptation to drought (McNeilly and Antonovics, 1968; Macnair *et al.*, 1989), or that a purely environmentally-induced difference in flowering time can lead to genetic differentiation because of the differential gene flow between populations (Stam, 1983).

Macnair and Christie (1983; Christie and Macnair, 1987) have shown that in *Mimulus guttatus* the copper tolerance gene is associated with a gene giving postmating reproductive isolation. When crossed to particular non-tolerant genotypes, all tolerant progeny are inviable. It is not clear whether this effect is a pleiotropic effect of the tolerance gene, or caused by another gene in tight linkage with it. Christie and Macnair (1987) showed that these sorts of genes were commonly polymorphic in this species. The importance of this result is that it gives a model of how these postmating isolating genes could be selected. If the tolerance gene (or any of the other genes involved in differentiating the ecotype) happens by chance to be linked to such a gene the isolating gene can spread by 'hitch-hiking'. The probability of this is quite high if these genes are common and if many genes giving a variety of different adaptations spread during the evolution of an ecotype.

A very common evolutionary route in plant speciation is the evolution of self-fertilizing species from outcrossing progenitors (Stebbins, 1970; Wyatt, 1988). An increase in self-fertility in mine ecotypes was demonstrated for both *A. capillaris* and *Anthoxanthum odoratum* by McNeilly and Antonovics (1968), and for *Armeria maritima* by Lefèbvre (1970), though there is no evidence that mine plants in the wild self-fertilize more than normal plants of these species. There are a variety of selective forces leading to the evolution of self-fertility (Jain, 1976), but in this context the most likely cause is the need for reproductive assurance in a marginal habitat.

The linking theme in each of the above potential partial isolating barriers is that they all could evolve as direct adaptations to the prevailing environment. Early flowering can be an adaptation to drought, and self-fertility for reproductive assurance. Postmating barriers could evolve through these genes 'hitch-hiking' with any adaptive gene. Macnair (1989) suggested that the difference between an ecotype and an edaphic endemic was that in the latter adaptations to the prevailing conditions included some that also acted as reproductive isolating barriers, while in the former they did not.

Conclusion

Heavy metal tolerance remains one of the clearest examples of microevolution, and one of the best systems to study the relationship between adaptation and ecology. It provides a model for the evolution of ecotypes and edaphic endemics. At the physiological and biochemical level, it is a model system for the study of the mechanisms of resistance to stress and pollution. Many questions remain unanswered, however, among the principal ones of which are:

(a) Why do some species evolve tolerance and others do not? Why do some species lack the genetic variability to adapt: is it because of a lack of the genetic potential to mutate to metal-tolerant forms, or the lack of ability to adapt to the other edaphic factors of the contaminated soil?
(b) What are the physiological, biochemical and molecular biological bases of tolerance? We still have a very imperfect understanding of these mechanisms.
(c) What are the 'down sides' to tolerance? Why is tolerance apparently disadvantageous in uncontaminated soils?

We can expect research into these and other questions to continue to provide further insights into this phenomenon in the next few years.

References

Al-Hiyaly, S. A., McNeilly, T., Bradshaw, A. D. (1989), *New Phytol. 110*, 571–580.
Allen, W. R., Sheppard, P. M. (1971), *Proc. Roy. Soc. Lond. Ser. B. 177*, 177–196.
Antoine, A. (1965), *Expl. Cell. Res. 40*, 570–584.
Antonovics, J., Bradshaw, A. D., Turner, R. G. (1971), *Adv. Ecol. Res. 7*, 1–85.
Baker, A. J. M. (1987), *New Phytol. 106*, 93–111.
Baker, A. J. M., Proctor, J. (1990), *Pl. Syst. Evol. 173*, 91–108.
Baker, A. J. M., Walker, P. L. (1989), *Chem. Spec. Bioavail. 1*, 7–17.
Baker, A. J. M., Walker, P. L. (1990), in: *Heavy Metal Tolerance in Plants: Evolutionary Aspects:* Shaw, A. J. (ed.) Boca Raton, Florida: CRC Press Inc.; pp. 155–177.
Baker, A. J. M., Grant, C. J., Martin, M. H., Shaw, S. C., Whitebrook, J. (1986), *New Phytol. 102*, 575–587.
Baker, A. J. M., Ewart, K., Hendry, G. A. F., Thorpe, P. C., Walker, P. L. (1990), *4th International Conference on Environmental Contamination, Barcelona, October 1990, Proceedings:* Barceló, J. (ed.) Edinburgh: CEP Consultants Ltd.; pp. 23–29.
Bradshaw, A. D. (1952), *Nature 169*, 1098.
Bradshaw, A. D. (1984) in: *Origins and Development of Adaptation, CIBA Foundation Symposium No. 102:* Evered, D., Collins, G. M. (eds.), London: Pitman; pp. 4–19.
Brown, M. T., Wilkins, D. A. (1985), *New Phytol. 99*, 101–106.
Butlin, R. (1987) *Trends Ecol. Evol. 2,* 8–13.
Christie P., Macnair, M. R. (1987), *Evolution 41*, 571–578.
Coyne, J. A., Orr, H. A. (1989), *Evolution 43*, 362–381.
Craig, C. C. (1977), *Trans. Rhod. Sci. Assoc. 58*, 9–16.

de Miranda, J. R., Thomas, M. A., Thurman, D. A., Tomsett, A. B. (1990), *FEBS Letts. 260*, 277–280.
de Vos, C. H. R., Vooijs, R., Schat, H., Ernst, W. H. O. (1989), *J. Plant Physiol. 135*, 165–169.
Ernst, W. H. O., Schat, H., Verkleij, J. A. C. (1990), *Evol. Trends Plants 4*, 45–51.
Gartside, D. W., McNeilly, T. (1974), *Heredity 32*, 335–348.
Gekeler, W., Grill, E., Winnacker, E.-L., Zenk, M. H. (1989), *Z. Naturforsch. 44c*, 361–369.
Grant, V. (1966), *Amer. Nat. 100*, 99–118.
Gregory, R. P. G., Bradshaw, A. D. (1965), *New Phytol. 64*, 131–143.
Grill, E., Winnacker, E.-L., Zenk, M. H. (1987), *Proc. Natl. Acad. Sci. 84*, 439–443.
Hickey, D. A., McNeilly, T. (1975), *Evolution 29*, 458–464.
Humphreys, M. O., Nicholls, M. K. (1984), *New Phytol. 95*, 177–190.
Ingram, C. (1988), PhD Thesis, University of Liverpool.
Jain, S. K. (1976), *Annu. Rev. Ecol. Syst. 7*, 469–495.
Kacser, H., Burns, J. A. (1981), *Genetics 97*, 639–666.
Kruckeberg, A. R. (1984), *California Serpentines: Flora, Vegetation, Geology, Soils, and Management Problems*, Davis: University of California Publications in Botany 78.
Kruckeberg, A. R., Rabinovitz, D. (1985), *Ann. Rev. Ecol. Syst. 16*, 447–479.
Lefèbvre, C. (1970), *Evolution 24*, 571–577.
Loeffler, S., Hochberger, A., Grill, E., Winnacker, E.-L., Zenk, M. H. (1989), *FEBS Letts. 258*, 42–46.
Lucas, P. H. (1981), BSc Thesis, University of Exeter.
Macnair, M. R. (1983) *Heredity 50*, 283–293.
Macnair, M. R. (1987), *Trends in Ecology and Evolution 2*, 354–359.
Macnair, M. R. (1989), *Bot. J. Linn. Soc. 100*, 1–14.
Macnair, M. R. (1990), in: *Heavy Metal Tolerance in Plants: Evolutionary Aspects:* Shaw, A. J. (ed.) Boca Raton, Florida: CRC Press Inc.; pp. 235–253.
Macnair, M. R. (1991), *Genetica 84*, 213–219.
Macnair, M. R., Christie, P. (1983), *Heredity 50*, 295–295.
Macnair, M. R., Watkins, A. D. (1983), *New Phytol. 95,* 133–137.
Macnair, M. R., Macnair, V. E., Martin, B. M. (1989), *New Phytol. 112*, 269–279.
Macnair, M. R., Cumbes, Q. J., Meharg, A. A. (1992), *Heredity 69*, 325–335.
McNeilly, T., Antonovics, J. (1968), *Heredity 23*, 205–218.
Morishima, H., Oka, H. I. (1977), *Jap. J. Genetics 52*, 357–372.
Nicholls, M. K., McNeilly, T. (1979), *New Phytol. 83*, 653–664.
Nicholls, M. K., McNeilly, T. (1985), *New Phytol. 101*, 202–217.
Outridge, P. M., Hutchinson, T. C. (1991), *New Phytol. 117*, 597–605.
Proctor, J., Woodell, S. R. J. (1975), *Adv. Ecol. Res. 9*, 255–366.
Rauser, W. E. (1990), *Ann. Rev. Biochem. 59*, 61–86.
Rauser, W. E., Curvetto, N. R. (1980), *Nature 287*, 563–564.
Robinson, N. J. (1990) in: *Heavy Metal Tolerance in Plants: Evolutionary Aspects:* Shaw, A. J. (ed.) Boca Raton, Florida: CRC Press Inc.; pp. 195–214
Robinson, N. J., Thurman, D. A. (1986), *Planta 169*, 192–197.
Schat, H., ten Bookum, W. M. (1992a), *Heredity 68*, 219–229.
Schat, H., ten Bookum, W. M. (1992b), in: *The Vegeation of Ultramafic (Serpentina) Soils:* Baker, A. J. M., Proctor, J., Reeves, R. D. (eds.) Andover: Intercept; pp. 337–352.
Seno, T. (1962), *Jap. J. Genet. 37*, 207–217.
Shaw, A. J. (ed.) (1990), *Heavy Metal Tolerance in Plants: Evolutionary Aspects* Boca Raton, Florida: CRC Press Inc. 350 pp.
Stam, P. (1983), *Heredity 50*, 105–118.
Stebbins, G. L. (1970), *Ann. Rev. Ecol. Syst. 1*, 307–326.

Steffens, J. C. (1990) *Ann. Rev. Pl. Physiol., Pl. Mol. Biol. 41*, 555–575.

Strange, J. (1988), PhD Thesis, University of Exeter.

Strange, J., Macnair, M. R. (1991), *New Phytol. 119,* 383–388.

Symeonidis, L., McNeilly, T., Bradshaw, A. D. (1985), *New Phytol. 101*, 309–315.

Verkleij, J. A. C., Schat, H. (1990), in: *Heavy Metal Tolerance in Plants: Evolutionary Aspects:* Shaw, A. J. (ed.) Boca Raton, Florida: CRC Press Inc.; pp. 179–193.

Verkleij, J. A. C., Koevoets, P., van't Reit, J., van Rossenberg, M. C., Bank, R., Ernst, W. H. O. (1989), in: *Metal Homeostasis: Molecular Biology and Biochemistry:* Winge, D., Hamer, D. (eds.) New York: Alan Liss; pp. 255–260.

Verkleij, J. A. C., Lolkema, P. C., De Neeling, A. L., Harmens, H. (1991), in: *Ecological Responses to Environmental Stresses:* Rozema, J., Verkleij, J. A. C. (eds.) Netherlands: Kluwer Academic Publishers; pp. 8–19.

Wainwright, S. J., Woolhouse, H. W. (1977), *J. Exp. Bot. 25*, 1025–1036.

Walker, P. L. (1990), PhD Thesis, University of Sheffield.

Walker, P. L., Baker, A. J. M. (1991), *8th International Conference on Heavy Metals in the Environment, September 1991, Edinburgh, Proceedings:* Farmer, J. G. (ed.) Edinburgh: CEP Consultants Ltd.; Vol. 1, pp. 98–102.

Walley, K. A., Khan, M. S. I., Bradshaw, A. D. (1974), *Heredity 32*, 309–319.

Watkins, A. J., Macnair, M. R (1991), *Heredity 66*, 47–54.

Wilkins, D. A. (1978), *New Phytol. 80*, 623–633.

Wilson, J. B. (1988), *Evolution 42*, 408–413.

Wyatt, R. (1988), in: *Plant Evolutionary Biology:* Gottlieb, L. D., Jain, S. K. (eds.) London: Chapman and Hall; pp. 109–132.

4 Plants that Hyperaccumulate Heavy Metals

Robert R. Brooks

Department of Soil Sciences, Massey University
Palmerston North, New Zealand

Contents

Introduction

The concentration of any element in dried plant material will usually reflect to some extent the content of this same element in the underlying substrate. This assumption indeed is the basis for a technique known as *biogeochemical prospecting* wherein plant material is analysed in order to give an indication of mineralisation of the analyte element in the substrate (Malyuga, 1964; Brooks, 1983; Kovalevsky, 1987).

In this chapter, the elements Co, Cu, Ni, and Zn will be covered. Before discussing the topic of *hyperaccumulation,* it will be appropriate to determine the normal abundances of the above elements in rocks, soils, and vegetation of the earth's crust (Table 4-1). From the table, it will be noted that except for Zn, the concentration of any element in the substrate is not exceeded in the plant material growing over it, at least as far as 'normal' vegetation is concerned.

Table 4-1. Mean elemental concentrations in (μg g^{-1}) in rocks, soils and vegetation (dry weight). After Brooks (1987).

	Co	Cr	Cu	Pb	Mn	Ni	Zn
Earth's crust	25	100	55	13	950	75	70
Granite	3	20	13	48	195	1	45
Basalt	47	114	110	8	1280	76	86
Ultramafic rocks	150	1600	10	1	1620	2000	50
Soils (non-ultramafic)	10	60	20	10	850	40	50
Soils (ultramafic)	250	2500	20	10	1000	2500	40
Vegetation (non-ultramafic)	1	1	10	5	80	2	100
Vegetation (ultramafic)[a]	10	10	10	5	100	80	100

[a] Not including hyperaccumulators

In an early study, Baumann (1885) reported about 1% zinc in dried leaves of *Thlaspi calaminare* growing over the well-known zinc deposits near Aachen, Germany. It was not until over 60 years later that the unusually high accumulation of another heavy metal was recorded, when Minguzzi and Vergnano (1948) reported over 1% nickel (10000 μg g^{-1}) in leaves of *Alyssum bertolonii* growing over ultramafic rocks in Tuscany. This find spurred further activity and resulted in the discovery of about 150 other *hyperaccumulators* of nickel (see below), of which one third were in the genus *Alyssum*. The term *hyperaccumulator* was first used by Brooks *et al.* (1977a) and was originally used to define plants containing more than 1000 μg g^{-1} (ppm) nickel in dry tissue. This threshold was not chosen arbitrarily but was based on histograms showing discrete plots of two different populations of even the same plant genus with no overlap of hyperaccumulators and non-accumulators at the 1000 μg g^{-1} Ni boundary. This is demonstrated in Fig. 4-1 (Brooks, 1987).

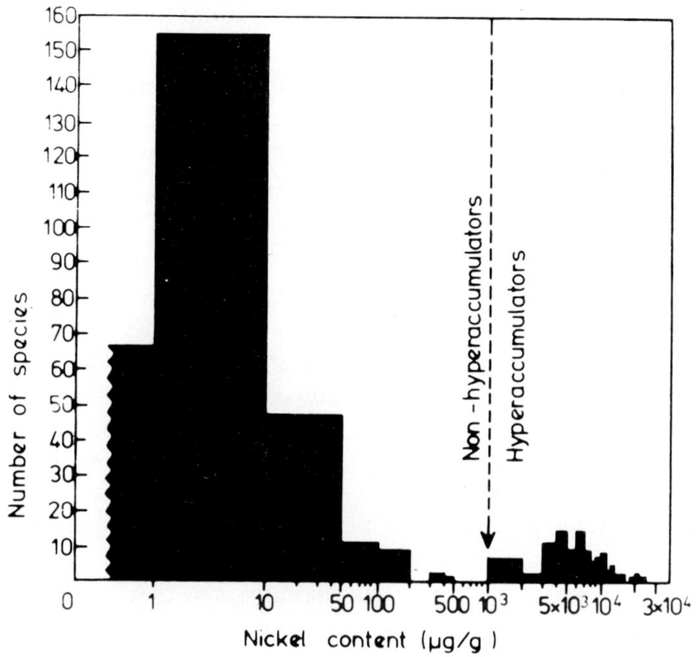

Fig. 4-1. Histogram showing nickel concentrations in the genus *Alyssum*. There are two distinct populations including hyperaccumulators with >1000 µg g^{-1} Ni in dried tissue.

Jaffré (1980) has modified the definition of hyperaccumulator by using the terms *hypermanganésiphores* and *hypernickelophores* to describe plants containing over 10000 µg g^{-1} (1%) Mn or Ni in dried plant material. It would be naive to suggest that the term *hyperaccumulator* should be applied to all plants for all trace elements with concentrations exceeding 1000 mg g^{-1} in dried tissue. The threshold of hyperaccumulation will vary considerably for different trace elements. Whereas for Ni a threshold of 1000 µg g^{-1} might seem appropriate, this would certainly not be the case for Zn for which the natural abundance in plants is quite high. For this element therefore, a threshold value of 10000 µg g^{-1} (1%) can be recommended. At the other extreme, a plant containing 1000 ng g^{-1} (1 µg g^{-1}) Au could be classed as a hyperaccumulator since the natural abundance of this element is only about 1 ng g^{-1} (Boyle, 1979).

The ultimate criterion as to the concentration of a given element in plant material that should accord the species hyperaccumulator status, cannot be based on arbitrary numbers alone. There must be distinct non-overlapping populations as shown in Fig. 4-1, before the threshold can be established. These limits will be discussed below under the individual elements.

Hyperaccumulating plant species represent perhaps the ultimate in plant tolerance to extremely hostile edaphic environments. They can thrive in soils that would kill many other species and have been the subject of several investigations involving such diverse subjects as: (1) geobotanical and biogeochemical prospecting, (2) colonisation of mine tips, (3) phytochemical investigations, (4) plant taxonomy, and (5) pollution studies. These topics are now discussed briefly.

Most hyperaccumulators are themselves endemic to the type of soil in which they are found. A classical case of this is the *Alyssum* genus that is widespread in Southern Europe and Asia Minor. Section Odontarrhena of the genus contains about 50 taxa that are endemic to ultramafic (serpentine) soils (Brooks, 1887; Brooks *et al.*, 1979). Due to their endemism they invariably indicate the presence of serpentine soils and thereby act as *geobotanical indicators* of potential mineralisation in the substrate. Another example is the presence of the Zn hyperaccumulators *Viola calaminaria* and *Thlaspi calaminare* confined to Zn deposits in Western Germany and Eastern Belgium.

Attention is now being paid to the potential use of hyperaccumulators to colonise mine dumps (Brooks, 1987). It is reasoned that a hyperaccumulator of Ni, for example, should be able to colonise a mine dump derived from Ni (or asbestos) mining and processing activities. In practice, metal-tolerant grasses are preferred for this purpose because they form a denser mat and grow faster than do herbs, but there are cases where even grasses find the task too difficult and recourse to nickel-accumulating herbs or shrubs might have to be made.

Hyperaccumulating plants have been the subject of several phytochemical studies because unlike most other plants, milligram rather than microgram quantities of organometallic complexes can be separated from the plant tissue. For the sake of brevity however, this topic is not addressed in this chapter (see Baker and Brooks, 1989 for further information).

Chemical analysis of the plant material can often be of assistance to the plant taxonomist. Cases of doubtful identity can sometimes be resolved. For example, the various species of *Alyssum* are rather similar morphologically. A simple colourimetric test of herbarium material can quickly confirm, or deny a given classification by the taxonomist.

Finally, because hyperaccumulators can accumulate relatively large quantities of heavy metals from the substrate, they could be used as decontaminating agents. A research group in the United States has already carried out trials with a Co hyperaccumulator in order to try to remove radioactive Co from contaminated soil.

Cobalt and Copper

Introduction

Copper and cobalt are considered together in this section because plant hyperaccumulators of these two metals are all derived from the same geographical region – Shaba Province, Zaïre. Copper is regarded as an essential element in plant nutrition and as with most essential elements, the Cu contents of most plants is internally rather than externally controlled (Timperley *et al.*, 1970). The Cu content of most plants therefore remains relatively constant at around 10 μg g^{-1}. This element is also very toxic to plants at high concentrations in the substrate and it is only in Shaba Province (see map in Fig. 4-2), that the evolutionary process has allowed for hyperaccumulators of this element. The Cu/Co deposits in this region are one of the world's largest and comprise over 100 separate outcrops spread over several thousand square kilometers of the province.

The Co content of most plants is around 1 μg g^{-1} and since this element is not normally considered to be essential for plant growth, its concentrations in plants tend to be externally controlled and reflect the Co content of the substrate to a better degree than does Cu. Despite the supposed non-essentiality of Co for plant growth, there

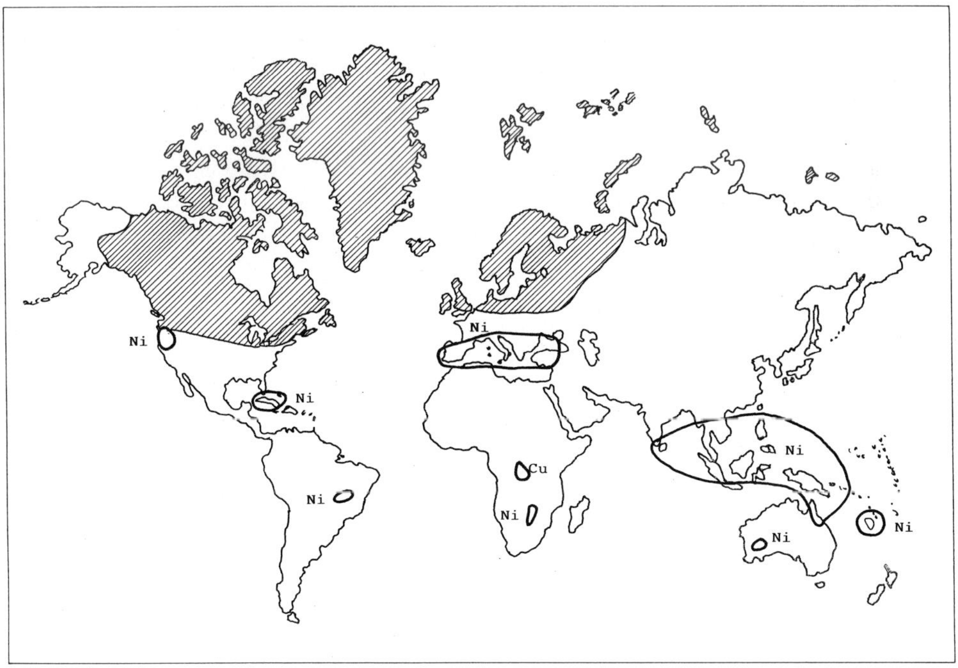

Fig. 4-2. World map showing the locations of hyperaccumulators of copper, cobalt, nickel, and zinc.

have been reports (eg, Marschner, 1986) that this element may be essential for the growth of legumes under symbiotic conditions.

To date, 41 hyperaccumulators of Cu and/or Co have been reported (Brooks and Malaisse, 1985; Brooks *et al.*, 1980, 1987) and of these, 15 accumulate Cu only, 17 Co only, and there are a further 9 taxa that hyperaccumulate both elements. These plants are listed in Table 4-2.

Table 4-2. Hyperaccumulators of copper and cobalt (maxium concentrations in $\mu g\ g^{-1}$ dry weight).

No.	Species	Copper	Cobalt	Refs[a]
1	*Aeollanthus biformifolius* De Wild. – Lamiaceae	3920	2820	9
2	*A. saxatilis* Duvign. et Plancke	–	1000	3
3	*Alectra sessiliflora* (Vahl) O. K.			
	var. *sessiliflora* – Scrophulariaceae	–	2782	4
4	var. *senegalensis* (Benth.) Hepper	1590	–	3
5	*A. welwitschii* Hemsl	–	1561	4
6	*Anisopappus* davyi S. Moore – Asteraceae	2889	2650	1
7	*A. hoffmanianus* Hutch	1065	–	7
8	*Ascolepis metallorum* Duvign. et Léon – Cyperaceae	1200	–	2
9	*Becium aureoviride* Duvign. subsp. *lupotoense* Duvign. – Lamiaceae	1135	–	4
10	*Buchnera henriquesii* Engl. – Scrophulariaceae	3520	2435	3, 4
11	*Bulbostylis mucronata* (Steud.) Cl. – Cyperaceae	7783	2130	3, 4
12	*Celosia trigyna* L. – Amaranthaceae	2051	–	4
13	*Commelina zigzag* Duvign. et Dewit. – Commelinaceae	1210	–	1
14	*Crassula alba* Forsk. – Crassulaceae	–	1712	4
15	*C. vaginata* Eckl. et Zeyh.	–	1405	1
16	*Crotalaria cobalticola* Duvign. – Fabaceae	–	3010	3
17	*Cyanotis longifolia* Benth. – Commelinaceae	–	4200	1
18	*Eragrostis boehmii* Hack. – Poaceae	2800	–	5
19	*Gutenbergia cupricola* Duvign. – Asteraceae	5095	2309	4
20	*Haumaniastrum homblei* (De Wild.) Duvign. et Denaeyer – Lamiaceae	–	2633	1, 4
21	*H. katangense* (S. Moore) Duvign. et Plancke	8356	2240	3, 4, 7
22	*H. robertii* (Robyns) Duvign. et Plancke	2070	10200	2, 4, 10
23	*H. rosulatum* (De Wild.) Duvign.	1089	–	4
24	*Hibiscus rhodanthus* Gurke – Malvaceae	–	1527	4
25	*Icomum tuberculatum* De Wild. – Lamiaceae	–	1429	4
26	*Ipomoea alpina* Rendle – Convolvulaceae	12300	–	1
27	*Lindernia damblonii* Duvign. – Scrophulariaceae	–	1113	4, 5
28	*L. perennis* Duvign.	9322	2300	4, 5
29	*Monadenium cupricola* Malaisse – Euphorbiaceae	–	1234	4
30	*Pandiaka metallorum* Duvign. et Van Bockst. – Amaranthaceae	6260	2139	1, 4

Tabble 4-2. Continued

No.	Species	Copper	Cobalt	Refs[a]
31	*Rendlia cupricola* Duvign. – Poaceae	1560	–	6
32	*Silene cobalticola* Duvign. et Plancke – Caryophyllaceae	1660	–	2
33	*Sopubia dregeana* Benth.	–	1767	4, 5
34	*S. metallorum* Duvign.	–	1742	4
35	*S. neptunii* Duvign. et Van Bockst.	–	2476	4
36	*Striga hermontheca* Benth.	1105	–	4
37	*Triumfetta digitata* (Oliv.) Sprague et Hutch. – Tiliaceae	1060	–	3
38	*T. welwitschii* Mast. var. *descampii*	–	2201	4
39	*Vernonia petersii* Oliv. et Hiern – Asteraceae	1555	–	4, 8
40	*Vigna dolomitica* Wilczek – Fabaceae	3000	–	5
41	*Xerophyta retinervis* Bak. var. *retinervis* – Velloziaceae	–	1520	4

[a] 1, Malaisse *et al.* (1979); 2, Duvigneaud and Denaeyer-De Smet (1963); 3, Brooks *et al.* (1980); 4, Brooks *et al.* (1980); 5, Malaisse and Grégoire (1978); 6, Brooks and Malaisse (1985); 7, Brooks *et al.* (1982); 8, Plaen *et al.* (1982); 9, Brooks *et al.* (1978); 10, Brooks (1977).

Hyperaccumulators of Cobalt

Among the hyperaccumulators of Co, *Haumaniastrum robertii* and *H. katangense* are rather unusual. Both will hyperaccumulate both Cu and Co and both have been known as 'copper flowers' (Duvigneaud and Denaeyer de Smet, 1963) and were used as prospecting guides in the 1950s. *Haumaniastrum robertii* has the capability of accumulating over 1% Co in dried tissue (Brooks, 1977) and is confined to the western part of the metalliferous Shaban Copper Arc. Its near relative, *H. katangense* is found in the eastern part of the Copper Arc and is very common near Lubumbashi (Mine de l'Etoile). It is more widespread and vigorous than *H. robertii.*

The first hyperaccumulator of Co to be discovered was *Crotalaria cobalticola.* Duvigneaud (1959) reported 354 $\mu g\ g^{-1}$ Co in this taxon but higher values of 970–3010 $\mu g\ g^{-1}$ were later reported by Brooks *et al.* (1980). This is thc only one of this genus that appears to have the ability to hyperaccumulate Co (Brooks *et al.*, 1977b).

Hyperaccumulators of Copper

Among the hyperaccumulators of copper listed in Table 4-2, *Aeollanthus biformifolius* has the greatest concentrating ability. Malaisse *et al.* (1978) reported as much as 13500 $\mu g\ g^{-1}$ (1.35%) Cu in the whole plant (dry weight). This is by far the highest copper content ever reported for any higher plant. Brooks *et al.* (1978) determined Cu (and Co) in 49 species of African *Aeollanthus* including all of those found in Shaba. The highest Cu value found, excluding *A. biformifolius*, was only 28 μg g^{-1} in *A. elongatum.*

Haumaniastrum katangense plays an important role in archaeology (Plaen *et al.*, 1982; Brooks and Malaisse, 1985; Brooks, 1989; Brooks and Johannes, 1990). During the precolonial period in Shaba Province, native artisans of the Kabambian culture smelted copper on termite mounds near rivers and brought the copper ore to the sites from mineralised areas elsewhere. After a few seasons, they would abandon the site, but left behind traces of ore that were sufficient to poison the soil and allow the sites to be colonised by metal-tolerant plants such as *H. katangense.* These plants formed a dense mat that could be recognised later by archaeologists. Excavation of these sites revealed the presence of numerous artifacts of the period, including copper crosses *(croisettes)* used as currency by the Kabambian culture (Brooks, 1989). Brooks and Malaisse (1985) coined the term *phytoarchaeology* to describe this technique.

Among the hyperaccumulators of Cu is *Silene cobalticola*. This taxon is of special interest because of tolerance tests by Baker *et al.* (1983) that established an ecophyletic series of Shaban plants in which the non-tolerant *S. burchelli* Otth. var. *angustifolia* Sond was the precursor and *S. cobalticola* was the ultimate development. The series included an intermediate ecotype of *S. burchelli,* with intermediate Cu/Co tolerance and ability to concentrate either element.

In any discussion of hyperaccumulation of heavy metals, the question immediately arises as to whether these taxa are *obligate* or *facultative* metallophytes: ie, requiring, or not requiring, the accumulated element for plant growth or development. From the many studies that have been carried out on these plants (summarised by Baker and Brooks, 1989), it is clear that all the species tested in the laboratory will grow quite well in soils containing 'normal' levels of heavy metals (eg, *Haumaniastrum robertii* and *H. katangense*). Nevertheless, in the course of laboratory experiments at this university, we have noticed that all the metallophytes tested, were very sensitive to fungal attack, and it is probably this factor, as well as an inability to withstand competitive pressure from other species, that restricts them to mineralised soils. To that extent they may be considered to have obligate characteristics, albeit related to be low abundance of fungi in their growth substrates.

Nickel

Introduction

By far the greatest volume of work and general interest in hyperaccumulators has been centred around the genus *Alyssum.* This followed the original discovery of 1% Ni in *A. bertolonii* by Minguzzi and Vergnano in 1948. Some years later, Doksopulo (1961) reported hyperaccumulation of Ni in *A. murale.* In the early 1970s a third nickel plant was discovered in Western Australia by Severne and Brooks (1972) and independently by Cole (1973). Shortly afterwards, Jaffré and Schmid (1974) reported hyperaccumulation in *Geissois pruinosa, Homalium guillaini,* and *Hybanthus austrocaledonicus.* Brooks *et al.* (1977 a) analysed some 2000 herbarium specimens of the genera *Homalium* and *Hybanthus* and detected five other hyperaccumulators of Ni, as well as reidentifying all previously known taxa with this capability.

Following this earlier work, nearly 150 hyperaccumulators of Ni have now been reported (Table 4-3), largely as the result of chemical tests on herbarium specimens. This has led to increased interest in these plants for the reasons outlined in the introduction of this chapter.

Table 4-3. Hyperaccumulators of nickel (maximum concentration in $\mu g\ g^{-1}$ dry weight).

No.	Genus	Species	Location	Ni concentration	Refs[a]
1	*Aguatea* (Violac.)	*deplanchei* Brongn. et Gris.	New Caledonia	2 500	36
2	*Alyssum* (Brassicaceae)	*akamasicum* Burtt.	Cyprus	9 090	1
3		*alpestre L.*	S. Europe	4 480	2, 3
4		*anatolicum* Hausskn. ex Nyar.	Anatolia	8 170	1
5		*argenteum* All.	Italy	29 400	1, 3, 4, 13
6		*bertolonii* Desv. subsp.	Italy	13 400	2, 3, 5
7		*scutarinum* Nyar.	Italy	10 200	13
8		*callichroum* Boiss. et Buhse	Anatolia	10 900	1
9		*caricum* T. R. Dudley et Huber-Morath	Anatolia	16 500	1, 13
10		*cassium* Boiss.	Anatolia	20 000	1
11		*chondrogynum* Burtt	Cyprus	16 300	1
12		*cicilicum* Boiss. et Balansa	Anatolia	13 500	1
13		*condensatum* Boiss. et Hausskn.	Syria, Iraq	4 990	1, 13
14		*constellatum* Boiss.	Anatolia	18 100	1

Table 4-3. Continued

No.	Genus	Species	Location	Ni concentration	Refs[a]
15		*corsicum* Duby	Anatolia, Corsica	13 500	1
16		*crenulatum* Boiss.	Anatolia	10 400	1
17		*cyprium* Nyar.	Cyprus	23 600	1
18		*davisianum* T. R. Dudley	Anatolia	19 600	1
19		*discolor* T. R. Dudley et Huber-Morath	Anatolia	11 700	1
20		*dubertretii* Gombault	Anatolia	16 500	1
21		*eriophyllum* Boiss. et Hausskn.	Anatolia	11 500	1
22		*euboeum* Halacsy	Euboa	4 550	2, 13
23		*fallacinum* Hausskn.	Crete	3 960	2
24		*floribundum* Boiss. et Balanse	Anatolia	7 700	1
25		*giosnanum* Nyar.	Anatolia	7 390	1
26		*heldreichii* Hausskn.	Greece	12 500	2
27		*hubermorathi* T. R. Dudley	Anatolia	13 500	1
28		*janchenii* Nyar.	Albania	9 610	1
29		*lesbiacum* (Candargy) Rech. f.	Lesbos	22 400	1
30		*malacitanum* T. R. Dudley	Spain	10 000	37
31		*markgrafii* O. E. Schulz	Albania	13 700	1, 13
32		*masmaenaeum* Boiss.	Anatolia	24 300	1
33		*murale* Waldst. et Kit.	Balkans	7 080	1, 2, 7
34		*obovatum* (C. A. Meyer)	Russia	4 590	1
35		*oxycarpum* Boiss et Balansa	Anatolia	7 290	1
36		*peltarioides* Boiss. subsp. *virgatiforme*	Anatolia	7 600	1, 13
37		*penjwinensis* T. R. Dudley	Iraq	7 860	1
38		*pinifolium* (Nyar.) T. R. Dudley	Anatolia	21 100	1, 13
39		*pintodasilvae* T. R. Dudley	Portugal	9 000	2, 8, 9
40		*pterocarpum* T. R. Dudley	Anatolia	22 200	1, 13
41		*robertianum* Bernhard ex Gren. et Godr.	Corsica	12 500	1
42		*samariferum* Boiss. et Hausskn.	Samar	18 900	1, 13
43		*singarense* Boiss. et Hausskn.	Iraq	1 280	1
44		*smolikanum* Nyar.	Greece	6 600	2
45		*syriacum* Nyar.	Syria	10 200	1
46		*tenium* Ha'ácsy	Tinos	3 430	2

Table 4-3. Continued

No.	Genus	Species	Location	Ni concentration	Refs[a]
47		*trapeziforme* Waldst. et Kit.	Anatolia	11 900	1
48		*troodii* Boiss.	Cyprus	17 100	1, 13
49		*virgatum* Nyar.	Anatolia	6 230	1
50	*Argophyllum* (Escalloniac.)	*grunowii* Zahlbr.	New Caledonia	1 375	36
51	*Berkheya*	*coddii* Roessl.	South Africa	11 637	40
52		*laxum* Schlecht.		1 900	36
53	*Blepharis* (Acanthaceae)	*acuminata* Oberm.	Zimbabwe	2 000	10
54	*Bornmuellera* (Brassicaceae)	*baldacci* (Degen) Heywood subsp. *baldacci*	Greece	21 300	13
55		subsp. *markgrafii* (Schulz) T. R. Dudley	Albania	27 300	13
56		subsp. *rechingeri* Greuter	Greece	12 000	13
57		*glabrescens* (Boiss. et Bal.) Cullen et Dudley	Anatolia	19 200	13
58		*tymphaea* (Hausskn.) Hausskn.	Greece	31 200	13
59		× *petri*	Greece	11 400	13
60	*Buxus-* (Buxaceae)	*flaviramea* (Britton) Howard	Cuba	4 500	11
61	*Cardamine* (Brassicaceae)	*resedifolia* L.	Italiy	1 050	4, 34
62	*Casearia* (Flacourtiaceae)	*silvana* Schlecht.	New Caledonia	1 490	12
63	*Chrysanthemum* (Asteraceae)	*alpinum* L.	Italy	3 200	34
64	*Cleidion* (Euphorbiac.)	*lasiophyllum* Pax et Hoffman	New Caledonia	9 900	36
65	*Dichapetalum* (Diachapetalac.)	*gelonoides* (Roxb.) Engl. subsp. *tuberculatum* Leenh.	Philippines	26 647	41
66	*Dicoma*	*niccolifera* Wild	Zimbabwe	1 000	38
67	*Geissois* (Cunoniaceae)	*hirsuta* Brongn. et Gris.	New Caledonia	4 000	14
68		*intermedia* Vieill, ex. Pampan		22 900	14
69		*magnifica* Bak. f.		3 250	14
70		*montana* Vieill. ex Brongn. et Gris.		5 740	14
71		*pruinosa* Brongn. et Gris.		34 000	14, 15
72		*racemosa* Labill.		1 000	14
73		*trifoliolata* Guill.		6 250	14

Table 4-3. Continued

No.	Genus	Species	Location	Ni concentration	Refs[a]
74	*Homalium* (Flacourtiac.)	*austrocaledonicum* Seemann	New Caledonia	1 805	12, 16
75		*deplanchei* (Vieill.) Warburg		1 850	12, 16
76		*francii* Guill.		14 500	12, 16
77		*guillainii* Vieill. Briq.		11 700	12, 15, 16, 36
78		*kanaliense* Vieill. Briq.		9 420	16–18
79		*mathieuanum* Vieill. Briq.		1 694	12, 16
80		*rubrocostatum* Sleumer		1 157	12, 16
81	*Hybanthus* (Violaceae)	*austrocaledonicus* Schinz et Guill.	New Caledonia	25 500	15, 18, 19, 36
82		*caledonicus* (Turcz.) Cretz.		17 500	16, 17, 36
		floribundus (Lindl.) F. Muell.	W. Australia		
83		subsp. *adpressus* Bennett		3 000	20
84		subsp. *curvifolius* Bennett		9 000	20
85		subsp. *floribundus*		9 800	20–22
86	*Lasiochlamys* (Flacourtiaceae)	*peltata* Sleumer	New Caledonia	1 000	12
87	*Leucocroton* (Euphorbiaceae)	*flavicans* Muell.	Cuba	7 700	11
88	*Linaria* (Scrophulariaceae)	*alpina* L.	Italy	1 990	34
89	*Luzula* (Juncaceae)	*lutea* (All.) DC.	Italy	2 050	34
90	*Merremia* (Asclepiacaceae)	*xanthophylla* Hall. f.	Zimbabwe	1 400	10
91	*Minuartia* (Caryophyllac.)	*laricifolia* Schinz et Thell.	Italy	1 910	34
92	*Myristica* (Myristicaceae)	*laurifolia* Spruce ex DC. var. *bifurcata*	Obi (Indonesia)	1 110	23
93	*Noccaea*	*aptera* (Velen.) F. K. Meyer	Jugoslavia	13 600	24
94	(Brassicaceae)	*boetica* F. K. Meyer	Greece	23 400	24
95		*firmiensis* F. K. Meyer	Greece	16 200	24
96		*tymphaea* F. K. Meyer	Greece	11 800	24
97	*Oncotheca* (Oncothecac.)	*balansae* Baill.	New Caledonia	2 500	36
98	*Pancheria* (Cunoniaceae)	*engleriana* Schlecht.	New Caledonia	6 300	36
99	*Pearsonia* (Fabaceae)	*metallifera* Wild	Zimbabwe	10 000	35

Table 4-3. Continued

No.	Genus	Species	Location	Ni concentration	Refs[a]
100	*Peltaria* (Brassicaceae)	*emarginata* Boiss. Hausskn.	Greece	34 400	25
101		*dumulosa* Post	Asia Minor	18 900	25
102	*Phyllanthus* (Euphorbiac.)	*aeneus* Baill	New Caledonia	2 100	26
103		*balansaeanus* Guill.		1 820	26
104		*cataractarum* Muell.		1 450	26
105		*chrysanthus* Baill.		1 180	26
106		*induratus* S. Moore		1 480	26
107		*kanalensis* Baill.		1 090	26
108		*maytenifolius* S. Moore		1 420	26
109		*ngoyensis* Schlecht.		9 550	26
110		*peltatus* Guill.		2 830	26
111		*serpentinus* S. Moore		38 100	26
112	*Planchonella* (Sapotaceae)	*oxyedra* Dubard	Indonesia	19 600	23
113	*Psychotria* (Rubiaceae)	*douarrei* (Beauv.) Dän.	New Caledonia	47 500	15, 19, 33, 36
114	*Rhus* (Anacardiaceae)	*wildii* R. & A. Fernandez	Zimbabwe	1 600	10
115	*Rinorea* (Violaceae)	*bengalensis* (Wall.) O. K.	S. E. Asia	17 500	27
116		*javanica* (Bl.) O. K.	Borneo	2 170	28
117	*Saxifraga* (Saxifragac.)	*aïzoon* Jacq.	Italy	3 840	34
118		*exarata* Vill.	Italy	2 970	34
119	*Sebertia* (Sapotaceae)	*acuminata* Pierre ex Baill.	New Caledonia	17 750	29, 36
120	*Stachys* (Lamiaceae)	*recta* L.	Italy	2 600	30
121	*Stackhousia*	*trysonii* Bailey	Queensland	13 900	39
122	*Streptanthus* (Brassicaceae)	*polygaloides* Gray	W. USA	14 800	31
123	*Thlaspi* (Brassicaceae)	*alpestre* L. subsp. *Virens* (Jord.) Hook f.	Central Europe	4 000	24
124			France	4 100	24
125		*alpinum* Crantz subsp. *sylvium* (Gaud.) Clapham	Central Europe	31 000	24
126		*bulbosum* Spruner ex Boiss.	Greece	2 000	24
127		*epirotum* Halácsy	Greece	3 000	24
128		*goesingense* Halácsy	Austria	12 000	24
129		*graecum* Jordan	Greece	12 000	24

Table 4-3. Continued

No.	Genus	Species	Location	Ni concentration	Refs[a]
130		*montanum* L. var. *californicum*	W. USA	7 940	32
131		var. *montanum*	W. USA	5 530	32
132		var. *siskiyouense*	W. USA	11 200	32
133		*ochroleucum* Boiss. ex Heldr.	Greece	4 000	24
134		*rotundifolium* (L.) Gaudin	Central Europe	18 300	24
135		var. *corymbosum* (Gay) Gaudin	Central Europe	2 000	24
136	*Trichospermum* (Tiliaceae)	*kjellbergii* Burret	Celebes	1 600	23
137	*Trifolium* (Fabaceae)	*pallescens* Schreber	Italy	2 000	24
138	*Xylosma* (Flacourtiac.)	*boulindae* Sleumer	New Caledonia	1 930	12
139		*confusum* Guill.		1 630	12
140		*dothioense* Guill.		1 780	12
141		*kaalense* Sleumer		1 900	12
142		*molestum* Sleumer		1 140	12
143		*pancheri* Guill.		1 130	12
144		*peltatum* (Sleumer) Lescot		1 000	36
145		*pininsulare* Guill.		1 280	12
146		*serpentinum* Sleumer		1 490	12
147		*tuberculatum* Sleumer		1 600	12
148		*vincentii* Guill.		3 750	12

[a] 1, Brooks *et al.* (1979); 2, Brooks and Radford (1978); 3, Vergnano Gambi *et al.* (1979); 4, Vergnano Gambi and Gabbrielli (1979); 5, Minguzzi and Vergnano (1948); 6, Brooks *et al.* (1981); 7, Doksopulo (1961); 8, Menezes de Sequeira (1969); 9, Dudley (1986 a); 10, Brooks and Yang (1984); 11, Berazain Iturralde (1981); 12, Jaffré *et al.* (1979 b); 13, Reeves *et al.* (1983 a); 14, Jaffré *et al.* (1979 a); 15, Jaffré and Schmid (1974); 16, Brooks *et al.* (1977 a); 17, Brooks *et al.* (1974); 18, Lee *et al.* (1977); 19, Kelly *et al.* (1975); 20, Severne (1972); 21, Severne and Brooks (1972); 22, Cole (1973); 23, Wither and Brooks (1977); 24, Reeves and Brooks (1983 b); 25, Reeves *et al.* (1980); 26, Kersten *et al.* (1979); 27, Brooks and Wither (1977); 28, Brooks *et al.* (1977 b); 29, Jaffré *et al.* (1976); 30, Lisanti (1952); 31, Reeves *et al.* (1981); 32, Reeves *et al.* (1983 b); 33, Baker *et al.* (1985); 34, Vergnano Gambi and Gabrielli (1979); 35, Wild (1970); 36, Jaffré (1980); 37, Dudley (1986 b); 38, Proctor *et al.* (1980); 39, Batianoff *et al.* (1990); 40, Morrey *et al.* (1989); 41, Homer *et al.* (1991).

The Distribution of Hyperaccumulators of Nickel

Hyperaccumulators of Ni ('nickel plants') are distributed throughout the world but only in regions that have not previously been glaciated. The main centres of distribution are shown in Fig. 4-2. The presence of these plants only in non-glaciated regions is probably a temporal factor. Presumably plant evolution and adaptation to the hostile environment of nickel-rich soils is a process considerably longer than the 10000 years that have elapsed since the last period of glaciation.

Nickel plants are confined to eight major regions as shown in Fig. 4-2: New Caledonia, Australia, Southern Europe and Asia Minor, The Malay Archipelago, Cuba, Western United States, Southern Africa, and Central Brazil. Some of these regions will be discussed below, but the reader is referred to Baker and Brooks (1989) for a more complete description and discussion.

Alyssum

The genus *Alyssum* contains by far the greatest number of individual nickel plants in any one genus. Some 48 members of section Odontarrhena (formerly described as a separate genus) have this hyperaccumulating ability. All of these plants are confined to a belt of ultramafic (serpentine) rocks stretching along Southern Europe from Portugal to Eastern Turkey. Anatolia is the centre of maximum multiplicity and diversity. The degree of endemism is also very high as some taxa are confined to outcrops of only a few hectares in area. It has been proposed by Brooks *et al.* (1979) that the greater the Ni content of the *Alyssum* species, the narrower its distribution range. For example, *A. troodii* (1.71% Ni) is confined to a small area of Cyprus, whereas *A. alpestre* (0.45% Ni) is found throughout the Eastern Mediterranean. There appears to be a relationship between high concentrations of Ni on the one hand, and a high degree of diversity, proliferation, and endemism on the other.

Hyperaccumulation of Ni appears to be a strategy where genera such as *Alyssum* have been able to evolve a physiological tolerance to phytotoxic nickel-rich soils, and avoid competition from other species by flourishing in environments so hostile, that often *Alyssum* is the only coloniser of the area.

Thlaspi

The only other genus with such a large number of hyperaccumulators of Ni is *Thlaspi*, a plant that seems to occupy the same ecological niches in Southern Europe as does *Alyssum*. Unlike the latter however, *Thlaspi* is much more widespread in Central and Western Europe, and has poor representation in Anatolia. Reeves and Brooks (1983 b) found 12 species with hyperaccumulator status.

Nickel Hyperaccumulators from New Caledonia and Brazil

Brief mention can be made of nickel plants from regions such as New Caledonia and Brazil. New Caledonia has perhaps the world's greatest concentration and diversity of these plants. About 50 hyperaccumulators from 14 genera and eight different families have been reported from this isolated Pacific island. These nickel plants belong to the following genera and families: *Agatea* (Violaceae), *Argophyllum* (Escalloniaceae), *Casearia* (Flacourtiaceae), *Cleidion* (Euphorbiaceae), *Geissois* (Cunoniaceae), *Homalium* (Flacourtiaceae), *Hybanthus* (Violaceae), *Lasiochlamys* (Flacourtiaceae), *Oncontheca* (Oconthecaceae), *Pancheria* (Cunoniaceae), *Phyllanthus* (Euphorbiaceae), *Psychotria* (Rubiaceae), *Sebertia* (Sapotaceae), and *Xylosma* (Flacourtiaceae). One of the most striking of these New Caledonian hyperaccumulators is *Sebertia acuminata* which possesses a blue-green sap with over 10% Ni on a wet-weight basis (Jaffré *et al.*, 1976).

Until recently, there had been no record of nickel plants from South America. However, Brooks *et al.* (1990) have now identified 12 hyperaccumulators from ultramafic soils in Goiás State. Among them is a *Jatropha* sp. (Euphorbiaceae) with a creamy latex containing 1.35% Ni on a dry weight basis. This is analogous to *Sebertia acuminata* from New Caledonia.

Zinc

Zinc is easily taken up by plants from the soil where it accumulates in the shoots. It is an essential element for plant nutrition (Marschner, 1986) and is important in both carbohydrate metabolism and protein synthesis.

The zinc-rich soils of Western Germany and Eastern Belgium support an unusual metal-tolerant flora (*Galmei*) flora dominated by the hyperaccumulators *Viola calaminaria* and *Thlaspi calaminare*. Baker and Brooks (1989) recommended a threshold of 10000 $\mu g\ g^{-1}$ (1%) Zn for defining hyperaccumulation of this element.

Reeves and Brooks (1983b) discovered several hyperaccumulators within the genus *Thlaspi*. These and other zinc plants are recorded in Table 4-4.

Conclusions

Hyperaccumulation of metallic elements in plants as reviewed above, is likely to draw increasing attention from scientists throughout the world for the reasons outlined in the introduction to this chapter. Studies on nickel plants could well involve scientists from such diverse fields as ecology, biogeography, mineral exploration, evolutionary biology, taxonomy, physiology, phytochemistry, and even archaeology (phytoarchaeology). The elements in this chapter are all common, but there is no reason why future studies should not extend to rarer elements including the noble metals.

Table 4-4. Hyperaccumulators of zinc (% dry weight).

No.	Species	Location	% Zn	Ref.[a]
1	*Cardaminopsis halleri* (L.) Hayek – Brassicaceae	Germany	1.36	1
2	*Haumaniastrum katangense* (S. Moore) Duvign. et Plancke – Lamiaceae	Zaire	1.98	2
3	*Noccaea eburneosa* F. K. Meyer – Brassicaceae	Switzerland	1.05	3
4	*Thlaspi alpestre* L. – Brassicaceae	Derbyshire	2.50	3, 4
5	*T. brachypetalum* Jordan	France	2.00	3
6	*T. bulbosum* Spruner ex Boiss	Greece	1.05	3
7	*T. calaminare* (Lej.) Lej. et Court	Germany	3.96	3
8	*T. caerulescens* J et C. Presl	Germany, Belg.	2.73	3, 5
9	*T. cepaeifolium* (Wulfen) Koch	Italy	2.10	3, 6
10	*T. limosellifolium* Reuter	France	1.10	3
11	*T. praecox* Wulfen	Bulgaria	2.10	3
12	*T. stenopterum* Boiss. et Reuter	Spain	1.60	3
13	*T. tatraense* Zapal	Czechoslovakia	2.70	3
14	*Viola calaminare* Lej. – Violaceae	Germany	1.00	7

[a] 1, Ernst (1974); 2, Brooks (unpub.); 3, Reeves and Brooks (1983b); 4, Shimwell and Laurie (1972); 5, Denaeyer-De Smet and Duvigneaud (1974); 6, Reeves and Brooks (1983a); 7, Von Linstow (1924).

References

Baker, A. J. M., Brooks, R. R. (1989). *Biorecovery, 1,* 81–126.

Baker, A. J. M., Brooks, R. R., Kersten, W. J. (1985). *Taxon, 34,* 89–95.

Baker, A. K. M., Brooks, R. R., Pease, A. J., Malaisse, F. (1983). *Pl Soil, 73,* 377–385.

Batianoff, G. N., Reeves, R. D., Specht, R. L. (1990). *Aust J Bot, 73,* 377–385.

Baumann, A. (1985). *Landwirt Versuch Sta, 31,* 1–53.

Berazain Iturralde, R. (1981). *Rev Jard Bot Nac (Cuba), 2,* 29–59.

Boyle, R. W. (1989). *Geol Surv Can Bull, 280,* 1–580.

Brooks, R. R. (1977). *Pl Soil, 48, ,* 541–544.

Brooks, R. R. (1983). *Biological Methods of Prospecting for Minerals,* New York: John Wiley.

Brooks, R. R., (1987). *Serpentine and its Vegetation*, Portland: Dioscorides Press.

Brooks, R. R. (1989). *Endeavour, 13,* 129–134.

Brooks, R. R., Johannes, D. (1990). *Phytoarchaeology,* Portland: Dioscorides Press.

Brooks, R. R., Malaisse, F. (1985). *The Heavy Metal-Tolerant Flora of Southcentral Africa,* Rotterdam: Balkema.

Brooks, R. R., Radford, C. C. (1978). *Proc Roy Soc (Lond) Sec B, 200,* 217–224.

Brooks, R. R., Wither, E. D. (1977). *J Geochem Explor, 7,* 295–300.

Brooks, R. R., Yang, X. H. (1984). *Taxon, 33,* 392–399.

Brooks, R. R., Lee, J., Jaffré, T. (1974). *J Ecol, 62,* 523–529.
Brooks, R. R., Lee, J., Reeves, R. D., Jaffré, T. (1977a). *J Geochem Explor, 7,* 49–77.
Brooks, R. R., McCleave, J. A., Malaisse, F. (1977b). *Proc Roy Soc (Lond) Sec B, 197,* 231–236.
Brooks, R. R., Reeves, R. D., Baker, A. J. M. (1990). *Nat Geogr Res, 6,* 205–219.
Brooks, R. R., Shaw, S., Aseni Marfil, A. (1981). *Physiol Plant, 51,* 167–170.
Brooks, R. R., Grégoire, J., Madi, L., Malaisse, F. (1982). *Geo-Eco-Trop 6,* 219–228.
Brooks, R. R., Morrison, R. S., Reeves, R. D., Malaisse, F. (1978). *Pl Soil, 50,* 503–507.
Brooks, R. R., Naidu, S. D., Malaisse, F., Lee, J. (1987). *Bull Soc Roy Bot Belg, 119,* 179–191.
Brooks, R. R., Reeves, R. D., Morrison, R. S., Malaisse, F. (1980). *Bull Soc Roy Bot Belg, 113,* 166–172.
Brooks, R. R., Morrison, R. S., Reeves, R. D., Dudley, T. R., Akam, Y., (1979). *Proc Roy Soc (Lond) Sec B, 203,* 387–403.
Cole, M. M. (1973). *J Appl Ecol, 10,* 269–320.
Denaeyer-De Smet, S., Duvigneaud, P. (1974). *Bull Soc Roy Bot Belg, 107,* 147–156.
Doksopulo, E. P. (1961). *Nickel in Rocks, Soils, Waters and Plants Adjacent to the Chorchanskaya Group* (in Russian), Tbilisi: Izdatel Tbiliskovo Universiteta.
Dudley, T. R. (1986a). *Fedd Rep, 97,* 135–138.
Dudley, T. R. (1986b). *Fedd Rep, 97,* 139–142.
Duvigneaud, R. (1959). *Bull Soc Roy Bot Belg, 91,* 111–134.
Duvigneaud, P., Denaeyer-De Smet, S. (1963). *Bull Soc Roy Bot Belg, 96,* 93–231.
Ernst, W. (1974). *Schwermetallvegetation der Erde,* Stuttgart: Fischer Verlag.
Homer, F. A., Reeves, R. D., Brooks, R. R., Baker, A. J. M. (1991). *Phytochemistry, 30,* 2141–2145.
Jaffré, T. (1980). PhD Thesis.
Jaffré, T., Schmid, M. (1974). *Compt Rend Acad Sci (Paris) Sér D, 278,* 1727–1730.
Jaffré, T., Brooks, R. R., Lee, J., Reeves, R. D. (1976). *Science, 193,* 579–580.
Jaffré, T., Brooks, R. R., Trow, J. M. (1979a). *Pl Soil, 51,* 157–162.
Jaffré, T., Kersten, W. J., Brooks, R. R., Reeves, R. D. (1979b). *Proc Roy Soc (Lond) Sec B, 205,* 385–394.
Kelly, P. C., Brooks, R. R., Dilli, S., Jaffré, T. (1975). *Proc Roy Soc (Lond) Sec B, 189,* 69–80.
Kersten, W. J., Brooks, R. R., Reeves, R. D., Jaffré, T. (1979). *Taxon, 28,* 529–534.
Kovalevsky, A. L. (1987). *Biogeochemical Prospecting for Mineral Deposits,* Utrecht: VNU Press.
Lee, J., Brooks, R. R., Reeves, R. D., Boswell, C. R., Jaffré, T. (1977). *Pl Soil, 46,* 675–680.
Lisanti, E. L. (1952). *Nuov G Bot Ital, 14,* 349–360.
Malaisse, F., Grégoire, J. (1978). *Bull Soc Roy Bot Belg, 111,* 252–260.
Malaisse, F., Grégoire, J., Brooks, R. R., Reeves, R. D. (1978). *Science, 199,* 887–888.
Malaisse, F., Grégoire, J., Morrison, R. S., Brooks, R. R., Reeves, R. D. (1979). *Oikos, 33,* 472–478.
Malyuga, D. P. (1964). *Biogeochemical Methods of Prospecting,* New York: Consultants Bureau.
Marschner, H. (1986). *Mineral Nutrition of Higher Plants,* London: Academic Press.
Menezes de Sequeira, E. (1969). *Agronom Lusit, 30,* 115–154.
Minguzzi, C., Vergnano, O. (1948). *Mem Soc Tosc Sci Nat Ser A, 55,* 49–74.
Morrey, D. R., Balkwill, K., Balkwill, M. J. (1989). *S Afr J Bot, 55,* 171–177.
Plaen, G. de, Malaisse, F., Brooks, R. R. (1982). *Endeavour, 6,* 72–77.
Proctor, J., Burrow, J., Craig, C. G. (1980). *Kirkia, 12,* 127–139.
Reeves, R. D., Brooks, R. R. (1983a). *Environ Pollut, 31,* 277–287.
Reeves, R. D., Brooks, R. R. (1983b). *J Geochem Explor, 18,* 275–283.

Reeves, R. D., Brooks, R. R., Press, J. R. (1980). *Taxon, 29,* 629–633.
Reeves, R. D., Brooks, R. R., McFarlane, R. M. (1981). *Am J Bot, 68,* 708–712.
Reeves, R. D., Brooks, R. R., Dudley, T. R. (1983). *Taxon, 32,* 184–192.
Severne, B. C. (1972). PhD Thesis, Massey University, New Zealand.
Severne, B. C., Brooks, R. R. (1972). *Planta, 103,* 91–94.
Shimwell, D. W., Laurie, A. E. (1972). *Environ Pollut, 3,* 291–301.
Timperley, M. H., Brooks, R. R., Peterson, P. J. (1970). *J Appl Ecol, 7,* 429–439.
Vergnano Gambi, O., Gabbrielli, R. (1979). *Ofioliti, 4,* 199–208.
Vergnano Gambi, O., Brooks, R. R., Radford, C. C. (1979). *Webbia, 33,* 269–277.
Von Linstow, O. (1924). *Fedd Rep, 31,* 1–151.
Wild, H. (1970). *Kirkia, 7 (suppl.),* 1–62.
Wither, E. D., Brooks, R. R. (1977). *J Geochem Explor, 8,* 579–583.

5 Plant Metalloenzymes

Robert W. Hay

Department of Inorganic Chemistry, University of St. Andrews
St. Andrews KY16 9ST, UK

Contents

Introduction

Enyzmes are present in all plant cells, and in greatest variety in seeds. Most plant enzymes are intracellular as they cannot diffuse through the cell wall. In a few cases the enzyme can be obtained from the plant juice, for example, the enzyme bromelain can be obtained from the stem of the pineapple. Otherwise, methods of extraction are used which depend on the rupturing of the cell wall, for example, grinding the plant with sand, either with or without freezing, or drying of the tissue.

In 1810, Planche showed that extracts of various plant roots contained oxidising enzymes, as they gave a blue colour with guaiacum tincture. This is an alcoholic extract of guaiacum gum, obtained from a species of Guaiacum trees, which contains guaiaconic acid and gives a blue oxidation product. A little later, several investigators (Robiquet and Boutron-Chalard, 1830; Liebig and Wöhler, 1837; Robiquet, 1838) examined the hydrolysis of amygdalin in bitter almonds and isolated emulsin. In 1830, Dubrunfaut prepared an extract of malt, and in 1833 Payen and Persoz isolated malt diastase by precipitation with alcohol. The study of plant enzymes thus has a long history which has developed over almost two centuries (Steel, 1934; Haldane, 1965). Typical examples of plant enzymes are shown in Table 5-1. A more extensive list is given by Dixon and Webb (1964). It should be recognised that enzymes can be obtained from a variety of sources – animal tissue, bacteria, moulds, plants, yeasts and fungi. A number of enzymes, for example, catechol oxidase can be obtained from plants, fungi and animals. Detailed investigations, including X-ray work, are frequently reported on a specific enzyme from an animal source although a similar enzyme occurs in plants. It is assumed that the enzymes from different sources operate by a similar overall mechanism. However, there will be differences in amino acid sequence and possibly secondary, tertiary and quaternary structures.

Roughly 30% of enzymes are metalloenzymes or require metal ions for activity and the present chapter will concentrate on the chemisty and structure of the plant metalloenzymes. As analytical methods have improved it has been possible to establish a metal ion requirement for a variety of enzymes which were initially considered to be pure proteins. A dramatic example is provided by the enzyme urease isolated from Jack beans and first crystallised by Sumner (1926) (the first enzyme to be crystallised). Sumner defined an enzyme as a pure protein with catalytic activity, however, Zerner and his coworkers (Dixon *et al.*, 1975) established that urease is in fact a nickel metalloenzyme. Jack bean urease contains two moles of nickel(II) per mole of active sites and at least one of these metal ions is implicated in its mechanism of action.

Table 5-1. Examples of some plant enzymes.

EC	Enzyme	Plant source	Substrate	Product
Hydrolases				
(A) Esterases				
3.1.1.3	Lipase	Oily seeds	Fats	Fatty acids + glycerol
3.1.1.11	Pectinesterase	Fruits; clover, etc.	Pectin	Pectic acid + methanol
3.1.1.1	Carboxylesterase	Plants, moulds	Carboxylic ester	Alcohol + carboxylic acid
3.1.1.6	Acetylesterase	Citrus fruits	Ester of acetic acid	Alcohol + acetic acid
3.1.3.2	Acid phosphatase	Yeast, seeds	Monoester of phosphoric acid	Alcohol + H_3PO_4
3.1.3.b	Glycerolphosphatase	Plants	2-Phosphoglycerol	Glycerol + H_3PO_4
3.1.3.6	3-Nucleotidase	Plants	3′-Ribonucleotide	Ribonucleoside + H_3PO_4
(B) Glycoside hydrolases				
3.2.1.1	α-Amylase	Plants especially Leguminosae	Starch, dextrins	Maltose
3.2.1.20	α-Glucosidase	Leaves, cereal grain	α-D-Glucoside	D-Glucose + an alcohol
3.2.1.20	β-Glusosidase	Leaves of cherry, etc.	β-D-Glucoside	D-Glucose + an alcohol
3.2.1.26	β-Fructofuranoside (invertase)	Leaves, yeast	β-D-Fructofuranoside	D-Fructose + an alcohol
3.2.1.4	Cellulase	Especially seeds	Cellulose	Cellobiose
3.2.1.7	Inulase	Yeast, artichokes, *Dahlia*	Inulin	Fructose
(C) Peptide hydrolases				
3.5.1.1	Asparaginase	Plants, germinating barley	L-Asparagine	L-Apartate + NH_3
3.4.4.10	Papain	*Carica papaya,* latex	Peptides, amides, esters	Carboxylic acid
3.4.4.c	Bromelain	Pineapple stems	Peptides	Carboxylic acid
3.5.1.5	Urease	Soy bean, Jack bean	Urea	NH_3 + CO_2
3.5.3.1	Arginase	Seedlings of vetch, etc.	Arginine	Ornithine + urea

Table 5-1. Continued

EC	Enzyme	Plant source	Substrate	Product
Lyases				
(A) Carbon-carbon lyases and carboxy-lyases				
4.1.2.f	Hydroxynitrile lyase	Sorghum, cherry leaves	Mandelonitrile	Benzaldehyde + HCN
4.1.1.31	Phosphopyruvate carboxylase	Plants	PO_4^{3-} + oxaloacetate	Phosphoenolpyruvate + CO_2
4.1.1.f	Ribulosediphosphate carboxylase	Plants	2,3-Phospho-D-glycerate	D-Ribulose-1,5-diphosphate
4.1.1.15	Glutamate decarboxylase	Plants, yeast	L-Glutamate	4-Aminobutyrate + CO_2
4.1.1.1	Pyruvate decarboxylase	Plants, yeast	2-Oxo-acid	Aldehyde + CO_2
(B) Carbon-oxygen lyases				
4.2.1.1	Carbonic anhydrase	Plants	CO_2 + H_2O	HCO_3^-
4.2.99.2	Threonine synthetase	Fungi	*o*-Phosphohomoserine	Threonine phosphate
(C) Carbon-nitrogen lyases				
4.3.1.1	Aspartase	Some plants	L-Aspartate	Fumarate + NH_3
4.3.1.a	Phenylalanine ammonia-lyase	Plants	L-Phenylalanine	*Trans*-cinnamate + NH_3
Oxidoreductases				
1.11.1.6	Catalase	All higher plants	$2H_2O_2$	$2H_2O$ + O_2
1.11.1.7	Peroxidase	All higher plants	Donor + H_2O_2	Oxidised donor + $2H_2O$
1.10.3.1	Catechol oxidase (Tyrosinase)	Plants, fungi, potato	*o*-Dihydroxy compounds	*o*-Quinones
1.10.3.3	Ascorbate oxidase	Squash	L-Ascorbate	Dehydroascorbate
1.10.3.2	*p*-Diphenol oxidase (Laccase)	Plants, fungi, Lac tree	*p*-Dihydroxy compounds	*p*-Quinones
1.2.3.2	Xanthine	Some plants eg lupin	Xanthine	Uric acid

An extensive listing of enzymes is given by Dixon and Webb (1964).
EC = Enzyme Code

Nickel Enzymes

An important development over the last decade has been the demonstration of a requirement for nickel in four enzymes: urease, hydrogenase, carbon monoxide dehydrogenase (CODH) and methyl-*S*-coenzyme M reductase. The origin of the unusually low Ni(III)/Ni(II) potentials in some of these enzymes has been studied, and it has been concluded that polarisable, electron-rich ligands are of importance (Kruger *et al.*, 1991). Recent spectroscopic studies indicate that the nickel-containing active site of the FeNi hydrogenase isolated from *Thiocapsa rosepersicina* consists of pentacoordinated nickel(II) with two sulphur and three nitrogen donors (Maroney *et al.*, 1990).

Urease

Urease (urea amidohydrolase) from Jack bean was the first enzyme shown to contain nickel. The enzyme catalyses the hydrolysis of urea to ammonia and carbon dioxide (Mobley and Hausinger, 1989). The protein has a multimeric structure with a molecular weight of 590000 Da and analysis indicates 12 nickel atoms per mole. Binding studies with inhibitors suggest an equivalent weight per active site of 105000 corresponding to 2 nickel atoms per active site (Dixon *et al.*, 1975). The nickel(II) can be removed with EDTA at pH 3.7, and the optical absorption and enzymic activity can be correlated with the nickel(II) content as the metal is removed. The enzyme is also sensitive to the chelating agents acetohydroxamic acid and phosphoramidate indicating that the nickel is essential for the activity of the enzyme (Dixon *et al.*, 1975).

The nickel is present as nickel(II) and appears to be diamagnetic. The spectrum obtained in the presence of the competitive inhibitor mercaptoethanol, after correction for Rayleigh scattering by the protein, has absorption bands at 324, 380 and 420 nm, with molar absorption coefficients of 1550, 890 and 460 $dm^3\ mol^{-1}\ cm^{-1}$, respectively (Fig. 5-1). These bands were assigned to sulphur-to-nickel charge transfer transitions. The spectrum is changed by the addition of other inhibitors, such as acetohydroxamic acid (Fig. 5-1B). Similar spectra have been observed in other nickel(II) proteins including Ni(II) carboxypeptidase, which also undergoes spectral changes on the addition of the inhibitor β-phenyl propionate (Rosenberg *et al.*, 1975). In the native zinc(II) carboxypeptidase the zinc is ligated by His-69, Glu-72 and His-196 but in the nickel(II) enzyme the nickel is believed to be coordinated by one cysteine sulphur, two histidine imidazoles and the carboxylate group of a glutamic acid residue.

EXAFS studies (Hasnain and Piggott, 1983; Alagna *et al.*, 1984) have shown that the spectrum of urease is similar to those of benzimidazole complexes (Fig. 5-2) suggesting that the nickel is coordinated by histidine nitrogen and oxygen ligands.

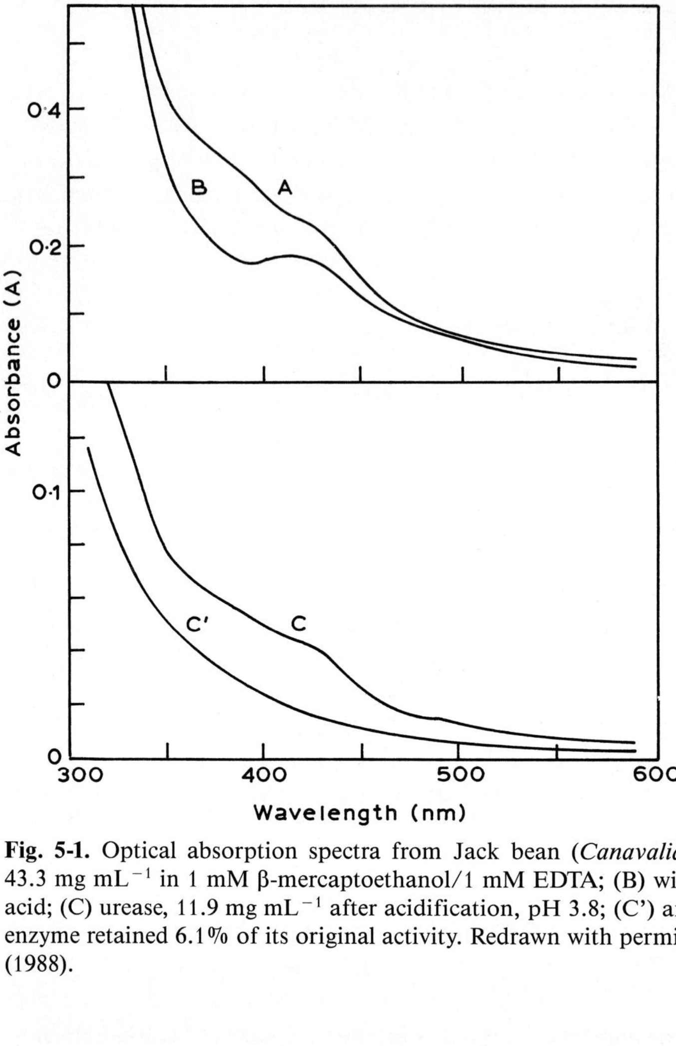

Fig. 5-1. Optical absorption spectra from Jack bean (*Canavalia ensiformis*). (A) Enzyme, 43.3 mg mL^{-1} in 1 mM β-mercaptoethanol/1 mM EDTA; (B) with 10 mM acetohydroxamic acid; (C) urease, 11.9 mg mL^{-1} after acidification, pH 3.8; (C') after 2 h at pH 3.8; the latter enzyme retained 6.1% of its original activity. Redrawn with permission from Carnmack *et al.* (1988).

Variable temperature magnetic circular dichroism (MCD) studies have revealed similar electronic and magnetic properties for the nickel centres in Jack bean and *Klebsiella aerogenes* urease. Native and 2-mercaptoethanol and acetohydroxamic inhibited forms of both enzymes have been investigated and in each case the energy of the temperature dependent MCD for the nickel(II) $d-d$ transitions indicates octahedral coordination with mainly oxygen donors.

The enzyme contains six identical subunits (M = 96600) arranged in the form of an octahedron. Each subunit contains two tightly bound nickel ions (% nickel = 0.12). After the electronic spectrum of the enzyme has been corrected for light scattering the $d-d$ bands associated with nickel(II) at 407, 747 and 1060 nm are consistent with nickel in an octahedral environment. In the presence of acetohydroxamic acid the

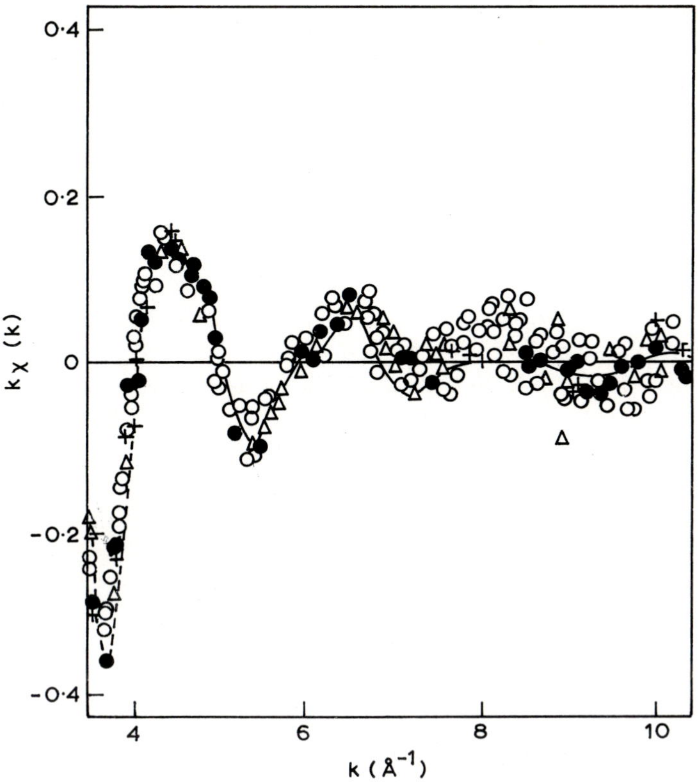

Fig. 5-2. Ni K-edge EXAFS spectrum of urease. The curve (+) is calculated for a single type of nickel site and the minimisation of parameters was based on those for the model complexes Ni(1-*n*-propyl-2-hydroxybenzylbenzimidazole)$_3$(ClO$_4$) and Ni(2-hydroxymethylbenzimidazole)$_3$Br$_3$. Atoms (with distances in nm given in parentheses) in the simulation were N (0.204), O (0.206), O (0.225), C (0.294), C (0.312), N (0.392) and C (0.394). Reproduced with permission from Blakely and Zerner (1984).

ligand field bands move to higher energy indicating direct binding to nickel(II) with retention of octahedral geometry. In addition there are profound changes in the magnetic properties of the nickel centre. In conjunction with magnetic susceptibility data for Jack bean urease (Clark and Wilcox, 1989) the results indicate that acetohydroxamic acid binding disrupts the exchange interaction in the binuclear Ni active site.

The *Klebsiella aerogenes* urease operon has been cloned, sequenced and shown to encode six genes (Mulrooney and Hausinger, 1990). The *ure*A, *ure*B and *ure*C genes encode the three urease subunits (M_r = 60300 (α), 11700 (β) and 11000 (τ)) which associate into an $\alpha_2\beta_4\tau_4$ complex containing two bi-Ni active sites (Todd and Hausinger, 1989). The precise functions of the *ure*E, *ure*F and *ure*G genes are unknown, but deletion of all three genes led to the synthesis of urease apo-enzyme. These ‘accessory genes’ appear to be involved in metal centre biosynthesis.

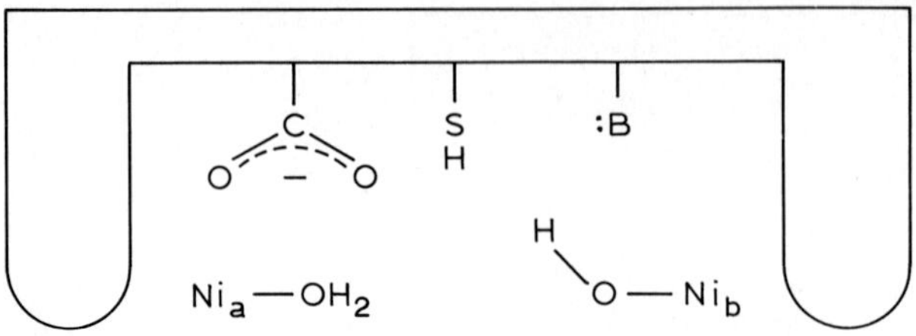

1

Biophysical studies of the urease metal centre in the presence and absence of inhibitors, in conjunction with kinetic data provide the model of the bi-Ni site shown in **1**. Certain inhibitors are thought to bridge the two nickel atoms consistent with a bridged transition state during urea hydrolysis. The ligands for nickel are believed not to contain sulphur, however, an essential cysteine is proximal to the active site. Comparisons of diethylpyrocarbonate reactivity for apo- and halo-enzyme are consistent with His as a ligand to nickel (Lee *et al.*, 1990).

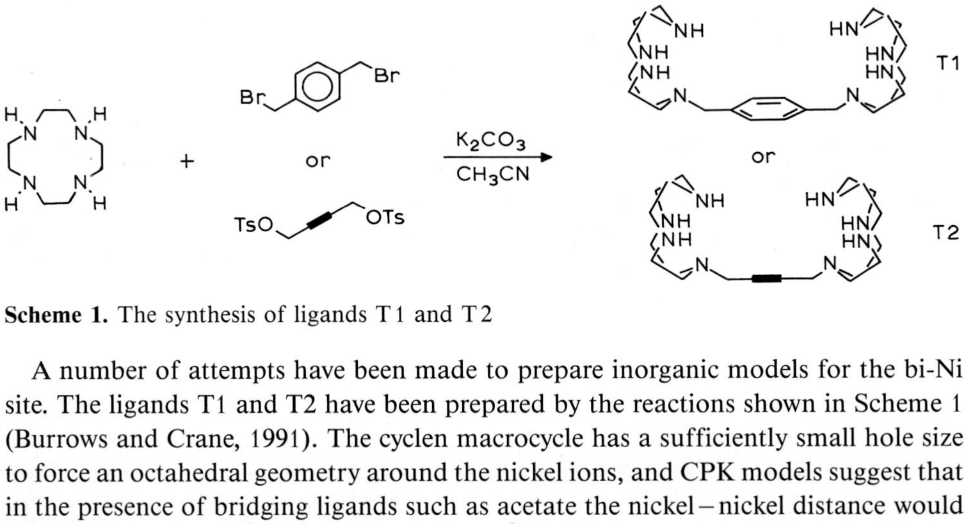

Scheme 1. The synthesis of ligands T1 and T2

A number of attempts have been made to prepare inorganic models for the bi-Ni site. The ligands T1 and T2 have been prepared by the reactions shown in Scheme 1 (Burrows and Crane, 1991). The cyclen macrocycle has a sufficiently small hole size to force an octahedral geometry around the nickel ions, and CPK models suggest that in the presence of bridging ligands such as acetate the nickel–nickel distance would be between 3.4–3.8 Å for T1 and 3.2–3.6 Å for T2. Preliminary kinetic studies on the effects of the dinickel complexes on the hydrolysis of 2,4-dinitrophenyl phosphate have been carried out. The dinickel complex of T1 led to a 5–6 fold rate enhancement. However, experiments using HEPES buffer gave no rate enhancement, suggesting that the buffer binds to the nickel sites.

Dixon *et al.* (1980) have proposed a mechanism for the action of urease (Fig. 5-3). The mechanism involves nucleophilic attack by hydroxide coordinated to one nickel, on urea which is coordinated to the other nickel(II) via the carbonyl oxygen.

A number of model studies have been carried out using cobalt(III) and rhodium(III) complexes of O- and N-bonded ureas (Dixon *et al.*, 1982, 1983; Curtis

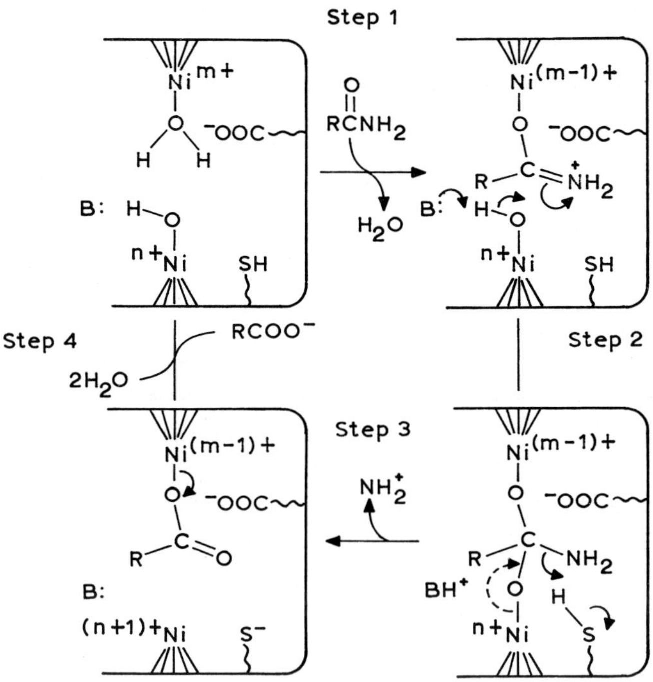

Fig. 5-3. Proposed reaction cycle for urease. For urea, R= NH_2. Step 1: urea is activated toward nucleophilic attack by O coordination to a nickel ion; the $=N^+H_2$ is stabilised by interaction with a protein carboxylate. Step 2: nucleophilic attack by an hydroxide ion, coordinated to the second nickel, to form a tetrahedral intermediate. Step 3: breakdown of the tetrahedral intermediate to form a coordinated carbamate ion. Step 4: hydrolysis releases carbamate ion, the initial product of urease on urea. Reproduced with permission from Blakely and Zerner (1984).

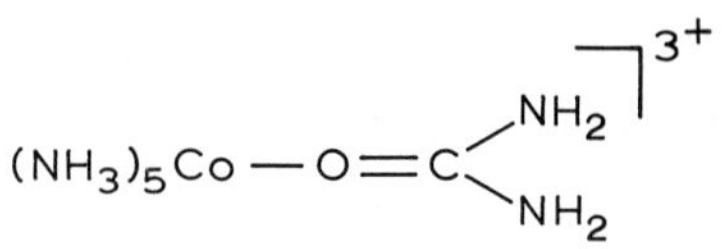

2

et al., 1983). In the base hydrolysis of the complex **2** it was found that O-coordination of urea to cobalt(III) did not provide sufficient activation of the carbonyl group to enable the anticipated urea hydrolysis to give the $[(NH_3)_5CoO_2CNH_2]^{2+}$ ion to compete with the facile base-catalysed (S_NICB) hydrolysis of the cobalt-oxygen bond (Dixon *et al.*, 1982). The deprotonated N-bonded dimethylurea complex $[(NH_3)_5CoNHCON(CH_3)_2]^{2+}$ has also been prepared (Dixon *et al.*,1983). This complex is stable in basic solution but rapidly isomerises to its O-bonded linkage isomer in basic solution.

Peroxidases, Catalases, Superoxide Dismutases

The presence of oxygen in the cell gives rise to toxic molecules such as superoxide ($O_2^{\bullet -}$), hydroxide radicals and hydrogen peroxide. Organisms eliminate superoxide using superoxide dismutase which catalyses the disproportionation or dismutation of superoxide to dioxygen and hydrogen peroxide ($2O_2^{\bullet -} + 2H^+ \rightarrow O_2 + H_2O_2$) which produce more hydrogen peroxide. Catalases convert hydrogen peroxide into water and dioxygen, whereas peroxidases reduce hydrogen peroxide to water and oxidise various substrates. Catalases of all living organisms have a similar structure, however, many peroxidases are structurally unrelated. Among the heme-containing peroxidases, animal peroxidases such as thyroid, eosinophil, lacto- and myeloperoxidases constitute a separate superfamily (Welinder, 1991; Fenna *et al.*, 1992). Similarly, the fungal chloroperoxidase from *Caldariomyces fumago* is structurally distinct, and the bacterial cytochrome c peroxidase from *Pseudomonas aeruginosa* contains a cytochrome c type heme. By contrast, yeast (*Saccharomyces cerevisiae*) cytochrome c peroxidase is the prototype structure for the plant peroxidase family of homologous proteins which are the subject of this discussion. The chemistry and biology of the plant peroxidase superfamily have been the subject of extensive recent reviews and surveys (Everse *et al.*, 1991; Lobarzewski *et al.*, 1991; Gaspar *et al.*, 1992).

Peroxidases

Peroxidases are enzymes catalysing the oxidation of a variety of organic and inorganic compounds by hydrogen peroxide or related compounds. An extensive list of the sources of peroxidase in plants and animals has been given (Saunders *et al.*, 1964). All of the peroxidases purified so far from plants contain the prosthetic group hemin or ferriprotoporphyrin IX, **3** (Dunford and Stillman, 1976; Hewson and Hager, 1979). Horseradish roots and the sap of fig trees are the richest source of plant peroxidases.

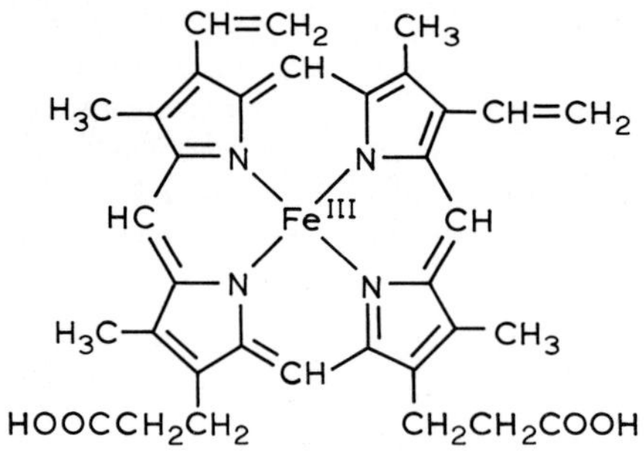

3

Peroxidases from animal sources which have been studied are thyroid peroxidase, lactoperoxidase, myeloperoxidase and glutathione peroxidase. Lactoperoxidase and myeloperoxidase are known as verdoperoxidases because of their green colour. Neither contain protoporphyrin IX as prosthetic group.

In addition to plant and animal sources, peroxidases are also found in mould, bacteria and microorganisms. A peroxidase from the mould *Caldariomyces fumago*, chloroperoxidase, has been isolated and characterised. Like the plant peroxidases it has ferriprotoporphyrin IX as the prosthetic group. In many of its chemical and physical properties chloroperoxidase is similar to horseradish peroxidase, but it has the unique ability amongst peroxidases to catalyse the oxidation of chloride ion (Hager *et al.*, 1966; Morris and Hager, 1966).

Table 5-2. Some properties of the peroxidases.

	Mol. weight	Prosthetic group	Carbo-hydrate content (%)	Crystallisation
Horseradish peroxidase (HRP)	40500[a] 39800[b]	Ferriprotoporphyrin IX	18.4 18	Microscopic needles
Cytochrome c peroxidase (CcP)	34100	Ferriprotoporphyrin IX	0	Long prisms
Chloroperoxidase (ClP)	40200[a] 42000[b]	Ferriprotoporphyrin IX	25–30	Brown needles
Lactoperoxidase (LP)	76500[a] 77500[b]	Derivative of mesoheme IX?	8	Needles
Thyroid peroxidase (ThP)	62000[c]	Not ferriprotoporphyrin IX?	–	–
Japanese radish peroxidase a (JRPa)	55700[f] 55500[d]	Ferriprotoporphyrin IX	28	Rhombic prisms
Japanese radish peroxidase c (JRPc)	41500[a]	Ferriprotoporphyrin IX		Tetragonal prisms
Myeloperoxidase (MP)	149000[b]	Two atoms of porphyrin bound iron	0	Needles
NADH peroxidase	12000[c]	FAD	–	–
Turnip peroxidase (TuP) A_1	49000[d]	Ferriprotoporphyrin IX	–	Fine needles
A_2	45000[d]			
B	65000[d]			
D	43000[d]			
Glutathione peroxidase	90000[d]	One atom of Se per sub-unit	–	–

[a] Heme content
[b] Hydrodynamic measurement
[c] Gel filtration
[d] Chemical analysis
[e] FAD content
[f] Osmotic pressure

Data from Dunford and Stillman (1976).

Most peroxidases are glycoproteins (exceptions are cytochrome c peroxidase and myeloperoxidase). The purpose of the carbohydrate component in these enzymes is uncertain. However, horseradish peroxidase has long been known to be a very stable enzyme, even under conditions of high temperature. In part this may be due to the carbohydrate component. Table 5-2 summarises some of the properties of the peroxidases and the spectral parameters are summarised in Table 5-3.

Table 5-3. Absorption spectra of peroxidases and derivatives (λ_{max} in nm)

	Soret	Visible	NIR
Horseradish peroxidase	403	498, 640	950, 1065, 1145
HRP-N_3	416	495 sh, 534, 565, 635	
HRP-CN	423	439, ~580 sh	
HRP-F	404	488, 530 sh, 560, 612	
Ferro-HRP	440	~510 sh, 557, ~580 sh	
Ferro-HRP-CO	423	542, 572	
Ferro-HRP-CN	432	536, 566	
HRP-I	400	~525 sh, 577, 622, 651	
HRP-II	420	527, 554	
HRP-III	416	546, 583, 673	
HRP-OH	416	542, 571, 633	
Chloroperoxidase	403	515, 542, 650	
CIP-N_3	432	549, 583, 665	
CIP-CN	437	555	

Data from Dunford and Stillman (1976). sh = shoulder.

Horseradish peroxidase (HRP) is a hemeprotein which catalyses the oxidation of a large variety of inorganic and organic substrates (Dunford and Stillman, 1976). Chloroperoxidase (CPO) is a versatile heme enzyme since it shares similar properties with classical peroxidases and P-450 monooxygenases and also catalyses the oxidative halogenation of organic substrates (Blake and Hager, 1990).

Vanadium bromoperoxidases have recently been discovered in brown and red seaweeds (Wever and Kustin, 1990) and constitute a new class of enzymes that contain vanadium as a prosthetic group. These enzymes have a molecular weight of ca. 90000 Da and in the presence of hydrogen peroxide are able to oxidise bromide to hypobromous acid. It is possible to remove the prosthetic group and to reconstitute the native enzyme by the addition of vanadate (V(V)). The affinity of the apoenzyme for vanadate is quite high with a dissociation constant of 55 nmol dm^{-3}. The dissociation of vanadate from the native enzyme is very slow requiring several weeks for completion. ESR measurements show that vanadium in the enzyme is present as vanadium(V). Reduction gives a vanadyl (VO^{2+}) type of ESR spectrum. Similarly, the K-edge X-ray absorption spectra indicate that vanadium in the native enzyme is in the +5 oxidation state which upon reduction is converted to the +4 state. EXAFS spectra for the native enzyme are consistent with the presence of one short vanadium–oxygen bond.

Reconstitution of the apobromoperoxidase is inhibited by tetrahedral ions such as phosphate, arsenate and molybdate and also very strongly by the fluoro complexes AlF_4^- and BeF_4^-. These observations suggest that these ions bind to the vanadium site of the apoenzyme. Steady state kinetic data (de Boer and Wever, 1988) show that during the catalysis hydrogen peroxide first binds to the active site and the bromide then reacts with the peroxo-complex, resulting in the formation of HOBr. When the vanadium in the enzyme was reduced to the +4 oxidation state, the enzyme lost its brominating activity. The role of the metal ion is the binding and activation of the hydrogen peroxide to allow subsequent reaction with bromide. Spectroscopic evidence for the existence of a peroxo-complex has recently been obtained (Tromp *et al.*, 1990). Mixing the enzyme with hydrogen peroxide gives changes in the absorption spectrum assigned to peroxo-complex formation. The peroxo-complex is stable for hours but rapidly decays in the presence of bromide.

Horseradish Peroxidase

Horseradish peroxidase (and urease) played an important role in the development of the modern concept of the nature of an enzyme and the role of metal ions (Sumner and Somers, 1943; Willstätter, 1965). The species now known as compound II (HRP-II) formed as a result of the reaction of HRP with H_2O_2, was discovered in 1937 (Keilen and Mann, 1937). Later compound I (HRP-I), formed prior to HRP-II was identified (Theorell, 1941). The spectra of HRP-I and HRP-II in the 400 nm (Soret band) region have been determined (Chance, 1949a, b) and measurements have also been extended to the visible region (Chance, 1952). Formation of HRP-I is first order in H_2O_2 and HRP (Chance, 1943) and the –OOH group is essential for the oxidation of HRP by peroxide. The enzymatic cycle can be summarised by the following equations (George, 1952),

$$\text{HRP} + H_2O_2 \rightarrow \text{HRP-I} \quad (1)$$
$$\text{HRP-I} + AH_2 \rightarrow \text{HRP-II} + \cdot AH \quad (2)$$
$$\text{HRP-II} + AH_2 \rightarrow \text{HRP} + \cdot AH \quad (3)$$

Reactions (1)–(3) are independent of pH. Detailed reviews of the enzyme are available (Dunford and Stillman, 1976; Dunford, 1991).

Crystal Structures of Peroxidases

Yeast cytochrome c peroxidase (CCP) is present in mitochondria and is induced under aerobic growth conditions. The enzyme preferentially oxidises cytochrome c, although small molecules can also act as substrates (De Pillis *et al.*, 1991). The 1.7 Å resolution coordinates for yeast CCP are currently available (Finzel *et al.*, 1984). As a result, CCP is a prototype structural model for peroxidases, although it shares less than 20% sequence identity with the secretory plant and fungal peroxidase (Welinder, 1985; Henrissat *et al.*, 1990). Morita and coworkers (1991) have recently presented a crystallographic model at 3.1 Å resolution of the first classical plant peroxidase, horseradish peroxidase isoenzyme E5 (HRP E5) (Fig. 5-4). The peroxidase structure contains 10 conserved helices (A–J) embedding a protochemin IX prosthetic group.

Conserved histidine-containing sequences were recognised at an early stage as the probable heme-linked sequences in classical plant peroxidases (Welinder and Mazza, 1977). Similar sequences were later found in CCP (Takio *et al.*, 1980) and recently in

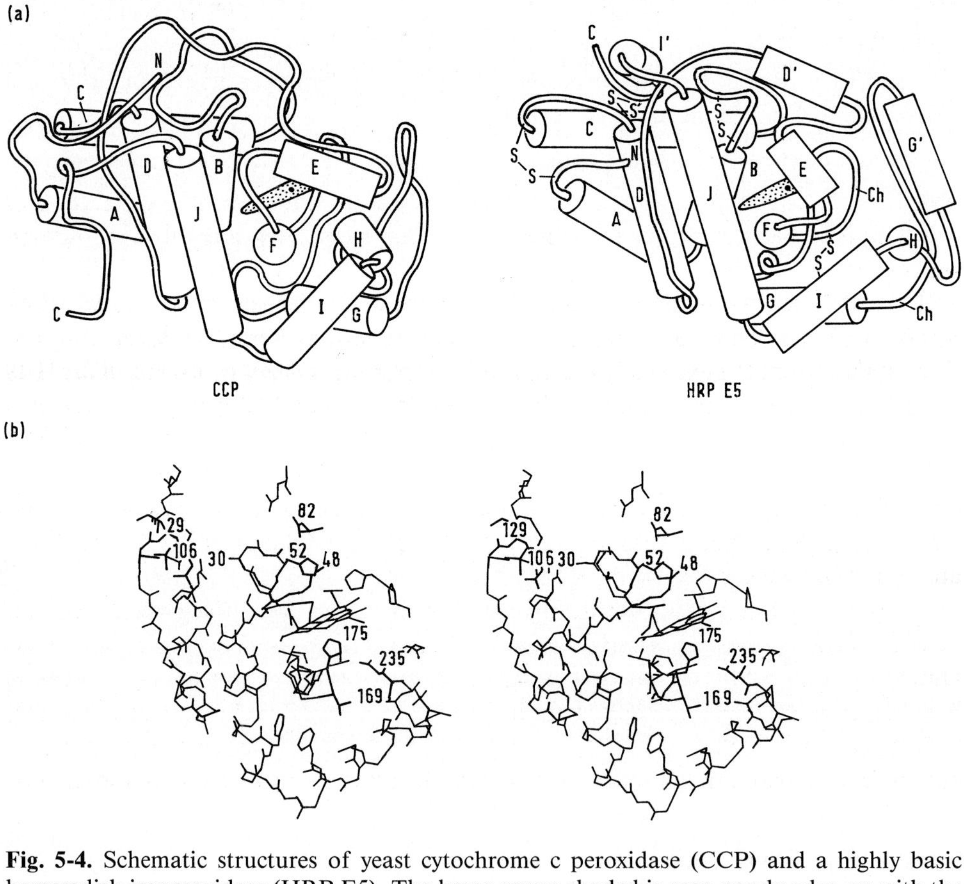

Fig. 5-4. Schematic structures of yeast cytochrome c peroxidase (CCP) and a highly basic horseradish isoperoxidase (HRP E5). The heme group shaded is seen nearly edge on with the iron atom indicated by a black dot. α-Helices are labelled A–J with a prime denoting additional helices present in HRP E5 but not in CCP. The position of cystine bridges (S–S), carbohydrate (Ch) and the amine (N) and carboxyl (C) termini are also indicated. Reproduced with permission from Welinder (1992).

fungal ligninase (Tien and Tu, 1987), bacterial peroxidase (Welinder, 1991) and ascorbate peroxidase (Mittler and Zilinkas, 1991). His-52 and His-175 (CCP numbers) are invariant. His-75 of helix F is coordinated to heme iron at the closed proximal site and is hydrogen bonded to the invariant Asp-275 of helix H as indicated in Fig. 5-4 (b). His-52, together with Arg-48 of helix B at the accessible distal end of heme, catalyses the heterolytic cleavage of hydrogen peroxide (Poulos, 1988; Dunford, 1991). In CCP, His-52 is hydrogen bonded to Asn-82.

Catalases

Catalases are enzymes which decompose H_2O_2 into H_2O and O_2 by the reaction $2H_2O_2 \rightarrow 2H_2O + O_2$. Catalases and peroxidases are related enzymes as they are both capable of promoting hydrogen peroxide oxidation by mechanisms which involve similar enzymatic intermediates. The peroxidatic activity of catalases is, however, low when compared with true peroxidases. Catalases are widely distributed among animals, plants and bacteria. All the catalases isolated so far have similar molecular weights (ca. 25000) and contain four iron atoms per molecule. The iron is present in four protoporphyrin IX prosthetic groups per molecule in all but the liver catalases where one or more prosthetic groups have been degraded to bile pigment hematins.

Recently manganese catalases have been discovered, the first being obtained by Kono und Fridovich (1983) from the lactic acid bacterium *Lactobacillus plantarum*. The catalase from the extremely thermophilic bacterium *Thermus thermophilus* HB8 was isolated by Barynin and Grebenko in 1984 (Barynin and Grebenko, 1986) and is very thermally stable, retaining 85% of its activity after heating at 95 °C for 10 min. The enzyme is inhibited by sodium azide and has a molecular weight of 210000 and consists of six identical subunits. Low temperature ESR spectroscopy (Barynin and Grebenko, 1986) indicates that the active centres of Mn-catalase from *T. thermophilus* are binuclear manganese complexes.

X-ray diffraction studies at 3.0 Å resolution have indicated the presence of two neighbouring manganese atoms (3.6 – 0.3 Å) in each subunit (Barynin *et al.*, 1986). Unlike the iron catalases, the Mn-catalases contain no heme and the active centres have an unusual structure. The subunit skeleton is a bundle of four nearly parallel helices, two of which are ca. 40 Å long. There are similarities to the structure of apoferritin, however, the polypeptide chain in Mn-catalase is twice that of apoferritin.

Superoxide Dismutases

Superoxide dismutase is an enzyme which catalyses the dismutation of superoxide into O_2 and H_2O_2 (Valentine and Pantoliano, 1981; Oberley, 1982; Bannister *et al.*, 1987): $2O_2^{\bullet -} + 2H^+ \rightarrow H_2O_2 + O_2$

The enzyme was first discovered in 1968 and plays an important role in the defences evolved to deal with the toxicity of dioxygen.

Superoxide dismutases have been isolated from a wide variety of eukaryotes including yeast, wheat germ, garden peas, chicken liver and erythrocytes. These enzymes contain copper and zinc. Copper containing proteins which also display this catalytic activity had been isolated from blood, brain and liver tissues many years previously and were known as *erythrocuprein, cerebrocuprein* and *hepatocuprein.*

Bovine superoxide dismutase (BSOD) is a dimeric enzyme with a molecular weight of 32000 containing two Cu(II) and two Zn(II) ions. The X-ray structure of the enzyme at 2 Å resolution is available (Tainer, 1982, 1983; Getzoff *et al.* 1982). The two metals share a common bridging imidazolate residue from the side chain of His-61 (Fig. 5-5). The coordination around copper is distorted square pyramidal, the other donors being three histidines and a water molecule. The coordination about zinc is pseudo-tetrahedral with the bridging imidazolate, two histidines and one aspartate carboxyl acting as donors. The enzyme can be regarded as a heterodinuclear metal complex (Fig. 5.5).

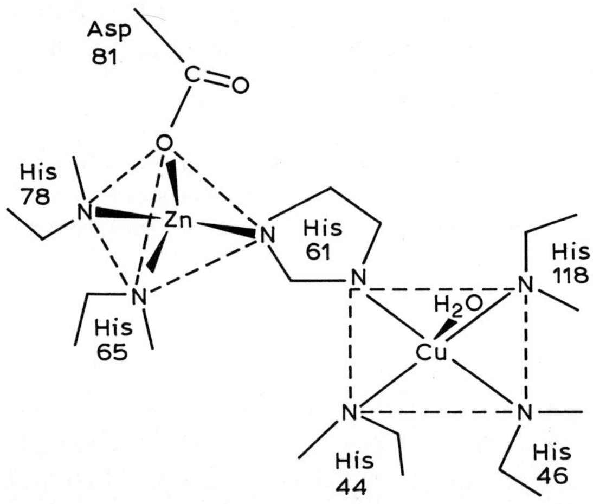

Fig. 5-5. Schematic drawing of the active site of bovine superoxide dismutase.

Cytochromes P-450

Cytochrome P-450 enzymes have been isolated from a variety of mammalian tissues, insects, plants, yeasts and bacteria. The P-450 cytochromes (Gunter and Turner, 1991) are membrane bound mono-oxygenase enzymes which catalyse oxygen atom transfer to entrapped non-polar substrates. The binding of carbon monoxide to the enzyme produces a split in the 420 nm Soret band to give bands at 364 and 450 nm. The absorption at 450 nm distinguishes the hemoprotein from all others and hence provides its name. Recently it has been established that dioxgen binding brings about a similar spectral change. These enzymes are not cytochromes as their main function is oxygen atom transfer rather than electron transfer.

The enzymes reductively activate dioxygen using NADPH as an electron source. One oxygen atom is then reduced to water and the other atom is transferred to a substrate, resulting in the hydroxylation of alkenes and arenes, the epoxidation of alkenes and the formation of N-oxides and S-oxides from amino and sulphur compounds. Other P-450 reactions include N-dealkylation, O-dealkylation and reductase-like dehalogenation of halocarbons. Typical P-450 reactions are summarised in Table 5-4.

Table 5-4. Typical P-450 reactions.

Reaction type	Simplified example	Typical substrate
Aliphatic hydroxylation	Cyclohexane → cyclohexanol	Pentobarbital
Aromatic hydroxylation	Benzene → phenol	Phenobarbital
Alkene epoxidation	Cyclohexene → cyclohexene oxide	Aldrin
N-Dealkylation	$CH_3N(H)CH_3 \rightarrow CH_3NH_2 + H_2C{=}O$	Methadone
O-Dealkylation	$C_6H_5OCH_3 \rightarrow C_6H_5OH + H_2C{=}O$	Codeine
Oxidative deamination	$(CH_3)_2CHNH_2 \rightarrow (CH_3)_2C{=}O + NH_3$	Amphetamine
S-Oxidation	$CH_3SCH_3 \rightarrow (CH_3)_2S{=}O$	Chlorpromazine
Reductive dehalogenation	$C_6H_5CH_2Br \rightarrow C_6H_5CH_3$	Halothane

Data from Dawson and Sono (1987).

Substrates processed by P-450 include endogenous compounds such as steroids, fatty acids, leukotrienes and prostoglandins, as well as exogenous drugs, pesticides, anaesthetics, solvents and chemical carcinogens. The bacterial camphor-hydroxylating enzyme from *Pseudomonas putida* was the first P-450 to be purified and crystal structures have recently been reported for the resting and substrate bound states of PC-450_{cam}. Cytochrome P-450_{cam} is an asymmetrically shaped hemoprotein resembling a triangular prism which is about 30 Å thick with a maximum dimension of around 60 Å. The structure is dominated by 12 helical segments that are arranged in three layers stacked one on top of the other with a protoporphyrin IX haem sandwiches between. The helical topography of the protein is very similar to that found in globins and cytochrome c peroxidase (Gunter and Turner, 1991). The camphor binding site is lined with hydrophobic residues and the substrate is buried about 4 Å from the porphyrin plane, directly adjacent to the O_2 binding site (Figs. 5-6 and 5-7). The iron porphyrin is a nearly independent entity, as in other hemoproteins, being attached to the protein by a single cysteine residue. The important features of the catalytic cycle (Fig. 5-8) can be summarised (Mansuy, 1987).

water ligand is displaced. The change of spin state facilitates the uptake of an electron, and subsequent dioxygen binding forms a stable low-spin hexacoordinate intermediate. A second electron is then accepted and heterolytic cleavage liberates water to form an active oxo-iron intermediate which is responsible for the rate determining step of substrate oxidation and product release. Crystal structures have been reported for iron porphyrin models of the first four intermediates of the reaction cycle. The reactive oxo-intermediate is the least understood component of the oxidation process.

Copper Proteins

Copper proteins are involved in a variety of biological functions, including electron transport, copper storage and many oxidase activities. A variety of reviews on this topic are available (Sykes, 1985; Chapman, 1991). Several copper proteins are easily identified by their beautiful blue colour and have been labelled 'blue copper' proteins. The blue copper proteins can be divided into two classes, the oxidases (laccase, ascorbate oxidase, ceruloplasmin) and the electron carriers (plastocyanin, stellacyanin, umecyanin, etc.).

Copper(II) sites in proteins can be classified into three types based on their spectral properties. The blue (Type I) copper proteins are characterised by a visible absorption

Table 5-5. Some characteristics of the blue copper proteins.

Protein (and source)	M_r	I_{max} (nm)	ε_{max} (mol^{-1} cm^{-1})	E° (mV)	g_{II}	A_{II} (cm^{-1} $\times 10^3$)
Plastocyanin (plants/algae)[a]	10000	597	4500	370	2.23	6.3
Azurin (bacteria)[b]	14000	626	4800	330	2.26	6.0
Stellacyanin (*Rhus vernicifera*)	20000[c]	609	4800	184	2.287	3.5
Umecyanin (*Armoracia lapatifolia*)	14600	610[d]	3400	283	2.317	3.5
Rusticyanin (*Thiobacillus ferrooxidans*)	16500	597	2240	680[e]	2.229	4.5
Basic blue protein (cucumber)	10100	597	3400	317	2.207	5.5

[a] Data are for spinach plastocyanin
[b] Data are for *P. aeruginosa* azurin
[c] 40% carbohydrate
[d] pH dependent
[e] Value determined at pH 2.0

band near 600 nm with an absorption coefficient of ca. 4000–5000 dm^3 mol^{-1} cm^{-1}, a small hyperfine coupling constant in the g_{II} region of the ESR spectrum and a relatively high (but variable) redox potential compared with values found for 'typical' copper complexes (Table 5-5). The 600 nm band is considered to arise from an L→M charge transfer transition involving a copper-cysteine bond. In 1978 with the publication of the X-ray crystal structure of poplar plastocyanin (Guss and Freeman, 1983) the first detailed structure of a Type I blue copper site became available (Figs. 5-9 and 5-10).

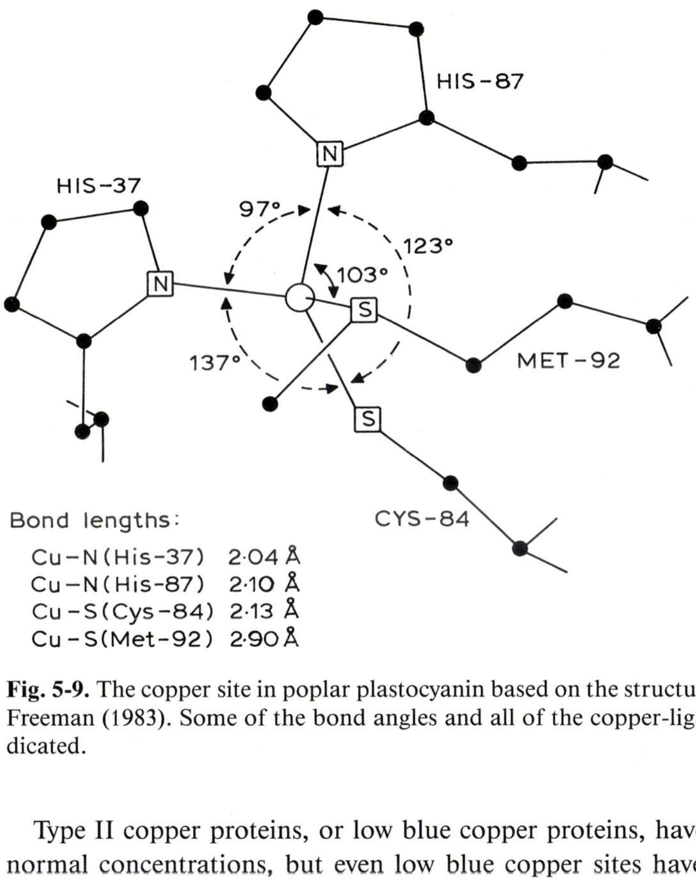

Fig. 5-9. The copper site in poplar plastocyanin based on the structure determined by Guss and Freeman (1983). Some of the bond angles and all of the copper-ligand bond distances are indicated.

Type II copper proteins, or low blue copper proteins, have less intense colours at normal concentrations, but even low blue copper sites have quite high absorption coefficients when compared with simple copper(II) coordination compounds. Bovine erthyrocyte superoxide dismutase (BSOD) is an example of a low blue copper(II) protein with λ_{max} 680 nm (ε = dm^3 300 mol^{-1} cm^{-1}).

Type III copper(II), found for example in *Rhus laccase*, is ESR inactive. Although copper(II) is present no ESR spectrum can be obtained. Recent magnetic susceptibility measurements on *Rhus laccase* indicate an antiferromagnetically coupled copper(II) dimer.

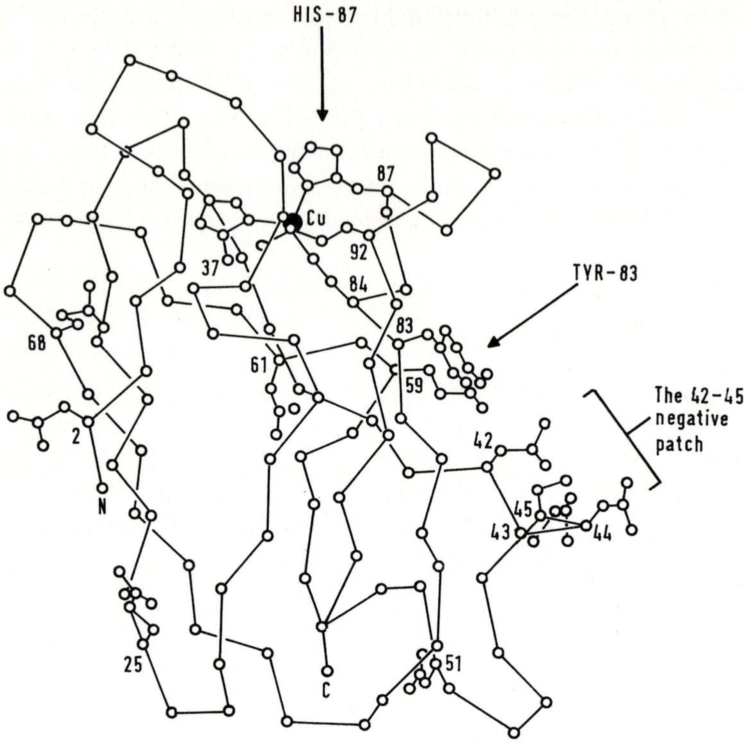

Fig. 5-10. The 3-D structure of poplar plastocyanin based on the structure determined by Guss and Freeman (1983). The entire α-carbon chain, the copper atom and ligand residues are shown. The 42–45 negative patch, Tyr-83 and His-87 are highlighted.

Plastocyanins

The plastocyanins are blue copper proteins found in the chloroplasts of higher plants and algae where they mediate electron transport between cytochrome f and P-700 (Barber, 1983; Haehnel, 1984, 1986; Cramer *et al.*, 1985; Sykes, 1985; Andersen *et al.*, 1987). Plastocyanins each contain one copper bound by a single polypeptide chain of molecular weight around 10500 (Sykes, 1985). The spectroscopic properties of the copper are those of a typical blue site. The properties of the plastocyanins have been the subject of detailed reviews (Sykes, 1985; Haehnel, 1986; Chapman, 1991).

Plastocyanins from 15 plants, three algae and a cyanobacterium (or blue-green alga) have now been fully sequenced (Chapman, 1991). The amino acid sequences show a high degree of conservation with around 50% identity amongst the higher plant plastocyanins. The 3-D crystal structure of poplar plastocyanin has now been refined to 1.6 Å resolution by Freeman and coworkers (Guss and Freeman, 1983) and is illustrated in Fig. 5-10. The protein is roughly cylindrical, with approximate dimensions of 40 Å × 32 Å × 28 Å. The peptide chain is folded into eight strands, seven of which have substantial β-sheet character. The β-strands are connected by loops at the end of the cylinder. The copper is located at one end of the cylinder, buried 6 Å below the protein surface. The imidazole ring of His-87, one of the ligands to copper, is surface accessible and is centred in a pocket of three loops formed by highly conserved non-polar residues. The region around His-87 has been proposed as a possible electron entry site and is often referred to as the 'north' site (Guss and Freeman, 1983), or simply the hydrophobic patch.

Recently the crystal structure of reduced, Cu(I), poplar plastocyanin has been reported (Guss *et al.*, 1986). Superposition of the structure of the reduced and oxidised proteins at near neutral pH shows no large differences in conformation. However, as the pH is lowered there are definite changes at the copper site. The copper(I) changes from a distorted tetrahedral coordination geometry at pH 7.8 to trigonal planar geometry at pH 3.8 (Guss *et al.*, 1986; Fig. 5-11). It appears that lowering the pH causes protonation of His-87 resulting in cleavage of the Cu(I)–N(His-87) bond. A trigonal geometry would be expected to stabilise Cu(I), making oxidation to Cu(II) difficult. It has been suggested that protonation of His-87 acts as part of a regulator mechanism preventing electron transfer to oxidants other than P-700 (Guss *et al.*, 1986).

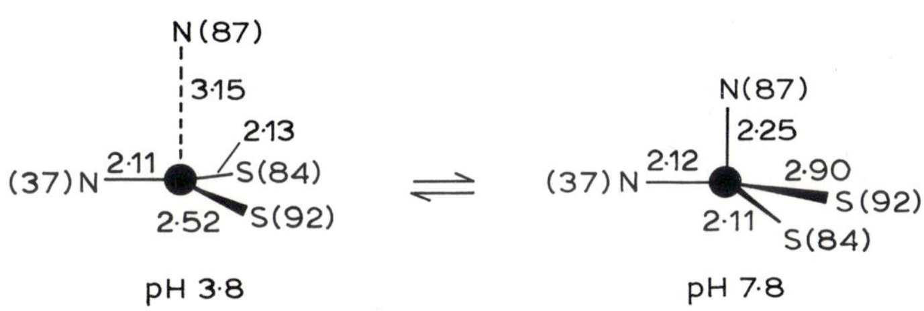

Fig. 5-11. The effect of pH at the Cu(I) site in poplar plastocyanin as determined by Guss *et al.* (1986). Bond distances are shown in Ångstroms and residue numbers are indicated in parentheses.

The crystal structure of apoplastocyanin has also been determined by Freeman and coworkers to 1.8 Å resolution (Garret *et al.*, 1984). The structure closely resembles that of the haloprotein and the positions of the ligand residues vary by only 0.1 to 0.3 Å between apo- and haloproteins. It appears that the folding of the protein imposes the unique geometry of the Type I site on copper. It as also noted that the imidazole ring of His-87 had flipped by 180° as observed in the reduced protein at low pH, and this observation led to the proposal that the imidazole acts like a 'revolving door' providing access to the copper site.

Cucumber Basic Blue Protein

Basic blue proteins have been isolated from a number of plant sources and have previously been referred to as 'plantacyanin' (Aikazyan and Nalbandyan, 1975, 1979; Sakurai *et al.*, 1982). The protein from cucumber has been the most extensively studied of the basic blue proteins and the crystal structure is now available (Guss *et al.*, 1988). The function of this basic blue protein is unknown, however, it is probably not involved in photosynthetic electron transport as it will not replace plastocyanin in that electron transport chain (Adman, 1985).

The cucumber basic blue protein is made up of a single polypeptide chain of 96 amino acids containing a single copper ion with a molecular weight of 10000 (Adman, 1985). The amino acid sequence (Bergman *et al.*, 1977) is very similar to those of stellacyanin and umecyanin. The copper in cucumber blue protein displays the characteristic spectroscopic properties of a typical blue copper site (Table 5-5).

The crystal structure of cucumber basic blue protein has now been refined to 3.0 Å resolution (Adman, 1985). The protein consists of eight strands, only five of which form a β-sandwich and the protein has less β-sheet character than plastocyanin or azurin. The ligands to copper are provided by the side chains of His-39, Cys-79, His-84 and Met-89. The copper site has the N_2SS' coordination seen in plastocyanin. The imidazole rings of the His-39 and His-94 residues are exposed to the solvent providing a likely entry site for electon transfer to the copper centre.

Umecyanin

The blue copper protein umecyanin may be isolated from horseradish roots (Paul and Stigbrand, 1970), where its function is as yet unknown. The protein is composed of a single polypeptide chain of molecular weight 14600 containing a single copper (Stigbrand, 1971). The amino acid sequence of the first 88 residues has been determined (Murata *et al.*, 1982) and shows considerable homology with stellacyanin and cucumber basic blue protein. It is known that there are no methionine residues after Met-74 and it is probable that unlike most blue copper proteins (which employ methionine in the C-terminal region of the molecule as a ligand to copper) umecyanin does not have methionine as a ligand and in this respect it resembles stellacyanin. The copper centre in the protein displays the properties of a normal blue copper site (Table 5-5). A variety of spectroscopic studies have been carried out (Strigbrand and Sjoholm, 1972; Chapman *et al.*, 1986), but they do not indicate the nature of the fourth ligand which may replace methionine. A rather limited number of studies dealing with electron transfer reactions with inorganic redox partners have been made (Chapman *et al.*, 1986) and there has been a brief investigation of the laccase catalysed oxidation of umecyanin by molecular oxygen (Stigbrand *et al.*, 1971).

Azurin

The structure of oxidised azurin has been determined and refined to 1.8 Å resolution (Baker, 1988). This work has now been extended by X-ray analysis of reduced azurin, apoazurin and cadmium substituted azurin (Shepard *et al.*, 1990). The metal site (Fig. 5-12) is essentially unchanged in all of these forms, showing that the ligand arrangement is relatively rigid, determined by constraints of protein structure. Removal of copper causes little or no change in the ligand positions. There are small but significant differences in the other two structures compared with oxidised azurin. Reduction causes a slight lengthening of the Cu-ligand bonds and concomitant shortening of the two N–H. . . S(Cys) hydrogen bonds. Such small changes are consistent with the requirements for rapid electron transfer. Replacement of copper by cadmium causes the axial carbonyl carbon to move closer, giving a Cd–O bond of ca. 2.6 Å, which is the only real deformation of the metal site. The result is also consistent with the presence of an oxygen in the metal site of stellacyanin, as the ^{113}Cd NMR spectrum is similar to that of azurin.

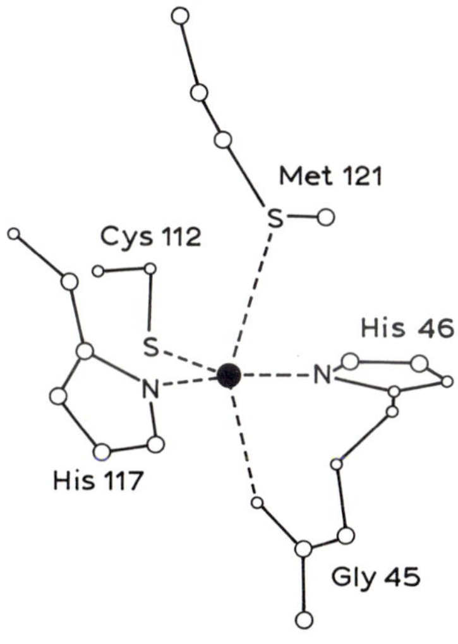

Fig. 5-12. Copper site in azurin.

The azurin structural gene has been cloned and expressed in large amounts in *E. coli* (Karlson *et al.*, 1989). The copper site in azurin is distorted-planar with two additional weakly interacting groups in axial positions. Site directed mutagenesis has been used to exchange His-46 for Met, Cys-112 for His and Met-121 for all other amino acids, in order to study the relationship between structure and function and to determine the prerequisites for the blue copper site. The Met-121 mutant proteins were characterised by their absorption and ESR spectra (Karlson *et al.*, 1991). At low pH, all mutants exhibit the characteristics of the blue (Type I) copper protein, indicating that Met is not needed to create a blue copper site. At high pH the Glu 121 and Lys 121 mutants constitute a new form of protein bound copper which is outside the range of Type I copper.

Nitrite Reductase

Conversion of NO_2^- to NO and/or N_2O is carried out by the nitrite reductases NiRs). The 2.3 Å X-ray structure of the NiR from *A. cyclocastes* has been determined (Fig. 5-13, Godden *et al.*, 1992). Hulse *et al.* (1989) have suggested that reduction occurs by the general scheme,

$$E{-}Cu \xrightarrow{NO_2^-} E{-}Cu{-}NO^- \xrightarrow{2H^+} E{-}Cu{-}NO^+ + H_2O$$

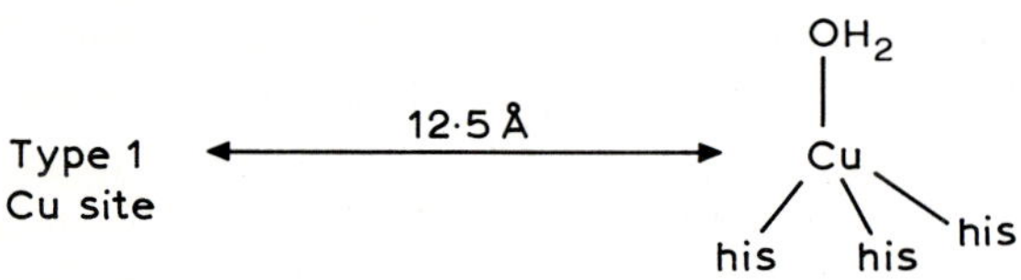

Fig. 5-13. Copper site of nitrite reductase form *A. cyclocastes.*

Recently several new active site model compounds have been prepared (Carrier *et al.*, 1991) using sterically hindered tris(pyrazoyl)hydroborate (L) and tris(thioethyl)amine (L^2) as ligands. The copper nitrite complex LCu^{II} (NO_2) models the enzyme substrate complex and X-ray studies confirm that the complex is tetrahedral. A mononuclear copper–nitrosyl complex similar to postulated NO adducts of copper proteins has also been prepared from LCu^I (MeCN) and NO, which has been tentatively identified as a $Cu^{II}-NO^-$ species on the basis of IR (ν NO 1711 cm^{-1}) and ESR evidence.

Ascorbic Acid Oxidase

The enzyme catalyses the oxidation of L-ascorbic acid, **4**, to dehydroascorbic, **5**. The blue protein belongs to the group of 'blue' oxidases together with laccase and ceruloplasmin. These are multicopper enzymes catalysing the electron reduction of molecular oxygen to water with concomitant one electron oxidation of the substrate.

Two crystalline forms of ascorbate oxidase from zucchini (Messerschmidt *et al.*, 1989) have been analysed at 2.5 Å resolution and a model of the polypeptide chain and the copper ions and their ligands has been prepared. The crystal forms M2 and M1 contain a dimer of 140000 M_r and a tetramer of 280000 M_r in the asymmetric unit. Each subunit of about 550 amino acid residues has a globular shape with dimensions of 49 Å × 53 Å × 65 Å. The subunit has three domains arranged sequentially

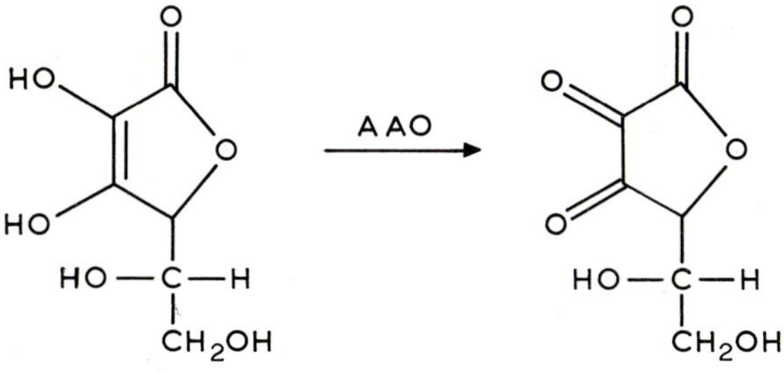

4 5

on the polypeptide chain. The folding of all three domains is of similar β-barrel type, and is distantly related to plastocyanin.

Each subunit has four copper atoms bound as mononuclear and trinuclear species. The mononuclear copper is Type I and has two histidines, a cysteine and a methionine as ligands and is located in the third domain. The trinuclear cluster has eight histidine ligands and may be divided into a pair of copper atoms arranged in a trigonal prismatic array. This pair probably represents the Type III copper. The remaining copper has two histidine ligands with the third site occupied by the pair of copper atoms, **6**, the fourth ligand may be hydroxide.

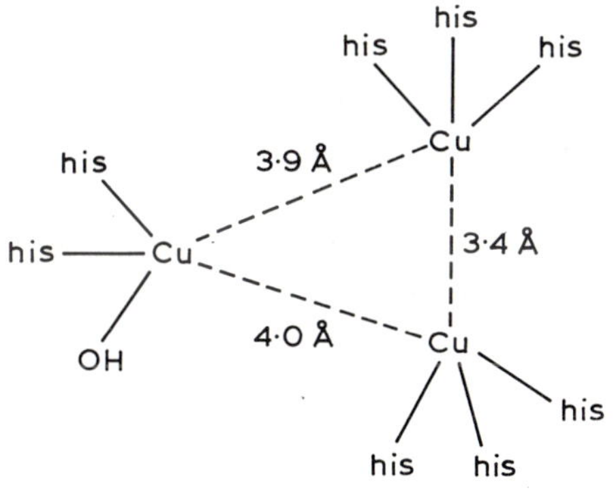

6

The enzyme obtained from summer crookneck squash (*Cucurbita pepo condensa*) is blue green in colour and has a molecular weight of ca. 150000. When a small amount of L-ascorbic acid is added to the blue solution of the enzyme in 0.1 M acetate buffer at pH 5.6, the colour is rapidly bleached to light yellow (Cu(II) → Cu(I); Fig. 5-14). Admission of dioxygen to the system slowly restores the blue colour due to re-oxidation to Cu(II).

The crystal structure of the resting form of ascorbate oxidase from zucchini has recently been refined to 1.9 Å resolution (Messerschmidt *et al.*, 1992) the subunits of 552 residues (70000 M_r) are arranged as tetramers with D2 symmetry. Asp 92 is the attachment site for one of the two N-linked sugar moieties, which has the defined elec-

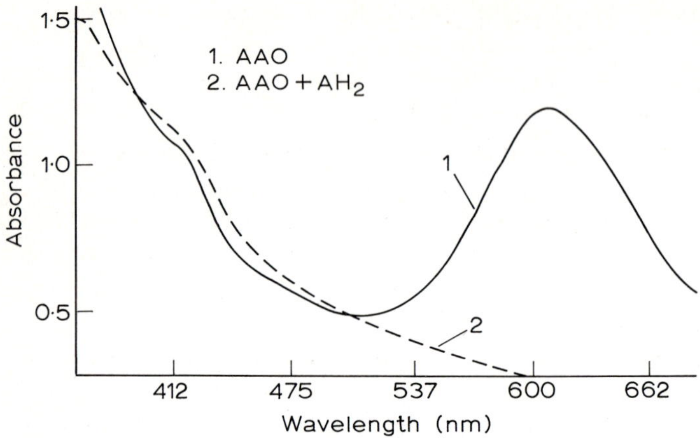

Fig. 5-14. The visible absorption spectrum of ascorbate oxidase (curve 1) and in the presence of a substrate AH_2 (curve 2). Reproduced with permission from Dawson (1960).

tron density for the N-linked N-acetyl-glucosamine ring. Each subunit is built up to three domains. The folding of all three domains is of a similar β-barrel type and is related to plastocyanin and azurin.

Each subunit has four copper atoms bound as mononuclear and trinuclear species. The mononuclear copper has two histidines, a cysteine and a methionine ligand and represents the Type I copper. The trinuclear cluster has eight histidine ligands symmetrically supplied from domains 1 and 3. It may be subdivided into a pair of copper atoms with six histidines whose ligating N-atoms (5 NE2 atoms and one ND1 atom) are arranged in a trigonal prismatic manner. The pair is the putative Type III copper. The remaining copper has two histidine ligands and is the specroscopic Type III copper. Two oxygen atoms are bound to the trinuclear species. The first one bridges the Type III copper pair and is either OH^- or O^{2-}. The second one is bound to the Type II copper *trans* to the copper pair as OH^- or H_2O.

The probable binding site for the reducing substrate is close to the Type I copper. Two channels can be identified providing access from the solvent to the trinuclear copper site which is the likely binding site for dioxygen.

Cellobiose Oxidase

The white-rot fungus *Sporotrichum pulverulentum* is one of the organisms able to degrade the polysaccharides and the lignin present in plant material. There have been many reports over the last few years of *S. pulverulentum* and other white-rot fungi producing extracellular enzymes capable of oxidising cellobiose to cellobionolactone (Ayers *et al.* 1978; Morpeth, 1984). In general, brown-rot fungi, which digest only the polysaccharides in wood do not produce cellobiose oxidoreductase (Ander and Eriksson, 1978).

One of the cellobiose oxidoreductases present in *S. pulverulentum* has been characterised and named 'cellobiose oxidase' (Ander and Eriksson, 1978). The enzyme contains both haem and flavin cofactors and binds irreversibly to concanavalin A-Sepharose, suggesting that it is a glycoprotein. Cellobiose oxidase from *S. pulverulentum* has now been purified to homogeneity by Morpeth (1985). The carbohydrate and amino acid compositions of the enzyme have been determined. The enzyme contains FAD and cytochrome b prosthetic groups and is a monomer with an M_r of 74400 determined by sedimentation equilibrium.

Disaccharides and even some insoluble polysaccharides are substrates, but not monosaccharides. Cellobiose oxidase is unusual among flavoproteins, as it stabilises the red anionic flavin semiquinone and forms a sulphite adduct, yet appears to produce the superoxide anion as its primary reduced oxygen product.

The visible absorption spectrum of oxidised cellobiose oxidase is typical of cytochrome b (Fig. 5-15). The flavin in cellobiose oxidase is weakly fluorescent, with emission maxima at 564 nm and excitation maxima at 380 and 444 nm. There are no obvious transient changes on reduction that can be readily ascribed to flavin semiquinone, but the strong absorbance of the cytochrome would make such changes difficult to detect.

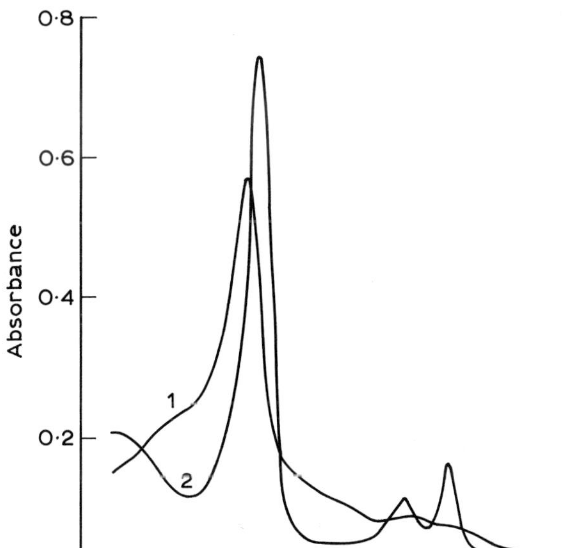

Fig. 5-15. Absorption spectra of oxidised (curve 1) and reduced (curve 2) cellobiose oxidase. Reproduced with permission from Morpeth (1985).

Galactose Oxidase

Galactose oxidase is an extracellular enzyme secreted by the fungus *Dactylium dendroides*. It is monomeric (M = 68000), contains a single copper site and catalyses the oxidation of a wide range of primary alcohols to the corresponding aldehydes. The two-electron transfer reaction $RCH_2OH \rightarrow RCHO + 2H^+ + 2e$ does not utilise a Cu(III)/Cu(I) couple, but a second redox site, involving a tyrosine radical which mediates the transfer of the second electron.

The crystal structure of galactose oxidase has been determined to 1.7 Å resolution and refined to a crystallographic *R* factor of 0.18 (Ito *et al.* 1991). The molecule is entirely composed of β-sheet structure, organised in three domains, two small domains (1 and 3) at the N- and C-termini, and a large central domain (2) containing the copper site. Domain (2) has approximately seven-fold symmetry, and consists of seven four-stranded antiparallel β-sheets, arranged like the petals of a flower with the copper site in the centre of one face. The copper has square planar coordination, with an axial tyrosine ligand (Tyr-272), and the fourth site is occupied by an acetate ligand. When acetate is not present in the buffer, the fourth site is vacant, and may be the binding site for the primary hydroxyl group of the substrate (Fig. 5-16).

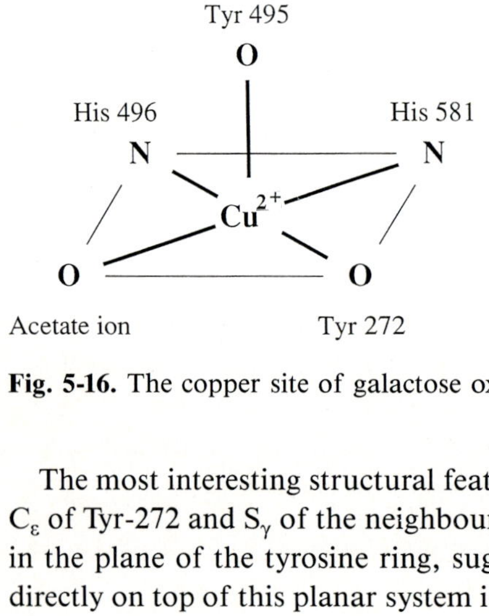

Fig. 5-16. The copper site of galactose oxidase.

The most interesting structural feature, however, is a covalent bond between a ring C_ε of Tyr-272 and S_γ of the neighbouring Cys-228. Both S_γ and C_β of the cysteine lie in the plane of the tyrosine ring, suggesting an extended aromatic system. Stacked directly on top of this planar system is the ring of Trp-290. It has been proposed that Tyr-272 is the site of the observed free radical in galactose oxidase. The protein has built a secondary cofactor from its polypeptide chain, thus avoiding dependence on an external cofactor.

The Ethylene Receptor Site in Plants

Ethylene is an important plant hormone which regulates plant growth, causing seed to sprout, flowers to bloom and fruit to ripen. The physiological activity of ethylene is also observed with acetylene, carbon monoxide and terminal olefines. Little is known of the receptor site, but it has been observed that the biological activity of the unsaturated compounds can be correlated with their affinity for copper(I) (Fig. 5-17). Ligands such as CN^-, SCN^-, I^- and 1,5-cyclooctadiene (cod) inhibit the activity of ethylene in carnation flowers and in pea epicotyl elongation suggesting that copper(I) is present at the ethylene receptor site in plants.

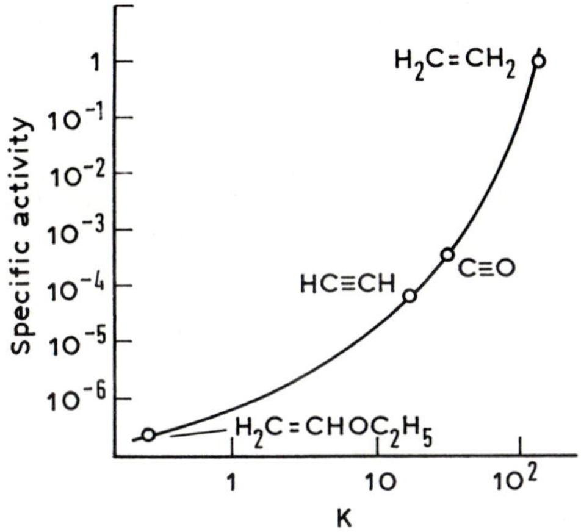

Fig. 5-17. A plot of specific activities of unsaturated compounds in pea epicotyl elongation against the formation constants (K) of their Cu(I) complexes.

Munakata and coworkers (1991) have recently characterised copper(I) complexes with ethylene, acetylene, carbon monoxide and cod and studied the structures by X-ray analysis to investigate the bonding involved. Ethylene and acetylene are sideways bonded to copper (Fig. 5-18). Sigma donation from ethylene to copper predominates and π-back donation is very weak.

Carbonic Anhydrase

Carbonic anhydrase is a zinc metalloenzyme present in animals, plants and certain microorganisms which catalyses the reversible hydration of carbon dioxide and the hydration of many aldehydes.

$$CO_2 + H_2O \rightleftharpoons HCO_3^- + H^+$$
$$CH_3CHO + H_2O \rightleftharpoons CH_3CH(OH)_2$$

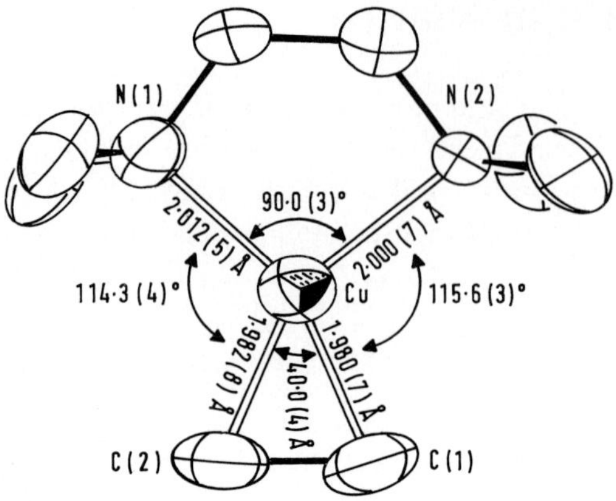

Fig. 5-18. Molecular structure of $[Cu(tmen)(C_2H_4)]ClO_4$.

The enzyme also catalyses the hydrolysis of various esters such as 4-nitrophenyl acetate and sultones. A number of slightly different enzymes, carbonic anhydrases A, B and C occur in different organisms. The most well characterised enzymes are the bovine and human carbonic anhydrases B, which are monomeric and contain one tightly bound zinc per 30000 molecular weight.

X-ray crystallographic studies (Kannen *et al.*, 1975) of the human enzyme have established that the molecule is roughly ellipsoidal (40 Å × 45 Å × 55 Å) with the zinc lying near the bottom of a deep cleft near the centre of the molecule. The zinc is ligated by three histidine residues (His-117, His-93 and His-95) in a distorted tetrahedral geometry, with the fourth site occupied by a water molecule.

Lindskog and Malmstrom (1962) first succeeded in removing the zinc under sufficiently mild conditions to avoid denaturation. The zinc was complexed with phenanthroline and the resulting complex removed by dialysis. The apoenzyme was catalytically inactive and the activity could be restored by addition of one mole of zinc per mole of apoenzyme. ORD studies have shown that the apoenzyme and the native enzyme have the same gross teriary structure, confirming that the function of the zinc is not to stabilise the tertiary structure, but it is directly involved in the catalytic activity of the enzyme.

Reaction of the apoenzyme with various metal ions allows the preparation of other metallocarbonic anhydrases and the Cu(II), Co(II) and Co(III) enzymes have been prepared. The cobalt(II) enzyme shows some catalytic activity. The $d-d$ spectrum of the cobalt(II) derivative, which is high spin, is markedly pH dependent (Fig. 5-19). A plot of the molar absorbance at 640 nm versus pH does not follow a simple pattern, as expected for a single ionising group, but is consistent with two ionising groups with pK values of < 6 and > 7.

The mechanisms of action of carbonic anhydrase have been the subject of considerable controversy. Kannen and coworkers (1977), as a result of X-ray investiga-

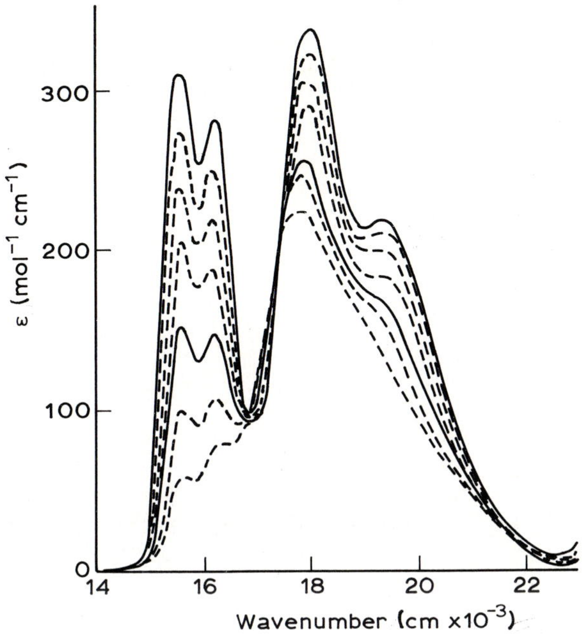

Fig. 5-19. Electronic spectra of CoHCAB as a function of pH in 10^{-2} M HEPES solutions at pH 6.1, 6.6, 7.1, 7.8, 8.3, 8.6 and 9.5 in order of increasing ε_{640}. The full lines represent the spectra obtained at the middle and at the end of the titration. Reproduced with permission from Bertini *et al.* (1980).

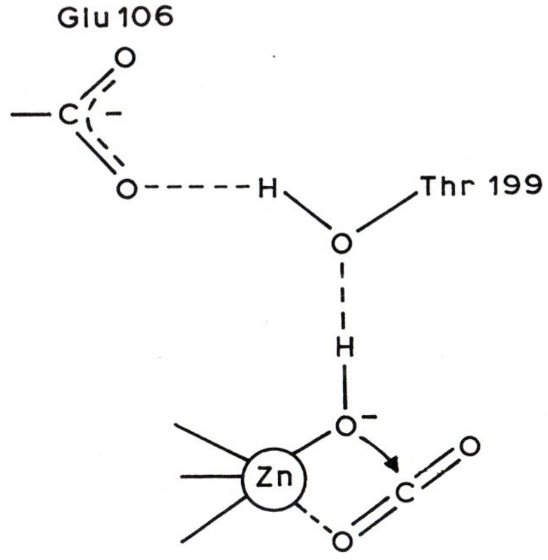

Fig. 5-20. Proposal for the hydration – dehydration mechanism of carbonic anhydrase. The pK of the zinc bound water is envisaged to be lowered to about 7.0 by the charge distribution on the metal ion and also helped by Glu-196 by the hydrogen bonding through Thr-199.

tions on the human carbonic anhydrase B-imidazole inhibitor complex, have proposed the hydration – dehydration mechanism shown in Fig. 5-20. Glutamic acid-106 acts as a general base. The zinc is five-coordinate with the zinc providing some Lewis acid catalysis in addition to providing an intramolecular reaction via the metal bound hydroxide.

The Purple Acid Phosphatases

The purple acid phosphatases (PAPs) are a class of phosphoprotein phosphatases which possess a μ-oxo(hydroxo)-bridged dinuclear iron centre. An enzyme has been isolated from beef spleen which is purple in colour, while a violet phosphatase has been characterised from red kidney beans (KBPase). This latter enzyme consists of two subunits with M = 58200 and contains two equivalents of Zn(II) and Fe(III) per dimer which are essential for catalytic activity. KBPase hydrolyses nucleosidetriphosphates as well as activated phosphomonoesters such as 4-nitrophenylphosphate or α-naphthyl phosphate (Beck *et al.,* 1986). As with the beef spleen enzyme, KBPase is inhibited by tetrahedral oxoanions such PO_4^{3-} and AsO_4^{3-}.

The inhibition of KBPase by oxoanions has been studied by ESR spectroscopy. The spectrum of the native enzyme has an intense signal at $g = 4.3$ which is typical of high-spin Fe(III) compounds. Additional signals have been detected at g values of 8.5, 5.6 and 2.9 (Beck, 1989). Phosphate converts the highly rhombic signal to a single species with $g = 4.3$. However, arsenate reduces the intensity of the 4.3 signal. Hyperfine structure becomes visible in the presence of arsenate (I = 3/2) at pH 7 indicating that the oxoanions bind to the Fe(III) centre under suitable conditions.

EXAFS studies indicate a Zn(II) – Fe(III) distance of 393 pm which decreases to 372 pm on binding phosphate. The change of Zn-edge structure and the appearance of a new peak in the Fourier transform of the EXAFS suggests that the inhibitor also binds to the Zn(II) site irrespective of pH.

Analyses of the XANES spectra of KBPase indicate a five-fold ligation of the zinc atom for the native enzyme and an increase of the coordination number towards six upon addition of phosphate (Priggenmeyer *et al.*, 1991) This result suggests a direct binding of the substrate-analogous inhibitor to the zinc atom. The intensities of the pre-edge peaks at the iron edges are within the range typical of six-coordinate complexes.

Various kinetic studies of the bovine spleen acid phosphatase have been reported (Vincent *et al.*, 1991). The results are consistent with a picture in which the oxyanions bind in a non-competitive fashion by bridging the two iron atoms in the PAP's dinuclear centre, with the smaller anions also able to bind in a competitive manner at a second site (Fig. 5-21).

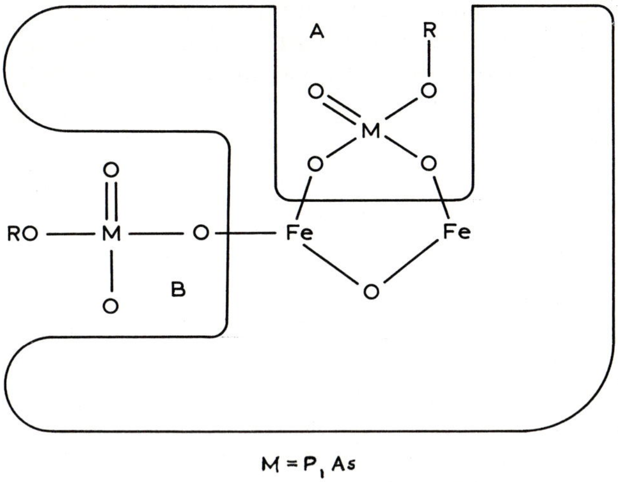

Fig. 5-21. Proposed model for oxyanion binding. A=non-competitive site, B=competitive site. Reproduced from Vincent *et al.* (1991).

The Photosynthetic Water Oxidation Centre

The enzyme responsible for the photolysis of water in plants is a multisubunit membrane protein (Fig. 5-22; Klein *et al.*, 1991). Four manganese ions, probably as a tetranuclear cluster, are thought to act as a charge accumulating system and as the active site for water oxidation. Both calcium and chloride ions are also required for activity (Babcock, 1987; Ghantokakis and Yocum, 1990). The water oxidation centre (WOC) contains a total of four Mn atoms and causes the oxidative coupling of two water molecules by a currently unknown mechanism.

Fig. 5-22. The water oxidation centre. Reproduced from Klein *et al.* (1991).

A search for inorganic model systems is being carried out by many groups. Christou and coworkers (Christou *et al.*, 1991; Libby *et al.*, 1991) have concentrated on Mn carboxylate chemistry and a variety of $Mn/O/RCO_2^-$ complexes of various nuclearities and oxidation states have been assessed. A high yield preparation of the $[Mn_4O_2(O_2CPh)_9(H_2O)]^-$ cluster has been developed. This complex contains a butterfly $[Mn_4O_2]^{8+}$ unit and a terminally bound water molecule.

Recent EXAFS results on oxygen evolving preparations from both spinach and cyanobacteria show three distinct Fourier peaks which can be modelled as interactions from bridging and terminal ligand atoms (O, N), Mn at 2.7 Å and Mn and/or Ca at 3.3 Å. A model consistent with the current results is shown in Fig. 5-22.

Cytochrome c Oxidase

Due to its central role in oxidative phosphorylation, cytochrome oxidase has a wide biological distribution. It is present in all animals and plants, in aerobic yeasts and in some bacteria. It is an integral membrane protein, being firmly associated with the inner membrane of mitochondria, the respiratory organelle of eukaryotic organisms, or, in bacteria the plasma membrane (Malstrom, 1990).

The reaction catalysed by cytochrome oxidase is

$$4c^{2+} + O_2 + 4H^+ \rightarrow 4c^{3+} + 2H_2O.$$

where c^{2+} and c^{3+} represent cytochrome c in its reduced and oxidised forms, respectively. This is the terminal reaction of the so-called respiratory chain, which provides most of the free energy needed for the life processes of aerobic organisms by coupling the electron transport of the synthesis of ATP.

It has long been established that the functional unit of cytochrome oxidase contains four redox-active metal centres. Two of these, cytochromes a and a_3, contain heme A (Fig. 5-23) coordinated in different ways in subunit I. The heme group is not covalently linked to the protein. The main structural features are the carbonyl group at position 8 and the isoprenoid chain at position 2 of the porphyrin ring.

The other two redox-active metals are two copper ions designated Cu_A and Cu_B also bound in different ways to the protein. The Cu_A is located in subunit II, and Cu_B appears to be in subunit I, together with the two cytochromes (Holm *et al.*, 1987). There is evidence that in some oxidase preparations there are three rather than two coppers, but the third copper appears to be non-functional.

Structural information about cytochrome oxidase is very limited beyond the level of primary structure as it has proved impossible to grow crystals of sufficiently high quality for high resolution X-ray studies. Cytochrome oxidase is a transmembrane protein which is inserted asymmetrically through the membrane. The overall shape is like a Y, with the stalk extending about 50 Å above the membrane plane on the

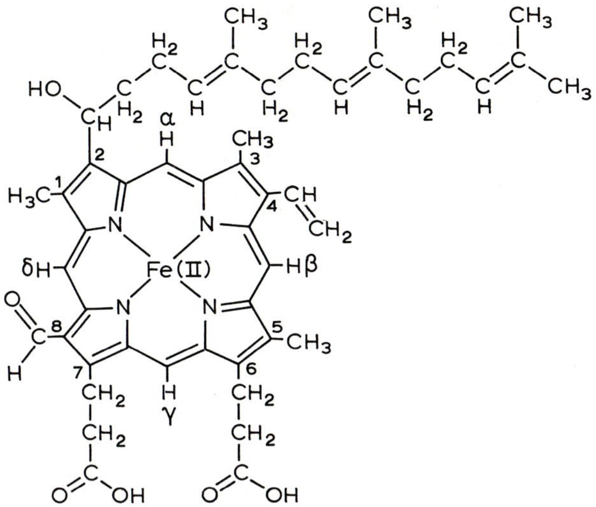

Fig. 5-23. Chemical structure of heme A.

cytoplasmic side with the arms crossing the lipid bilayer and extending about 10 Å on the side of the mitochondrial matrix. The distance between the centres of the two arms has been estimated to be about 40 Å.

Various spectroscopic studies allow the assignment of probable ligand structures to the four redox centres. Coupled with information about conserved amino acids, it is possible to establish in which subunits the centres are located. Thus, subunit II is the only one that contains conserved cysteines and provides the binding site for Cu_A. This subunit has only two conserved histidines which are probably also both ligands to Cu_A. All other metal centres must be in subunit I.

Subunit I contains nine conserved histidine residues, which are sufficient to provide the six ligands suggested by the spectroscopic data. Two for cytochrome a, one for cytochrome a_3 and three for Cu_B. Holm and coworkers (1981) have developed a schematic model of subunit II of cytochrome oxidase (Fig. 5-24). Cytochrome a is coordinated to two histidine residues close to the membrane surface in separate transmembrane helices. Cytochrome a_3-Cu_B is located between the same helices, with a third helix providing two of the Cu_B ligands. This arrangement places the two heme groups very close together (ca. 14 Å centre to centre) which seems to be at variance with the relatively slow electron transfer from cytochrome a to the bimetallic site. In addition, electron spin relaxation measurements indicate that the distance is about 20 Å (Brudvig *et al.*, 1984). As a result Malmstrom (1989) has suggested an alternative model which retains the structure of cytochrome a but places cytochrome a_3 bound to one of the transmembrane helices on the opposite side from cytochrome a.

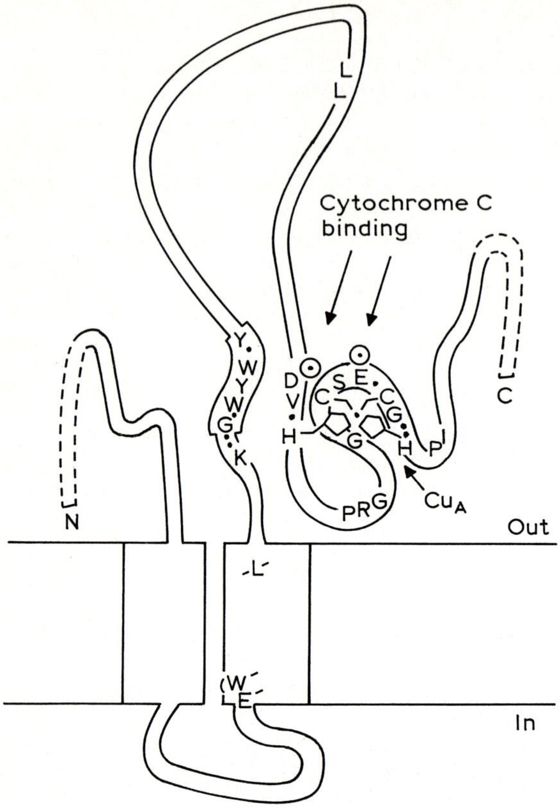

Fig. 5-24. Schematic model of sub-unit II of cytochrome oxidase. Two hydrophobic helices anchor the sub-unit to the membrane. The Cu_A binding site is indicated in a location outside the membrane close to the COOH-terminal end of the peptide. Invariant amino acid residues are shown. Reproduced from Holm *et al.* (1987).

References

Adman, E. T. (1985), in: *Metalloproteins*: Harrison, P. M. (ed.) Weinheim: Verlag Chemie; Vol. 1, pp. 1–42.

Aikazyan, V. Ts., Nalbandyan, R. M. (1975). *FEBS Lett.*, *55*, 272.

Aikazyan, V. Ts., Nalbandyan, R. M. (1979). *FEBS Lett.*, *104*, 127.

Alagna, L., Hasnain, S. S., Piggot, B., Williams, D. J. (1984). *Biochem. J.*, *220*, 591.

Ander, P., Eriksson, K.-E. (1978). *Prog. Ind. Microbiol.*, *14*, 1–58.

Andersen, G. P., Sanderson, D. G., Lee, C. H., Durrell, S., Andersen, L. B., Gross, E. L. (1987). *Biochim. Biophys. Acta*, *894*, 386.

Ayers, A. R., Ayers, S. B., Eriksson, K.-E. (1978). *Eur. J. Biochem.*, *90*, 171.

Babcock, G. T. (1987), in: *New Comprehensive Biochemistry, Vol. 15, Photosynthesis*: Amesz, J. (ed.) Amsterdam: Elsevier.

Bannister, J. V., Bannister, W. H., Rotilio, G. (1987). *Crit. Rev. Biochem.*, *22*, 111–180.

Baker, E. N. (1988). *J. Mol. Biol.*, *203*, 1071.
Barber, J. (1983). *Plant Cell Environ.*, *6*, 311.
Barynin, V. V., Grebenko, A. (1986). *Dokl. Akad. Nauk USSR*, *286*, 461.
Barynin, V. V., Vagin, A. A., *et al.* (1986). *Dokl. Akad. Nauk USSR*, *288*, 877.
Beck, J. L., McConachie, L. A., Summors, A. C., Arnold, W. N., de Jersey, J., Zerner, B. (1986). *Biochim. Biophys. Acta*, *869*, 61.
Beck, J. L., de Jersey, J., Zerner, B., Hendrich, M. P., Debrunner, P. G. (1988). *J. Am. Chem. Soc.*, *110*, 3317.
Bergman, C., Ganvick, E. K., Nyman, P. O., Strid, L. (1977). *Biochem. Biophys. Res. Commun.*, *787*, 1052.
Bertini, I., Luchinat, C., Scozzafava, A. (1980). *Inorg. Chim. Acta*, *46*, 85.
Blake, R. S., Hager, L. P. (1990). *J. Biol. Chem.*, *265*, 12454.
Blakely, R. L., Zerner, B. (1984). *J. Mol. Catal.*, *23*, 263.
Brudvig, G. W., Blair, D. F., Chan, S. I. (1984). *J. Biol. Chem.*, *259*, 11001.
Burrows, C. J., Crane, C. G. (1991). *J. Inorg. Biochem.*, *43*, 661.
Carnmack, R., Fernandez, V. M., Schneider, K. (1988), in: *Bioinorganic Chemistry of Nickel*: Lancaster, J. R. (ed.) Deerfield Beach: VCH Publishers.
Carrier, S. M., Ruggiero, C. E., Tolman, W. B. (1991). *J. Inorg. Biochem.*, *43*, 199.
Chance, B. (1943). *J. Biol. Chem.*, *151*, 553.
Chance, B. (1949a). *Arch. Biochem. Biophys.*, *21*, 416.
Chance, B. (1949b). *Science*, *109*, 204.
Chance, B. (1952). *Arch. Biochem. Biophys.*, *41*, 404.
Chapman, S. K. (1991), in: *Perspectives on Bioinorganic Chemistry*: Hay, R. W., Dilworth, J. R., Nolan, K. B. (eds.) Connecticut: JAI Press; Vol. 1, pp. 95–140.
Chapman, S. K., Orme-Johnson, W. H., McGinnis, J., Sinclair-Day, J. D., Sykes, A. G., Ohlsson, P.-I., Paul, K.-G. (1986). *J. Chem. Soc. Dalton Trans.*, 2063.
Christou, G., Libby, E., Wang, S., Blackman, A. G., Huffman, J. C. (1991). *J. Inorg. Biochem.*, *43*, 364: see also Christou, G. (1989). *Acc. Chem. Res.*, *22*, 328.
Clark, P. A., Wilcox, D. E. (1989). *Inorg. Chem.*, *28*, 1326.
Cramer, W. A., Widger, W. R., Herrmann, R. G., Trebset, A. (1985). *Trends Biochem. Sci.*, *10*, 125.
Curtis, N. J., Nixon, N. E., Sargeson, A. M. (1983). *J. Am. Chem. Soc.*, *105*, 5347.
Dawson, C. R. (1960). *Ann. NY Acad. Sci.*, *88*, *Art 2*, 353.
Dawson, J. H., Sono, M. (1987). *Chem. Rev.*, *87*, 1255.
de Boer, E., Wever, R. (1988). *J. Biol. Chem.*, *263*, 12326.
De Pillis, G. D., Sishta, B. P., Mauk, A. G., Montellano, P. R. O. (1991). *J. Biol. Chem.*, *266*, 19334.
Dixon, M., Webb, E. C. (1964). *Enzymes*, 2nd edn, London: Longmans.
Dixon, N. E., Gazzola, C., Blakeley, R. L., Zerner, B. (1975). *J. Am. Chem. Soc.*, *97*, 4131.
Dixon, N. E., Riddles, P. W., Gazzola, C., Blakeley, R. L., Zerner, B. (1980). *Can. J. Biochem.*, *58*, 1335. See also Dixon, N. E., Blakeley, R. L., Zerner, B. (1980). *Can. J. Biochem.*, *5*, 469; Dixon, N. E., Gazzola, C., Asher, C. J., Lee, D. S. W., Blakeley, R. L., Zerner, B., (1980). *Can. J. Biochem.*, *58*, 474.
Dixon, N. E., Jackson, W. G., Marty, W., Sargeson, A. M. (1982). *Inorg. Chem.*, *21*, 688.
Dixon, N. E., Fairlie, D. P., Jackson, W. G., Sargeson, A. M. (1983). *Inorg. Chem.*, *22*, 4038.
Dunford, H. B., Stillman, J. S. (1976). *Coord. Chem. Rev.*, *19*, 187.
Dunford, H. B. (1991), in: *Peroxidases in Chemistry and Biology*: Everse, J., Everse, K. E., Grisham, M. B. (eds.) Boca Raton: CRC Press; Vol. 2, pp. 1–23.
Everse, J., Everse, K. E., Grisham, M. B. (1991). *Peroxidases in Chemistry and Biology*, Boca Raton: CRC Press; p. 2.

Fenna, R. E. *et al.* (1992). *J. Mol. Biol.* (in press).
Finzel, B. C., Poulos, T. L., Kraut, J. (1984). *J. Biol. Chem.*, *259*, 13027.
Garret, T. P. J., Clingelleffer, D. J., Guss, J. M., Rodgers, S. J., Freeman, H. C. (1984). *J. Biol. Chem.*, *259*, 2822.
Gaspar, T., Penel, C., Grepin, H. (eds.) (1992). *Plant Peroxidases 1980–1990: Progress and Prospects in Biochemistry and Physiology*, Geneva: University of Geneva.
Ghantokakis, D. F., Yocum, C. F. (1990). *Ann. Rev. Plant Physiol. Plant Mol. Biol.*, *41*, 255.
George, P. (1952). *Advan. Cat.*, *4*, 367.
Getzoff, E. D., Tainer, J. A., Weiner, P. K., Kollman, P. A., Richardson, J. S., Richardson, D. C. (1983). *Nature*, *306*, 287.
Godden *et al.* (1992). *Science* (in press).
Gunter, M. J., Turner, P. (1991). *Coord. Chem. Rev.*, *108*, 115.
Guss, J. M., Freeman, H. C. (1983). *J. Mol. Biol.*, *169*, 521.
Guss, J. M., Hanswell, P. R., Murata, M., Norris, V. A., Freeman, H. C. (1986). *J. Mol. Biol.*, *192*, 361.
Guss, M. J., Merrit, E. A., Phizackerly, R. P., Hedman, B., Murata, M., Hodgson, K. O., Freeman, H. C. (1988). *Science*, *241*, 806.
Haehnel, W. (1984). *Ann. Rev. Plant Physiol.*, *35*, 659.
Haehnel, W. (1986), in: *Photosynthesis III- Encyclopedia of Plant Physiology*: Staelin, L. A., Arntzen, C. J. (eds.) Berlin: Springer-Verlag; Vol. 196, pp. 547–559.
Hager, L. P., Morris, D. R., Brown, F. S., Eberwein, H. (1966). *J. Biol. Chem.*, *241*, 1769.
Haldane, J. B. S. (1965). *Enzymes*, Connecticut: The MIT Press.
Hasnain, S. S., Piggott, B. (1983). *Biochem. Biophys. Res. Commun.*, *112*, 279.
Henrissat, B., Saloheimo, M., Lavaitte, S., Knowles, J. K. C. (1990). *Proteins*, *8*, 251.
Hewson, W. D., Hager, L. P. (1979), in: *The Porphyrins*: Dolphin, D. (ed.) New York: Academic Press; Vol. 7, pp. 295–332.
Holm, L., Saraste, M., Wikstrom, M. (1987). *EMBO J.*, *6*, 2819.
Hulse, C. L., Averill, B. A., Tiedje, J. M. (1989). *J. Am. Chem. Soc.*, *111*, 2322.
Ito, N., Phillips, S. E. V., Stevens, C., Ogel, Z. B., McPherson, M. J., Keen, J. N., Yadav, K. D. S., Knowles, P. F. (1991). *Nature*, *350*, 87.
Kannen, K. K., Nostrand, B., Fridborg, K., Lovgren, S., Petet, M., (1975). *Proc. Natl. Acad. Sci. USA*, *72*, 51.
Kannen, K. K., Petet, M., Fridborg, K., Cid-Dresdner, H., Lovgren, S. (1977). *FEBS Lett.*, *73*, 115.
Karlson, B. G., Pascher, T., Nordling, M., Arvidson, R. H. A., Lundberg, L. G. (1989). *FEBS Lett.*, *246*, 211.
Karlson, B. G., Pascher, T., Nordling, M., Tsai, L.-C., Sjolin, L., Lundberg, L. G. (1991). *Prot. Eng.*, 343.
Keilen, D., Mann, T. (1937). *Proc. Roy. Soc. Ser. B*, *122*, 119.
Klein, M. P., Yachandra, V. K., DeRose, V. J., Muckerji, I., Latimer, M. J., Sauer, K. (1991). *J. Inorg. Biochem.*, *43*, 363. See also Klein, M. P. *et al.* (1991). *Biochemistry*, *30*, 1335.
Kono, Y., Fridovich, I. (1983). *J. Biol. Chem.*, *258*, 6015.
Kruger, H. J., Peng, G., Holm, R. H. (1991). *Inorg. Chem.*, *30*, 734.
Lee, M. H., Mulrooney, S. B., Hausinger, R. P. (1990). *J. Bacteriol.*, *172*, 4427.
Libby, E., McCuster, J. K., Schmitt, E. A., Folting, K., Henrickson, D. N., Christou, G. (1991). *J. Am. Chem. Soc.*, *30*, 3486.
Liebig, J. V., Wöhler, F. (1837). *Ann. Chem.*, *22*, 1.
Lindskog, S., Malmstrom, B. (1962). *J. Biol. Chem.* *237*, 1129.
Lobarzewski, J., Greppin, H., Penel, C., Gaspar, T. (eds.) (1991). *Biochemical, Molecular and Physiological Aspects of Plant Peroxidases*, Geneva: University of Geneva.

Loew, G. H., Collins, J., Luke, B., Walsh, A., Pudzianowski, A. (1986). *Enzyme*, *36*, 54.
Malmstrom, B. G. (1989). *FEBS Lett.*, *250*, 9.
Malmstrom, B. G. (1990). *Chem. Rev.*, *90*, 1247.
Mansuy, D. (1987). *Pure Appl. Chem.*, *59*, 759.
Maroney, M. J., Colpas, G. J., Bagyinka, C. (1990). *J. Am. Chem. Soc.*, *112*, 7067.
Messerschmidt, A., Rossi, A., Ladenstein, R., Huber, R., Bolognesi, M., Gatti, G., Marchesini, A., Petruzzelli, R., Finazzi-Agro, A., (1989). *J. Mol. Biol.*, *206*, 513.
Messerschmidt, A., Ladenstein, R., Huber, R., Bolognesi, M., Avigliano, L., Petruzzelli, R., Rossi, A., Finazzi-Agro, A. (1992). *J. Mol. Biol.*, *224*, 179.
Mittler, R., Zilinskas, B. A. (1991). *FEBS Lett.*, *289*, 257. (First complete sequence of an ascorbate peroxidase). The protein is described by Mittler, R., Zilinskas, B. A. (1991). *Plant Physiol.*, *97*, 962.
Mobley, H. L. T., Hausinger, R. P. (1989). *Microbiol. Rev.*, *53*, 85.
Morita, Y., Mikami, B., Yamashita, H., Lee, J. Y., Aibara, S., Sato, M., Katsube, Y., Tanaka, M. (1991), in: *Biochemical, Molecular and Physiological Aspects of Plant Peroxidases*: Lobarzewski, J., Greppin, H., Penel, C., Gaspar, T. (eds.) Geneva: University of Geneva; pp. 81–88.
Morpeth, F. F. (1984), in: *Flavins and Flavoproteins*, Bray, R. C., Mayhew, S. G., Engel, P. C. (eds.) Berlin: Walter De Gruyter and Co.; pp. 679–682.
Morpeth, F. F. (1985). *Biochem. J.*, *228*, 557.
Morris, D. R., Hager, L. P. (1966). *J. Biol. Chem.*, *241*, 1763.
Mulrooney, S. B., Hausinger, R. P. (1990). *J. Bacteriol.*, *172*, 587.
Munakata, M., Kutagawa, S., Maekawa, M. (1991). *J. Inorg. Biochem.*, *43*, 198.
Murata, M., Begg, G. S., Lambrou, F., Leslie, B., Simson, R. J., Freeman, H. C., Morgan, F. J. (1982). *Proc. Natl. Acad. Sci. USA*, *79*, 6434.
Oberley, L. W. (ed.) (1982). *Superoxide Dismutase*, Boca Raton: CRC Press; Vols. I and II.
Paul, K.-G., Stigbrand, T. (1970). *Biochim. Biophys. Acta*, *221*, 255.
Payen, A., Persoz, J. (1833). *Ann. Chim. Phys.*, *53*, 73.
Planche, L. (1810). *Bull. Pharm.*, *82*, 578.
Poulos, T. L. (1988). *Adv. Inorg. Biochem.*, *7*, 1.
Priggenmeyer, S., Rompel, A., Eggers-Borkenstein, P., Krebs, B., Henkel, G., Nolting, H.-F., Hermes, C., Korner, M., Witzel, H. (1991). *J. Inorg. Biochem.*, *43*, 542.
Robiquet (1838). Mentioned in: *Extrait du procès verbal de la Société de Pharmacie de Paris*, Séance du 7 mars.
Robiquet, Boutron-Chalard (1830). *Ann. Chim. Phys.*, *44*, 352.
Rosenberg, R. C., Root, C. R., Grey, H. B. (1975). *J. Am. Chem. Soc.*, *97*, 21.
Sakurai, T., Okamoto, H., Kawahara, A. (1982). *FEBS Lett.*, *147*, 220.
Saunders, B. C., Holmes-Siedel, A. G., Stark, B. P. (1964). *Peroxidase*, London: Butterworth.
Shepard, W. E. B., Anderson, D. A., Lewandowski, D. A., Norris, G. S., Baker, E. N. (1990). *J. Am. Chem. Soc.*, *112*, 7817.
Steele, C. C. (1934). *An Introduction to Plant Biochemistry*, London: G. Bell and Sons.
Stigbrand, T. (1971). *Biochim. Biophys. Acta*, *236*, 246
Stigbrand, T., Sjoholm, I. (1972). *Biochim. Biophys. Acta*, *262*, 244.
Stigbrand, T., Malmstrom, B. G., Vangard, T. (1971). *FEBS Lett.*, *12*, 260.
Sumner, J. B. (1926). *J. Biol. Chem.*, *69*, 435.
Sumner, J. B., Somers, G. F. (1943). *Chemistry and Methods of Enzymes*, New York: Academic Press; pp. 121–122, 181–182.
Sykes, A. G. (1985). *Chem. Soc. Rev.*, *14*, 283–315.
Tainer, J. A., Getzoff, E. D., Beem, K. M., Richardson, J. S., Richardson, D. C. (1982). *J. Mol. Biol.*, *160*, 181.

Tainer, J. A., Getzoff, E. D., Richardson, J. S., Richardson, D. C. (1983). *Nature*, *306*, 284.
Takio, K., Titani, K., Ericsson, L. H., Yonetani, T. (1980). *Arch. Biochem. Biophys.*, *203*, 615.
Theorell, H. (1941). *Enzymologia*, *10*, 250.
Tien, M., Tu, C.-P. D. (1987). *Nature*, *326*, 520.
Todd, M. J., Hausinger, R. P. (1989). *J. Biol. Chem.*, *264*, 15835.
Tromp, M. G. M., Olafson, G., Kren, B. E., Wever, R. (1990). *Biochim. Biophys. Acta*, *1040*, 192.
Valentine, J. S., Pantoliano, M. W. (1981), in: *Metal Ions in Biology*: Spiro, G. T. (ed.) New York: Wiley-Interscience, Vol. 3; Chapter 8, pp. 291–358.
Vincent, J. B., Crowder, M. W., Averill, B. A. (1991). *J. Inorg. Biochem.*, *43*, 542.
Welinder, K. G. (1985). *Eur. J. Biochem.*, *151*, 497.
Welinder, K. G. (1991a). *Biochim. Biophys. Acta*, *1080*, 215.
Welinder, K. G. (1991b), in: *Biochemical, Molecular and Physiological Aspects of Plant Peroxidases*: Lobarzewski, J., Greppin, H., Penel, C., Gaspar, T. (eds.) Geneva: University of Geneva; pp. 3, 13.
Welinder, K. G. (1992). *Current Opinions in Structural Biology*, *2*, 388.
Welinder, K. G., Mazza, G. (1977). *Eur. J. Biochem.*, *73*, 353.
Wever, R., Kustin, K. (1990). *Inorg. Chem.*, *35*, 81.
Willstätter, R. M. (1965). *From My Life*, New York: Benjamin; pp. 382–385.

6 Toxic Effects of Metals

Jaco Vangronsveld and Herman Clijsters

Limburgs Universitair Centrum, Department SBG, Universitaire Campus
B-3590 Diepenbeek, Belgium

Contents

Introduction

Heavy metals are natural elements, ubiquitous in trace concentrations in the different compartments of the terrestrial ecosystem. Many of them are essential micronutrients for plants, animals, and man. High natural soil levels of metals originating from geological processes are occasionally found. Generally, however, high metal concentrations are the result of human industrial and agricultural activities. Metalliferous mining and smelting industries and other industrial emissions, waste disposal, vehicle emissions, metal contaminated sludge, pig slurry, etc., all are important sources of metal dispersion and enrichment in the environment (Freedman and Hutchinson, 1981). Therefore, heavy metal environmental contamination is not a recent phenomenon, but its intensity and frequency highly increased during the last decades.

Plants easily accumulate zinc, copper, cadmium, nickel and other metals. General physiological processes, eg, transpiration, respiration, and photosynthesis, are affected and plant development in general can be visibly inhibited (Wallnöfer and Engelhardt, 1984). Stunted growth, leaf epinasty and chlorosis are striking symptoms of strong metal toxicity. However, at lower degrees of soil pollution, these visible symptoms are less pronounced or can even be absent, whereas at the cellular level several processes are affected, due to increased local metal concentrations. Several mechanisms of metal action at the physiological and biochemical level were described (Foy *et al.*, 1978; Van Assche and Clijsters, 1990b).

Whatever the ulitmate site(s) of action in the plant, the metals have to cross the plasma membrane of the root cells before reaching the intracellular compartment. Therefore, this barrier (including its associated transport systems) can be considered as the first target for metal action. Only after passing the membrane, metals can interact with other cellular components and processes.

In this chapter, four different physiological aspects of metal phytotoxicity in higher plants will be discussed at the cellular level:

(1) Direct or indirect effects of metals on the alteration of the plasma membrane permeability, which can lead to leakage of ions and other solutes
(2) inhibitory action of metals on enzymes
(3) increase in capacity of enzymes
(4) defence mechanisms against increased metal phytotoxicity (antioxidative intermediates and processes, superoxide dismutase, peroxidase, catalase, phytochelatins).

In a further section, a brief review of possible interactions between metals and nucleic acids is presented. Finally, attention will be paid to the application of enzymatic changes as diagnostic criteria for the phytotoxicity of plant substrates.

When metal interaction with subcellular components (membranes, chloroplasts, mitochondria) and enzymes is discussed, it is a prerequisite to take into account the experimental approaches used. According to the method of application the studies can generally be classified in two main categories:

(a) *in-vitro* metal application:
(a1) Organelles or proteins originate from plants grown on a non-contaminated substrate and metal ions are supplied to the assay medium. Measurements are performed after a short incubation period and compared with a control without metal application. Only direct effects of metals can be detected by this method.
(a2) Metals are supplied to excised plant organs (leaves, roots) or organ fragments (leaf discs, root segments) before extraction and measurements. In this approach only short distance metal transport is included; selectivity or regulation of the metal uptake by the root system and long distance translocation processes throughout the plant are not taken into account.
(b) *in-vivo* metal application: comparative observations are performed on membranes, organelles or proteins isolated either from controls or from plants grown on a nutrient medium or substrate contaminated with heavy metals. There is no metal addition to the assay medium. This experimental approach adequately reflects the situation as for plants growing on a contaminated substrate *in situ*.

It is indicative to distinguish between the experimental results obtained with the main approaches cited, since their significance for understanding the metal action in the intact plant is obviously different. Therefore for each result cited the method of metal application will be mentioned as far as possible, and attention will be preferentially paid to the data obtained by *in-vivo* metal application.

Membrane Damage

An important harmful effect of metals at the cellular level is the alteration of the plasma membrane permeability, leading to leakage of ions like potassium and other solutes (Passow and Rothstein, 1960; Wainwright and Woolhouse, 1978; De Filippis, 1979; De Vos *et al.*, 1988, 1991). After supply of copper ions Ohsumi *et al.* (1988) demonstrated for yeast cells and De Vos *et al.* (1989) for root cells of *Silene cucubalus* that the permeability barrier (controlled by means of potassium leakage) of the plasma membrane was almost immediately lost. Oshumi *et al.* (1988) also reported a quick release of amino acids, especially glutamate and aspartate. After McBrien and Hassall (1965) and Overnell (1975), who studied potassium release from algal cells, the increased permeability of the plasma membrane may be considered to constitute the primary toxic effect of copper.

Two main mechanisms are generally proposed for copper induced changes to the plasma membrane function: direct effects on sulphydryl groups of membrane constituents and direct or indirect free radical mediated lipid peroxidation.

Metals are known to demonstrate relatively high affinity for sulphydryl and carboxyl groups, depending on the physicochemical properties of the cation. Salhany *et al.* (1978) suggested that the oxidation and cross-linking of membrane-protein sulphydryls of erythrocytes play an important role in the mechanism of copper mediated damage and subsequent haemolysis. In cells of the green alga *Chlorella*, the ability of the metals Zn, Hg, and Cu to induce potassium leakage was strongly correlated to the strength of the metal–sulphydryl bond (De Filippis, 1979). In higher plants, ATP-ases of the plasmalemma, responsible for electrogenic pumping and involved in the cellular permeability barrier, may be primarily affected by metals, since they are very sensitive to various sulphydryl reagents, including metals (Beffagna *et al.*, 1979; Kennedy and Gonsalves, 1989). Veltrup (1982) demonstrated a copper and zinc induced decrease of ATP-ase activity in excised roots of *Agrostis tenuis* and Lindberg and Wingstrand (1985) reported that *in vitro* cadmium ions competitively inhibit Mg^{2+}-ATP-ase by forming a complex with ATP thereby decreasing the substrate (Mg–ATP) concentration. Kennedy and Gonsalves (1987, 1989) showed that supply of the cations zinc, cadmium, mercury, copper and lead to maize roots causes an inhibition of Mg^{2+}–ATP-ase activity and also affects ATP-ase related processes such as H^+-ion efflux and transmembrane and transroot potentials in maize. On the contrary, Ros *et al.* (1990) found an increase in the Mg^{2+}-ATP-ase activity of plasma membranes of rice shoots also after *in-vivo* application of cadmium and nickel. They associated this increase with changes in lipid composition of the plasma membrane, namely reductions in the campesterol/sitosterol ratio.

Lipid peroxidation initiated by the formation of reactive free radicals may severely affect the functioning of biomembranes, including the plasma membrane. Radicals can be defined as species possessing one or more unpaired electrons. They may be either rather small and diffusible ('free' radicals) or large and by consequence rather immobile as part of subcellular constituents such as membranes. Under natural conditions, oxygen-free radicals are the most important free radical species formed in aerobic cells. They are generated as intermediates of the reduction of oxygen to water by single electron transfer. These intermediates are respectively the superoxide anion (O_2^-), hydrogen peroxide (H_2O_2), the hydroxyl radical ($OH^\bullet$). Reactive oxygen species can be highly toxic since they attack several cell constitutents, such as amino acids, proteins, carbohydrates, nucleic acids, and lipids. At the biomembrane level, lipid peroxidation is a very important destructive reaction. This process induces lipid phase transition and increases membrane permeability (Pauls and Thompson, 1981) leading to cell decompartmentation. The hydroxyl free radical ($OH^\bullet$) can initiate the membrane peroxidation process. Unsaturated membrane fatty acids are transformed into radical and hydroperoxy products which in turn promote free radical chain reactions with other cell constituents (Kappus, 1985). This cellular damage caused by reactive oxygen radicals is called 'oxidative stress' (Sies, 1985).

Under natural conditions, reactive oxygen species are formed at low rates. However, the cells dispose of an efficient antioxidative protection system consisting of enzymatic and non-enzymatic reactions. These reactions are extensively described in the section on defence mechanisms.

Metals can induce membrane lipid peroxidation and oxidative stress directly or indirectly. Indirect effects of metals on membrane damage originate from interactions with biochemical redox processes. Especially redox active metals as copper and iron, may interact with cellular redox processes or may even directly generate reactive free radicals, such as O_2^- (Kumar *et al.*, 1978), singlet oxygen ($^{\bullet}O_2$) (Ding and Chan, 1984), hydrogen peroxide (Freedman *et al.*, 1989) and lipid free radicals (Sandmann and Böger, 1980a; Chan *et al.*, 1982). By consequence, the toxicity of these metals can be ascribed to an increased production of free radicals in cells, which results in oxidative stress (Sandmann and Böger, 1980a, b; Freedman *et al.*, 1989).

Ions of several metals such as zinc, cadmium, lead, nickel, and mercury form very stable complexes with glutathione, disturb the interconversion of oxidised reduced glutathione, which results in a lowering of the level of available antioxidants in cells (Christie and Costa, 1984).

De Vos *et al.* (1989) suggest that the copper-induced damage to the permeability barrier in roots of *Silene cucubalus* is caused by a direct metal action on both membrane lipids and thiols. They propose that the first damaging effects of copper ions is the oxidation and cross-linking of membrane protein sulphydryls. However, they also adjudge an important role to the copper induced membrane lipid peroxidation, possibly due to direct free radical formation in the membrane; this effect could be enhanced by a depletion of thiols such as glutathione which results in a concomitant decrease of the cellular defence system against free radicals.

Lipid peroxidation may also be driven enzymatically. The iron containing enzyme lipoxygenase (EC 1.13.11.12) catalyses the hydroperoxidation of *cis,cis*-1,4-pentadiene groups of polyunsaturated fatty acids. Most lipoxygenases are specific for free fatty acids as substrate and show little reactivity toward esterified fatty acids (Todd *et al.*, 1990), but there exist also isozymes which react with complex lipids (Galliard and Chan, 1980; Hildebrand *et al.*, 1988). Gora and Clijsters (1989) found a stimulation of both lipoxygenase activity and lipid degradation products (thiobarbituric acid reactive products and ethane) in primary leaves of *Phaseolus vulgaris* seedlings, whereas De Vos (1991) concluded that lipoxygenase was absent in roots of *Silene cucubalus*. The last author suggested that the copper-induced increase of lipoxygenase-like activity observed in his plant material probably results from enhanced activity of aspecific peroxidases. This enzyme activity is unlikely to play a role in lipid peroxidation *in vivo*, since it is mainly localised in the extracytoplasmatic space (Schloß *et al.*, 1987) and therefore probably not in contact with the fatty acids in the hydrophobic lipid bilayer of biomembranes.

Disintegration of biomembranes by lipid peroxidation is not only limited to metal toxicity, but is a general process also caused by various other stress factors as air pollution (Chia *et al.*, 1984; Elstner, 1984) and herbicides (Youngman and Elstner, 1984; Schmidt and Kunert, 1986).

Inhibition of Enzymes

Metal inhibition is reported for many enzymes. The high affinity of metals for sulphydryl groups is suggested to be one of the main mechanisms of enzyme inhibition. *In vitro* it was shown (Mathys, 1975) that nitrate reductase, which catalyses the initial step of nitrate assimilation, is one of the most affected enzymes by metals. For many enzymes different responses were described depending on the method of application: after *in-vitro* metal supply, malate dehydrogenase activity was reported to be inhibited by zinc and cadmium in *Silene cucubalus* (Mathys, 1975) but was not affected at all by copper in non-tolerant clones of *Agrostis stolonifera* (Wu *et al.*, 1975). After *in-vivo* metal pretreatment however the same enzyme showed no effect from zinc in *Silene cucubalus* (Mathys, 1975), but a copper-induced reduction in activity in *Agrostis stolonifera* (Wu *et al.*, 1975) and a cadmium-induced stimulation in soybean (Lee *et al.*, 1976a).

The inhibitory effect of zinc on soluble and cell wall-bound acid phosphatases *in vitro* was studied in an attempt to find parameters distinguishing between tolerant and non-tolerant ecotypes of grasses (Wainwright and Woolhouse, 1978; Cox and Thurman, 1978).

In soybean seedlings *in-vivo* supply of cadmium and lead was reported to stimulate the respiration rate (Lee *et al.*, 1976a, b); this effect was ascribed to a demand for ATP production through oxidative phosphorylation because photophosphorylation was reduced. *In vitro*, several metal ions were effective inhibitors of the mitochondrial respiratory electron transport chain (Kleiner, 1974; Koeppe, 1977); Koeppe (1977) considered the inhibition of the electron transfer at the terminal NADH-oxidase to be specific for cadmium.

Sublethal concentrations of zinc, cadmium and mercury in the culture medium inhibit both respiration and photosynthesis in the green alga *Euglena* (De Filippis *et al.*, 1981b). Photosynthesis probably is one of the most metal-sensitive processes of plant metabolism. In consequence, this aspect will be treated extensively.

Metal Inhibition of Enzymes Related to Photosynthesis

Photosynthesis can be affected in many ways. Metals can influence biosynthesis of biomembranes and photosynthetic pigments, especially chlorophyll. They may inactivate enzymes by oxidising SH-groups necessary for catalytic activity or by substitution for other divalent cations in metalloenzymes. They finally can also interact with the photosynthetic electron transport and with the related photophosphorylation.

Inhibition of Chlorophyll Biosynthesis

The inhibition of chlorophyll biosynthesis by metals was described for higher plants (Bazinsky *et al.*, 1980; Prassad and Prassad, 1987) as well as for algae (De Filippis and Pallaghy, 1976; De Filippis *et al.*, 1981 a). Sensitivity to metals was found for two enzymes of this pathway: δ-aminolaevulinic acid (ALA)-dehydratase (EC 4.2.1.24) and protochlorophyllide reductase. Stobart *et al.* (1985) reported that the synthesis of δ-ALA is also an important site of inhibition in barley (*Hordeum vulgare*).

δ-ALA-dehydratase: This chloroplast enzyme, which is not membrane bound (Prassad and Prassad, 1987), catalyses the conversion of δ-ALA into porphobilinogen in both the synthesis of chlorophyll and haemoglobin. In *Pennisetum typhoideum* seedlings treated with toxic amounts of lead and mercury, this enzyme was significantly inhibited (Prassad and Prassad, 1987). The authors ascribed this effect to the interaction of the metals with functional sulphydryl groups of the enzyme, as lead and mercury possess a strong affinity for the SH-ligand (Passow *et al.*, 1961; Nieboer and Richardson, 1980). The results obtained in plants are in agreement with the effects described for human blood: the lead-induced inhibition of δ-ALA-dehydratase is considered to be the major cause of anemia in lead-exposed people (Vallee and Ulmer, 1972). In mammalian biochemistry, two different possibilities were proposed for the interaction of lead with this enzyme: (1) the most obvious inhibitory mechanism is metal–SH interaction (Vallee and Ulmer, 1972; Wilson *et al.*, 1972); (2) because δ-ALA-dehydratase contains zinc (Vallee, 1976) essential for its catalytic activity (Cheh and Neilands, 1973) and since the lead-induced δ-ALA-dehydratase inhibition in rats can be relieved by zinc (Rehman, 1984), it has been suggested that in blood lead-induced zinc deficiency could be responsible for the inhibitory effect (Rehman, 1984). A similar mechanism could also explain the decrease of chlorophyll content in *Phaseolus vulgaris* (Austenfeld, 1979) induced by cobalt. This metal has only little affinity for SH-groups but inhibits zinc uptake. Consequently, the substitution of zinc by cobalt in δ-ALA-dehydratase could affect the enzyme activity and explain the reduction in chlorophyll content observed. In this respect, it is noticeable that a treatment with toxic amounts of zinc did not change the chlorophyll content of *Phaseolus vulgaris* (Van Assche *et al.*, 1979).

Protochlorophyllide reductase: In the green alga *Chlorella vulgaris* (De Filippis and Pallaghy, 1976) grown in media containing sublethal mercury concentration, chlorophyll biosynthesis was strongly inhibited and protochlorophyll was accumulated. Zinc had a similar but less pronounced effect. A more metal sensitive alga, *Euglena gracilis* (De Filippis *et al.*, 1981 a), showed also a decrease of chlorophyll and a marked enhancement of protochlorophyllide levels in the presence of sublethal concentrations of zinc, cadmium and mercury. In these reports it was speculated that the metals inhibit the final reductive steps of chlorophyll formation. This hypothesis was confirmed by the data obtained with leaf discs of barley (*Hordeum vulgare*) floating on cadmium-containing solutions (Stobart *et al.*, 1985). This metal was found to inhibit protochlorophyllide reductase activity. Cadmium – SH interaction was proposed as the mechanism of action since SH-groups are involved in the formation of the NADPH – substrate – enzyme complex (Griffiths, 1975) and since similar results were obtained with other sulphydryl inhibitors, as *N*-methylmaleimide.

It should be noted that in most of these studies, especially in those on the barley leaf discs, the metal concentrations applied are significantly higher than those found in the field. It remains, therefore, to be proven that the cadmium effects described occur at the metal concentrations present in intact plants growing *in situ* on a cadmium-contaminated substrate.

Inhibition of Photosynthetic Electron Transport and Photophosphorylation

In-vitro approach: Data are available in abundance concerning metal effects on isolated chloroplasts (for a review, see Clijsters and Van Assche, 1985). All the metals studied were found to be potential inhibitors of photosystem 2 (PS 2); photosystem 1 (PS 1) was reported to be less sensitive. From the *in-vitro* experiments, at least two potential metal-sensitive sites can be derived in the photosynthetic electron transport chain: the water-splitting enzyme at the oxidising side of PS 2, and the NADPH-oxidoreductase (an enzyme with functional SH-groups) at the reducing side of PS 1 (Clijsters and Van Assche, 1985). Moreover, *in vitro*, non cyclic photophosphorylation was very sensitive to lead (Hampp *et al.*, 1973 b) and mercury (Honeycutt and Korgmann, 1972). Both cyclic and non-cyclic photophosphorylation were proven to be inhibited by excess of copper (Uribe and Stark, 1982) and cadmium (Lucero *et al.*, 1976).

In contrast to the previous results, Weigel (1985 a, b) reported that in mesophyll protoplasts of *Valerianella locusta* and in intact chloroplasts of *Spinacea oleraceae* cadmium affects photosynthesis by inhibition of several reaction steps of the Calvin cycle and not by interaction with the electron transport or photophosphorylation (cf. section on photosynthetic CO_2 fixation).

Interaction of metals with functional SH-groups was generally proposed as the mechanism of inhibition at several metabolic levels (Shioi *et al.*, 1978; Uribe and Stark, 1982). Sandmann and Böger (1980 a) suggested that the copper-involved effects on photosynthetic processes are the results of loss of chloroplast membrane function due to copper-induced lipid peroxidation.

In-vivo metal application: Reports of metal effects on electron transport and photophosphorylation after application of toxic amounts of metals to intact plants are less frequent. Cadmium and zinc inhibited PS 2 activity in *Lycopersicon esculentum* (Bazinsky *et al.*, 1980) and *Phaseolus vulgaris*, respectively (Van Assche and Clijsters, 1983). In the green alga *Euglena gracilis* (De Filippis *et al.*, 1981 b), PS 2 was sensitive to cadmium, copper and zinc. In the three species mentioned above, the water-splitting enzyme was the site of action.

Manganese ions are essential for an efficient water-splitting activity; for optimal functioning, 5 to 8 Mn ions per 400 chlorophyll molecules are required (Cheniae, 1980). Bazinsky *et al.* (1980) reported the restoration of the cadmium-inhibited electron transport by additional application of manganese. A sharp increase of the thylakoid-bound Zn/Mn ratio after zinc intoxication of *Phaseolus vulgaris* (Van Assche and Clijsters, 1986a) suggests a partial substitution of manganese by zinc in the water-splitting enzyme, causing inhibition of photolysis of water. For these higher plants, no metal inhibition of PS 1 electron transport was reported. In *Euglena* grown in a metal-containing medium (zinc, cadmium or mercury), the reduction in photosynthetic capacity was attributed to an inhibition of both the water-splitting enzyme and the NADPH-oxido-reductase (De Filippis *et al.*, 1981b). SH-interaction was proposed as the mechanism of action on the latter enzyme.

Although photophosphorylation was reported to be very sensitive to metals *in vitro*, the reduction of photosynthetic ATP production was shown to be related to an inhibition of the electron flow rate in *Euglena gracilis* treated with toxic concentrations of zinc, cadmium and mercury (De Filippis *et al.*, 1981b), in cadmium-treated *Lycopersicon esculentum* (Bazinsky *et al.*, 1980) and in zinc-treated *Phaseolus vulgaris* (Van Assche and Clijsters, 1986a).

Photosynthetic CO_2 Fixation

Cadmium (10 μM) inhibited CO_2 fixation of *Chlorella pyrenoidosa* by 62% (Hart and Scaife, 1977); a comparable decrease (74%) was found for spinach chloroplasts (Hampp *et al.*, 1976). *In-vivo* external application of 450 μM cadmium reduced CO_2 fixation by only 34%, indicating that a large percentage of cadmium never reached the chloroplasts (Huang *et al.*, 1974).

The primary inhibitory effect of cadmium on photosynthesis of excised leaves was proposed to be metal-induced stomatal closure (Bazzaz *et al.*, 1974; Lamoreaux and Chaney, 1978). In epidermal peels floating on a metal-containing solution, stomatal closure was reported (Bazzaz *et al.*, 1974). Inhibition of transpiration by several metals, especially by cadmium, is well documented (Poschenrieder *et al.*, 1989), but is not necessarily at the stomatal level. Metal-induced stomatal closure can indirectly be responsible for a decrease of photosynthetic CO_2 fixation indeed. However, several enzymes of the Calvin cycle are directly affected by metals.

For cadmium, Weigel (1985a, b) concluded that, *in vitro*, this metal inhibits photosynthesis mainly by interaction with several sites in the Calvin cycle and not by interaction with photochemical reactions located on the thylakoid membrane. *In vitro* studies showed a 90% inhibition of phosphoribulokinase (EC 2.7.1.18) by cadmium ions (Hurwitz *et al.*, 1956). This element also inhibited light activation of the Calvin cycle enzymes glyceraldehyde-3-phosphate kinase (EC 1.2.1.13) and ribulose-5-phosphate kinase (EC 2.7.1.19) in mesophyll protoplasts of *Valerianella locusta* (Weigel,

1985a). The author suggested that upon light activation, functional SH-groups become exposed. Ribulose-5-phosphate kinase was reported to be also very sensitive to lead applied to the enzyme *in vitro* (Hampp *et al.*, 1973b).

The majority of the metal effects on enzymes related to the Calvin cycle is described on the key enzyme ribulose-1,5-diphosphate carboxylase/oxygenase (RubisCo; EC 4.1.1.39). Bivalent cations (Mg^{2+}) play a major role in the activation of the RubisCo. Van Assche and Clijsters (1983, 1986b) found that zinc inhibits the carboxylase activity, but the oxygenase activity is unaffected. Experimental arguments were developed for an *in-vivo* substitution of the loosely bound magnesium by zinc in the active ternary RubisCo – CO_2 – metal complex. Similar concentrations of competing ions in the enzyme environment are a prerequisite for metal substitution (Clarkson and Hanson, 1980). After metal intoxication of the plant only zinc reached a concentration comparable to that of magnesium in the chloroplast; therefore, *in-vivo* substitution of bivalent cations for magnesium in RubisCo of higher plants might be restricted to zinc (Van Assche and Clijsters, 1990b).

The highly reversible binding of magnesium in the enzyme – CO_2 – metal complex was originally demonstrated in experiments where the purified enzyme was incubated with micromolar concentrations of cobalt, nickel or manganese; these cations substituted for magnesium, resulting in a general loss of carboxylation activity and in a decrease of the carboxylase/oxygenase ratio (Robinson *et al.*, 1979; Wildner and Henkel, 1979). Stiborova *et al.* (1988) demonstrated a slight reactivation of the carboxylase activity (0.5 – 1% of the values of the magnesium activated enzyme) after *in-vitro* incubation with stoichiometric amounts of copper, cobalt and zinc. Prolonged incubations resulted in inhibition.

In barley, cadmium, copper and, to a lesser degree, lead inhibited RubisCo *in-vivo*; zinc had almost no effect. After *in-vitro* application, all the metals tested inhibited the RubisCo (Stiborova *et al.*, 1986). Two SH-groups of cysteine (cys 172 and cys 456) are located in the RubisCo active centre (Lorimer, 1981) and these groups are essential for the enzyme activity. The inhibition observed after the *in-vitro* incubation of RubisCo with the four metals cited was explained by metal interaction with these functional SH-groups (Stiborova *et al.*, 1986). Stiborova (1988) further demonstrated that the irreversible cadmium inhibition of barley leaf RubisCo *in-vitro* can be ascribed to irreversible binding of cadmium to SH-groups involved in the stabilisation of the quaternary structure of the enzyme. This modification of the SH-groups by cadmium results in an irreversible dissociation and, by consequence, loss of enzyme activity. Since this effect already occurs at very low cadmium concentrations, the author speculated that this mechanism might explain the cadmium inhibition of RubisCo *in-vivo*. Measurement of the cadmium concentration in the chloroplast could deliver a decisive argument for this hypothesis.

In the brown alga *Lamionaria saccharina*, RubisCo is also inhibited by cadmium (Kremer and Markham, 1982). *In-vivo*, no direct interaction of cadmium with the enzyme was found, but *de novo* protein synthesis was inhibited. It was concluded that cadmium in the culture medium inhibits one or several steps in the protein biosynthesis and thus leads to enzyme deficiency. This is in contradiction with Stiborova *et al.* (1986) who found that the *in-vivo* inhibition of RubisCo in barley could not be ascribed to a reduction of *de novo* protein synthesis.

Summarising, three possible mechanisms of action were proposed for the observed metal inhibition of RubisCo: (1) SH-interaction (Cd, Cu and Pb; Stiborova *et al.*, 1986; Stiborova, 1988), (2) substitution of zinc for magnesium (Van Assche and Clijsters, 1986b) and (3) a general cadmium-induced enzyme deficiency (Kremer and Markham, 1982).

Carbonic Anhydrase

Carbonic anhydrase (CA; EC 4.2.1.1) is a zinc-metalloprotein. It catalyses the reversible hydration of carbon dioxide to bicarbonate and hydrogen ions and might be important in providing CO_2 for photosynthesis, although its role is somewhat controversial (Sandmann and Böger, 1983).

Lee *et al.* (1976a) reported a decreased CA activity in soybean (*Glycine max*) seedlings grown in cadmium-containing mineral solutions. In cotton (*Gossypium hirsutum*), CA activity was not detectable in leaves showing severe zinc deficiency (Okhi, 1976). CA activity increased curvilinearly when the zinc status improved from deficiency to optimal. Since CA activity is zinc-dependent, Lee *et al.* (1976a) supposed that cadmium interacts antagonistically with zinc, probably substituting for it and reducing the enzyme activity. This substitution is possible since in the CA enzyme, zinc binding involves two imidazole-N-residues of histidine and two carboxyl side chains of glutamate (Woolhouse, 1983).

Superoxide Dismutase

Superoxide dismutases (SOD; EC 1.15.1.1) constitute a group of metalloisoenzymes that neutralise the very reactive superoxide radical, generated in the cell as a by-product of the reduction of molecular oxygen (Salin, 1988). SOD is a highly efficient enzyme system involved in the cellular antioxidant system (cf. section on defence mechanisms).

In higher plants, three types of SOD were distinguished. The first is a copper-zinc-containing protein, sensitive to CN^- and predominantly associated with the chloroplast (Salin, 1988). The group of the CN^--insensitive enzymes consists of a manganese-containing and an iron-containing SOD-isozyme. In plants, the Mn–SOD is found mainly associated with the mitochondria. The iron-containing SOD-form was observed only in some plant families. It is associated with chloroplasts and is absent in mitochondria (Salin, 1988).

In leaf extracts from *Phaseolus vulgaris* grown on a substrate enriched with toxic concentrations of zinc or cadmium, no change in total SOD-activity was observed (Cardinaels *et al.*, 1984). Careful examination of these metal effects revealed that only cadmium almost completely inhibited the Cu–Zn–SOD activity and that this inhibition was fully compensated by stimulation of the CN^--insensitive SOD. This inhibitory effect was ascribed to a cadmium-induced zinc deficiency; leaf zinc content decreased from 1.13 μmol per gram dry weight for the control to 0.23 μmol for the cadmium treated plants; the cadmium content on the contrary increased from 0.85 nmol per gram dry weight to 0.13 μmol (Cardinaels *et al.*, 1984). De Vos (1991) suggested that in copper tolerant *Silene cucubalus* copper deficiency induces decrease of Cu–Zn–SOD activity, which was also compensated by a relatively high Mn–SOD activity.

The importance of zinc for a normal functioning of the Cu–Zn–SOD was shown in *Lemna gibba*. In zinc-deficient culture media the activity of Cu–Zn–SOD was strongly inhibited whereas in copper-deficient media little change was found in the enzyme activity (Vaughan *et al.*, 1982). In extracts of zinc-deficient plants, restoration of enzyme activity was possible by supplying zinc to the enzyme assay medium.

Concluding Remarks

Two main inhibitory mechanisms of metal action on enzymes can be distinguished:

(1) binding of metals to functional groups (eg, SH-groups involved in the catalytic action and/or in the structural integrity of the enzyme protein)
(2) induction of metal deficiency in metalloenzymes and substitution of the toxic for the essential metal in the enzyme complex.

Several conditions have to be fulfilled to make interference of metals with enzymes possible *in-vivo*: the metal should be accumulated (1) in an active form (permitting free interactions with the enzyme) and (2) at an inhibitory concentration in the same cellular compartment where the enzyme is present; (3) the affinity of the metal for the SH-group should be high, when interaction with this ligand is the mechanism of action and (4) the metal concentration should be comparable to that of the active metal when enzyme inhibition is due to substitution or competition for the same binding site.

It is clear that for many of the enzymes discussed, conflicting results are described which do not always allow a conclusive interpretation of the mechanism of interaction. Several of these discrepancies can be attributed to differences in the mode of metal application. It is obvious that extrapolation of data obtained in *in-vitro* experiments to the *in-vivo* situation can easily lead to misinterpretation.

Induction of Enzymes

A variety of responses can be initiated by the direct interaction of metals with cellular components. Membrane damage and enzyme inhibition are examples of such a metal effect. Above a certain threshold concentration of metals in the cell, its physiological state is irreversibly changed (Van Assche *et al.*, 1988). This response is reflected by an increase in capacity (activity under non-limiting substrate and coenzyme concentrations) of certain enzymes. This increase in capacity is generally called enzyme induction. These secondary, indirect effects of metals are considered to play an important role in the stress metabolism induced by toxic metal concentrations.

Peroxidases

Peroxidase (POD) induction was observed in leaves and roots of various plant species after application of toxic amounts of cadmium, zinc, copper, nickel, lead and mercury (Table 6-1). This induction of the POD capacity is strongly correlated with the plant tissue level of copper (Mukherji and Das Gupta, 1972; *Lactuca sativa*), zinc and cadmium (Van Assche *et al.*, 1988; *Phaseolus vulgaris*) and with shoot growth and biomass reduction.

Table 6-1. Metal induction of peroxidase in different organs of higher plant species (after Van Assche and Clijsters, 1990b, adapted).

Metal applied	Plant species	Organ	Reference
Cadmium	*Glycine max*	Leaves	Lee *et al.* (1976a)
	Phaseolus vulgaris	Leaves	Van Assche *et al.* (1988)
Copper	*Lactuca sativa*	Seedlings, not specified	Mukherji and Das Gupta (1972)
	Phaseolus vulgaris	Roots	Van Assche *et al.* (1986)
Lead	*Glycine max*	Leaves	Lee *et al.* (1976b)
	Medicago sativa	Leaves	Maier (1978a)
	Zea mays	Leaves and roots	Maier (1978a)
Nickel	Nickel-sensitive *Silene italica*	Not specified	Gabrielli *et al.* (1987)
	Phaseolus vulgaris	Leaves	Van Assche *et al.* (1986)
Zinc	Zinc-sensitive *Silene cucubalus*	Leaves	Mathys (1977)
	Phaseolus vulgaris	Leaves	Van Assche *et al.* (1988)
Mercury	*Lemna minor*	Fronds	Subhadra *et al.* (1991)
	Allium cepa	Roots	Subhadra *et al.* (1991)
Mixture in soil	*Phaseolus vulgaris*	Leaves, roots	Van Assche and Clijsters (1990a)

Induction of POD capacity is not restricted to heavy metals only but was found for a variety of other stress factors: chilling (Levitt, 1972; Kacperska-Palacz and Uliasz, 1974), wounding (Gaspar *et al.*, 1982) or pathogenic infection (Van Loon, 1986), air pollution by SO_2, ozone, NO_X or fluoride (Horsman and Wellburn, 1976; Rabe and Kreeb, 1979; Endress *et al.*, 1980; Khan and Malhotra, 1982; Decleire *et al.*, 1984). POD-capacity of tree leaves was even utilised as a parameter for monitoring and mapping air pollution (Keller, 1974).

In higher plants, different isoenzymes of POD can be distinguished. The electrophoretic pattern of leaf extracts of *Oryza sativa* treated with toxic amounts of zinc, copper and mercury was changed as compared to controls: some new isoperoxidases became visible, while other bands disappeared (Nag *et al.*, 1981). As compared with the control, the activity of one particular anionic band was very pronounced. Also in *Phaseolus vulgaris*, the isoperoxidase pattern was shown to be affected by metals (Van Assche *et al.*, 1986, see Fig. 6-1). In the primary leaf, zinc induced two anionic bands, that were absent in the controls, but the isoperoxidase pattern of the root was not affected. With copper, the same two bands appeared in the roots only, while cadmium induced the same two isoperoxidases in both primary leaves and roots.

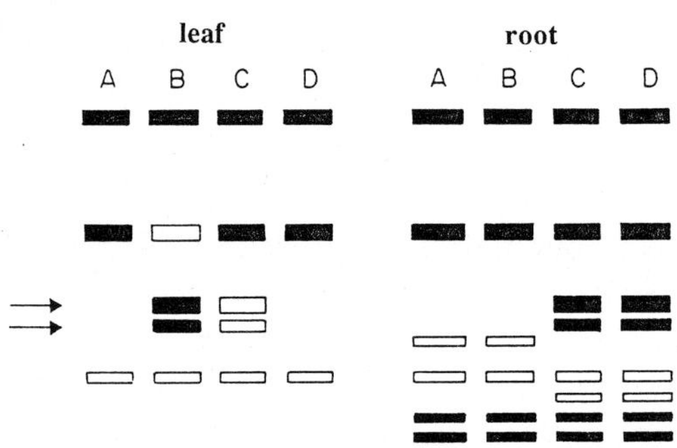

Fig. 6-1. Electrophoretic pattern of anionic isoperoxidases in leaf and root of *Phaseolus vulgaris*, treated with a toxic concentration of zinc (B), cadmium (C) and copper (D), as compared to the control (A) (staining with benzidine/H_2O_2 (after Van Assche et al., 1986, and Van Assche and Clijsters, 1990)). The isoperoxidases, induced by the metals, are indicated with arrows. Black symbols: high activity; open symbols: moderate activity.

In *Sedum album*, an interesting evolution in the isoperoxidase pattern was found after ozone fumigation (Castillo and Greppin, 1986). In a first phase, some new cationic isoperoxidases appeared. After 24 h these isozymes disappeared but some anionic bands, involved in lignification, were induced. Supply of a toxic concentration of zinc to 8 day old seedlings of *Phaseolus vulgaris* also resulted in a quick (2h), transient appearance of new cationic PODs, while only 2 days after metal application the anionic D-bands already described became detectable (Van Assche *et al.*, 1990). These results suggest that different POD-isoforms might be involved in different physiological phenomena, occurring as a response to the stress factor.

The general occurrence of POD induction by different stress factors raises the hypothesis of a basic stress metabolism involving formation of H_2O_2 or organic peroxides. Elstner *et al.* (1988) proposed oxygen activation as a general stress response. The very reactive intermediary forms of oxygen could be quenched by an enhanced capacity of specific enzymes like POD, SOD and catalase. This type of defence mechanism, which could protect biomembranes against metal-mediated peroxidation, will be discussed in the following section.

Involvement of POD was also described for the degradation of indolacetic acid (IAA; Salin, 1988). Increase of IAA-oxidase activity was found in *Lactuca sativa* seedlings (Mukherji and Das Gupta, 1972) and *Hordeum vulgare* roots (Coombes *et al.*, 1976) treated with toxic amounts of copper. This response was suggested to be responsible for the growth inhibition of the seedlings and roots described.

Enhancement of POD-capacity and appearance of new isoforms is generally considered as an important criterion for senescence (Hazell and Murray, 1982). Lee *et al.* (1976a) suggested that the cadmium-induced capacity increase of POD and several hydrolytic enzymes (ribonuclease, deoxyribonuclease, acid phosphatase) in *Glycine max* should constitute an accelerated senescence response. In *Zea mays*, an induction of leaf acid phosphatase was also reported for toxic concentrations of lead (Maier, 1978b).

Catalase

A transient increase of catalase capacity was found after application of a toxic lead concentration to *Zea mays* seedlings (Hoxha *et al.*, 1985). Evidence was presented for a protective role of catalase after intoxication with heavy metals. De Vos (1991) reported a copper-induced increase of catalase capacity in copper-sensitive *Silene cucubalus*. In *Lemna minor* fronds and *Allium cepa* roots, the same enzyme induction was observed with mercury (Subhadra *et al.*, 1991).

Esterases

Karataglis *et al.* (1988) reported changes in the number of isoesterase bands and quantitative changes of band activity after treatment of *Triticum aestivum* with toxic concentrations of zinc, cadmium, copper, lead, and nickel. In nickel treated *Silene vulgaris*, a change of the isoesterase pattern was also found (Pandolfini and Maier, 1985).

Superoxide Dismutase

As already cited, under conditions of cadmium stress, a significant stimulation of the CN^--insensitive SOD occurred in primary leaves of *Phaseolus vulgaris* while Cu – Zn – SOD was almost completely inhibited. The net result was an unchanged total SOD capacity (Cardinaels *et al.*, 1984). Palma *et al.* (1987) demonstrated that the specific activity of peroxisomal Mn – SOD in the leaves of *Pisum sativum* was higher in copper-tolerant than in non-tolerant plants and suggested that this increased activity could play a role in the copper tolerance mechanism of this plant. De Vos (1991) found an increase of both Cu – Zn – SOD and Mn – SOD in roots of copper sensitive *Silene cucubalus* after supply of toxic concentrations of copper. For copper-tolerant plants of the same species, the author reported a copper-induced increase of Cu – Zn – SOD and a corresponding decrease of Mn – SOD, resulting in an unchanged total-SOD activity.

An increased SOD capacity after copper supply was also found in other organisms (Shatzman and Kosman, 1978, in the fungus *Dactylium dendroides*; Naiki, 1980, *Saccharomyces*; Ljutakova *et al.*, 1984, rat liver; Lee and Hassan, 1985, *Saccharomyces*). The increased biosynthesis of SOD in copper-treated yeast was suggested to be the result of an enhanced intracellular O_2^- flux (Lee and Hassan, 1985).

It is interesting to mention that SOD was also reported to increase after treatment of plants with ozone (Lee and Bennett, 1982) and SO_2 (Tanaka and Sugahara, 1980), as was also observed for POD (cf. above).

Enzymes of the Intermediary Metabolism

The capacity of several enzymes involved in or closely related to the Krebs cycle is significantly enhanced after assimilation of toxic amounts of metals by the plant: isocitrate dehydrogenase (ICDH; EC 1.1.1.42), malate dehydrogenase (MDH; EC 1.1.1.37), glutamate dehydrogenase (GDH; EC 1.4.1.2) and malic enzyme (ME; EC 1.1.1.40). A summary of the data from literature is presented in Table 6-2. Ernst (1980) suggested that the capacity of enzymes involved in dark respiration is possibly stimulated to compensate for the decrease of ATP and NADPH normally provided by the metal-sensitive photosynthetic reactions. A similar argument could be proposed to explain the induction of ME and glucose-6-phosphate dehydrogenase (G6PDH; EC 1.1.1.49), the key enzyme of the pentose-phosphate pathway. The capacity of G6PDH was significantly induced by zinc in zinc-sensitive ecotypes of *Silene cucubalus* (Mathys, 1975) and by zinc and cadmium in *Phaseolus vulgaris* (Van Assche *et al.*, 1988). The capacity of these enzymes, catalysing NADPH delivering reactions could be increased to compensate for a possible shortage of reducing power in the cell, which results from the limited electron transport in the chloroplast.

GDH plays a key role in amino acid metabolism. Several authors (Mukherji and Das Gupta, 1972; Lee *et al.*, 1976b; Van Assche *et al.*, 1979; Weigel and Jäger, 1980a)

Table 6-2. Enzymes of the intermediary metabolism induced after application of toxic concentrations of metals. Most of the experiments were performed on *Phaseolus vulgaris* cultivars, except those who are specially marked, which were performed on *Glycine max* (1) and on a zinc resistant (2) and a zinc sensitive (3) clone of *Silene cucubalus* (after Van Assche and Clijsters, 1990b, adapted).

Metal applied	Enzyme induced	Organ	Reference
Cadmium	ICDH	Leaves, roots	Weigel and Jäger (1980a)
		Leaves	Van Assche *et al.* (1988)
	GDH	Leaves	Weigel and Jäger (1980a)
		Roots	Weigel and Jäger (1980a, b)
		Leaves	Van Assche *et al.* (1988)
	MDH	Leaves (1)	Lee *et al.* (1976a)
		Leaves, roots	Weigel and Jäger (1980a)
		Leaves	Van Assche *et al.* (1988)
	ME	Leaves	Van Assche *et al.* (1988)
	G6PDH	Leaves	Van Assche *et al.* (1988)
Copper	ICDH, MDH	Roots	Weigel and Jäger (1980a)
Zinc	ICDH, GDH, ME, G6PDH	Leaves	Van Assche *et al.* (1988)
	ICDH	Leaves (2)	Mathys (1975)
	G6PDH	Leaves (3)	Mathys (1975)

reported a metal-induced increase of the water-soluble protein content. After inhibition of the ribosomal function using cycloheximide, it was shown at least for one of these enzymes that the zinc-induced increase of enzyme capacity was due to *de novo* synthesis of proteins (Clijsters *et al.*, 1984). Under metal stress conditions, the stimulated Krebs cycle enzyme capacity might be linked to *de novo* protein synthesis by GDH.

Concluding Remarks

Enzyme induction is an indirect effect of metal toxicity. In consequence, it only appears after *in-vivo* metal application. Increase in capacity in the presence of toxic concentrations of metals implies that the enzyme involved is insensitive to or well masked from direct metal action. In *Silene cucubalus* POD and to a lesser extent ICDH were shown to be very tolerant to zinc, copper and cadmium applied *in-vitro* (Mathys, 1975). MDH extracted from roots of both copper tolerant and non-tolerant clones of *Agrostis stolonifera* was found to be insensitive *in-vitro* to copper concentrations up to 180 μM. However, in the literature, conflicting results are reported about the effects of *in-vitro* and *in-vivo* application of metals on enzymes. In *Phaseolus vulgaris*, GDH was inhibited by cadmium *in-vitro*, while an induction was found after *in-vivo* application of a toxic dose of the same metal (Weigel and Jäger, 1980b).

Obviously, discrepancies in observed enzyme responses may also result from differences in concentration of the active free metal in the direct environment of the enzyme. Phosphoenolpyruvate carboxylase (EC 4.1.1.31) for instance, which is extremely sensitive to *in-vitro* zinc application, was also inhibited in intact 7 day old *Zea mays* seedlings, treated with very high amounts of zinc (Stiborova *et al.*, 1988). In *Phaseolus vulgaris* grown on a substrate containing toxic amounts of zinc, on the contrary, the same enzyme showed a significant capacity increase as compared to the control (Clijsters *et al.*, 1984).

All the data cited clearly demonstrate how important it is to know the *in-vivo* free metal concentration for understanding the metal effect on enzyme capacity. They also warn against to thoughtless interpretation and extrapolation of results.

Defence Mechanisms Against Metal Phytotoxicity

Various mechanisms have been developed by plants to alleviate stress imposed by exposure to toxic metal concentrations. Although the biochemical and molecular bases of adaptation and resistance of plant communities to high metal levels will be discussed extensively in Chapter 3, a survey summarising the possible anti-metal stress defence strategies and mechanisms forms an essential part of this chapter on toxicology. Every plant species, whether it is resistant or not to high metal concentrations, possesses defence responses against increased metal concentrations in the cell indeed.

The metal defence mechanisms can be classified in two main categories:

(a) Mechanisms which prevent interaction between metals and their site(s) of actions.

(a1) at the cellular level:

– Binding of the metals to the cell wall complex: the wall of root cells is directly exposed to the metals available in the soil water phase. From a few to about 80% of the total root metal content was found to be associated with the root cell walls in various dicotyledonous plants (Ernst, 1972, 1974; Farago *et al.*, 1980). Metals can be electrostatically bound to carboxylic functions of pectins and to charged groups of wall proteins; even more complex chemical bounds can occur. Moreover, amorphous metal precipitates as phosphate, carbonate and silicate have been reported (Ernst *et al.*, 1990).

– Reduction of active uptake in the cell: cadmium resistance of *Euglena gracilis* strains was suggested to be the result of a membrane transport mechanism (inhibition of the membrane cadmium ion transport or increase of the cadmium exclusion mechanism; Bariaud *et al.*, 1985).

– Translocation to cell compartments where direct interaction with essential metabolic processes is impossible: accumulation in the vacuole was proposed as an important intracellular mechanism for the sequestration of zinc (Ernst, 1969), cadmium (Ernst, 1980) and nickel (Ernst, 1974). Vacuolar deposits of zinc, cadmium, lead and copper were reported in both metal sensitive and resistant plants (Mullins *et al.*, 1985; Rauser and Ackerley, 1987).

– Sequestration in the cytoplasm: in many plant species, increased levels of metals appear to induce the production of metal-binding polypeptides, generally classified as phytochelatins (for a review, see Rauser, 1990 and Steffens, 1990). These polypeptides probably play a central role in the homeostatic control of metal ions in plants (Steffens, 1990). They may also be involved in the mechanism of metal tolerance of selected cell lines and intact plants (Rauser, 1984; Tomsett *et al.*, 1989). The similarity of phytochelatins to glutathione indicates that they cannot be primary gene products and that their biosynthesis shares a common enzymology with glutathione (Steffens, 1990). They can be viewed as linear polymers of the γ-glutamylcysteine portion of glutathione. Evidence for the linkage of glutathione metabolism to phytochelatin synthesis has been found by many authors in both *in-vivo* and *in-vitro* studies (Rauser, 1987; Scheller *et al.*, 1987; Grill *et al.*, 1989; Rüegsegger *et al.*, 1990). In *in-vitro* studies, Grill *et al.* (1989) showed that phytochelatins were formed from glutathione in the presence of cadmium ions by an enzyme (purified from *Silene cucubalus*) they characterised as γ-glutamylcysteine dipeptidyl transpeptidase (phytochelatin synthase).

(a2) at the level of the plant:

– Changes in plant metal uptake can be the consequence of a modification of metal fluxes through the plasma membranes of root cells or of mycorrhizal fungi. The latter might be of special importance in several species since they can modify the host's metal uptake.

– Changes in metal translocation within the plant: reduction of root to shoot translocation of metals leads to the protection of sensitive essential metabolic processes of the shoot (eg, photosynthesis). Accumulation of metals in old leaves just prior to abcission results in a removal of excessive metals from the plant body.

(b) Mechanisms which counteract metal-induced damaging processes.

The formation of free radical species, which can be initiated directly or indirectly by metals, can cause severe damage to different cells components. Cells are provided with an efficient antioxidative system, consisting of several non-enzymatic and enzymatic mechanisms (Table 6-3). Molecules such as glutathione, ascorbic acid (vitamin C), α-tocopherol (vitamin E), hydroquinones, β-carotene and flavonoids serve as reductants and radical scavengers. Glutathione, vitamins C and E and hydroquinones react directly or via enzyme catalysis with O_2^-, H_2O_2 or $OH^\bullet$, while carotenes and flavonoids can directly interact with singlet oxygen. POD, SOD and catalase, enzymes showing changes in capacity as a result of increased metal levels in

Table 6-3. Antioxidant systems in plants (after Elstner, 1982; Salin, 1988; Cadenas, 1989; De Vos, 1991).

Antioxidant system	
Enzymatic:	
	Superoxide dismutase
	Peroxidase
	Catalase
Non-enzymatic:	
	RSH (eg, glutathione)
	Ascorbic acid (vitamin C)
	α-Tocopherol (vitamin E)
	β-Carotene
	Flavonoids
	Hydroquinones
Ancillary enzymes:	
	Glutathione reductase (uses NADPH)
	NADPH-delivering enzymes

the cell (cf. section on enzyme induction), can function as effective quenchers of reactive intermediary forms of oxygen. Superoxide dismutase catalyses the dismutation of two superoxide molecules according to the following equation:

$$O_2^- + O_2^- + 2H^+ \rightarrow H_2O_2 + O_2$$

The reaction mechanism involves alternate reduction and oxidation of the metal (M) associated with the enzyme:

$$O_2^- + M^{2+} \rightarrow O_2 + M^+$$

$$O_2^- + M^+ + 2H^+ \rightarrow H_2O_2 + M^{2+}$$

The heme-containing enzymes catalase and peroxidase are responsible for the removal of long-lived H_2O_2 according to the following reactions:

$$H_2O_2 + H_2O_2 \rightarrow 2H_2O + O_2 \text{ (catalase)}$$

$$H_2O_2 + R(HO)_2 \rightarrow 2H_2O + RO_2 \text{ (peroxidase)}$$

Glutathione reductase and several $NADP^+$ reducing enzymes may play a more indirect role. When glutathione (GSH) participates in a reduction reaction it becomes oxidised to oxidised glutathione (GSSG). To provide the cell again with 'active' reduced glutathione, GSSG has to be transformed into GSH by glutathione reductase with NADPH as an electron and proton donor. This mechanism illustrates the importance of a high cellular concentration of reduced coenzymes. Enzymes such as glucose-6-phosphate dehydrogenase, malic enzyme and isocitrate dehydrogenase catalyse reactions involving $NADP^+$ reduction. In consequence, the increase in capacity of these enzymes as a response to toxic metal concentrations in the cell might play a key role in the cellular defence mechanism against metal stress. In roots of copper-sensitive *Silene cucubalus* the capacity of glutathione reductase was strongly increased after an exposure to 20 μM copper for 3 days or more (De Vos, 1991). Elevated total glutathione levels and glutathione reductase activities have been found as a result of several other forms of environmental stress, such as air pollutants (Grill *et al.*, 1979; Miszalski and Ziegler, 1979; Chiment *et al.*, 1986; Melhorn *et al.*, 1986; Chen *et al.*, 1991), water stress (Dhinsa, 1987) and chilling (Levitt, 1972).

Interaction of Metals with Nucleic Acids

Most of the reports about interactions of metals with nucleic acids and their metabolism are based on studies with bacteria and animals. Information from higher plants is rather poor. Ernst (1980) cited an increase of the number of structural chromosome aberrations after cadmium treatment of *Crepis capillaris* seeds.

In general metals can interact directly or indirectly with nucleic acids. The numerous nucleophilic centres in nucleic acids are favourite binding sites for metal ions. The type and localisation of binding apparently depend on the metal considered. Induction of crosslinks between both DNA strands, single-strand DNA breaks and chelation or formation of complexes between DNA and metals were described (see Gebhart and Rossman, 1991, for a review). Similar reactions can also be expected between metals and RNA, thereby affecting the specific functions of the latter.

Besides these direct effects, metals can indirectly induce breaks in the DNA strand: depurination of DNA which can have mutagenic effects was reported to be induced by copper(II), nickel(II) and chromium(IV) (Schaaper *et al.*, 1987). Generation of oxygen free radicals can also cause DNA damage. Exposure of DNA to oxygen free radical generating systems was found to induce extensive strand breakage and degradation of DNA (Halliwell and Gutteridge, 1984).

Effects of metals on the function of various enzymes involved in the nucleic acid metabolism can also indirectly contribute to alterations in the genetic information by affecting replication fidelity.

Finally, the activity of hydrolytic enzymes as ribonuclease and deoxyribonuclease was found to increase after *in-vivo* cadmium supply to *Glycine max* seedlings. These effects were considered as symptoms of cadmium accelerated senescence (Lee *et al.*, 1976a; see section on enzyme induction).

Enzymatic Changes in Plants as Diagnostic Criteria of a Biological Test for the Evaluation of Phytotoxicity of Metal-Polluted Soils

Enzyme induction and/or isoenzyme variation have been used to characterise widely different phenomena. These enzymatic parameters were shown to be useful criteria for monitoring air pollution (Keller, 1974; Flückiger *et al.*, 1978; Rabe and Kreeb, 1979), for ecogenetic characterisation of plant populations (Verkleij *et al.*, 1980; Triest, 1991), for the study of callus tissue differentiation (Coppens and Gillis, 1987) and plant development and differentiation (Scandalios, 1974). The evaluation of phytotoxicity of metal-polluted soils and other substrates can also be realised by enzymatic parameters.

Generally, monitoring of the potential phytotoxicity of metal-polluted soils is performed by chemical analysis. These results can give rise to misinterpretations since plant availability in and consequently metal uptake from the substrate are functions of the chemical form of the element in the soil and of several soil parameters, eg, pH, organic matter and cation exchange capacity. Moreover, soils are frequently contaminated by a mixture of metals. Each of these materials separately can be phytotoxic, or they can mutually interact in a synergistic, antagonistic or cumulative way. The physiological and metabolic responses of test plants can be considered as a biological criterion for the total phytotoxic effect, since it is the result of the mutual interactions of the metals present in the soil with other soil factors and with the plant. Phytotoxicity responses of test plants grown under controlled environmental conditions only reflect further the interference with metabolic processes of metals assimilated through the roots. The correlation between enzyme induction and the metal concentration is so strict that the precise toxic threshold value for zinc and cadmium in the primary leaf of *Phaseolus vulgaris* could be calculated from the enzyme induction curves in this organ (Van Assche *et al.*, 1988). Due to the metal specific responses in the different organs (cf. section on enzyme induction) isoperoxidase analysis appears to be particularly useful for the evaluation of the phytotoxicity of metal-polluted soils.

The quantitative relation observed in *Phaseolus vulgaris* between enzyme induction and metal assimilation (Van Assche *et al.*, 1988) in combination with the metal-specific changes in the isoperoxidase pattern (*vide supra* and Van Assche *et al.*, 1986) and data of morphological parameters (length, weight, leaf area) delivers a reliable biological test system for the evaluation of soil phytotoxicity (Van Assche and Clijsters, 1990a; Vangronsveld and Clijsters, 1992). Cultivation of the test plants under controlled environmental conditions is very important for these studies since several stress factors can affect enzyme activities. Attention must also be paid to the test plant species as the sensitivity towards metals varies between species and even between cultivars of the same species (Marschner, 1983). For routine screening of soil phytotoxicity, a *Phaseolus vulgaris* cultivar (Limburgse vroege) with moderate sensitivity to zinc, cadmium and copper was used. Other more sensitive species as *Spinacea oleracea* and *Lycopersicon esculentum* are now included in the test system.

The values obtained for each parameter examined can be 'transposed' into phytotoxicity classes: each parameter is used to classify the substrate in a given phytotoxicity class. A classification system was developed for the parameters studied on *Phaseolus vulgaris* cv. Limburgse vroege and is presented in Table 6-4. The 'phytotoxicity index' for a given substrate represents the mean value of the toxicity class numbers obtained for each parameter separately.

Table 6-4. Biological evaluation of phytotoxicity of metal contaminated substrates with *Phaseolus vulgaris* as the test plant. Classification of biological data into phytotoxicity classes, depending on the magnitude of the effect (% of control for morphological parameters and enzymes capacities) (after Van Assche and Clijsters, 1990; Vangronsveld and Clijsters, 1992).

Parameters	Degree of phytotoxicity (class number)			
	Not toxic (class 1)	Slightly toxic (class 2)	Moderately toxic (class 3)	Strongly toxic (class 4)
Shoot length	>85	85–70	70–50	<50
Primary leaf area	>85	85–70	70–50	<50
Root weight	>85	85–70	70–50	<50
Enzyme capacity				
Leaf				
POD	<150	150–325	325–500	>500
ME, ICDH	<125	125–175	175–250	>250
Root				
POD, ME, GDH	<125	125–175	175–250	>250
D-isoperoxidases/ total isoperoxidases (X 100)				
Leaf	0	0–25	35–50	>50
Root	0	0–15	15–30	>30

Conclusions

Uptake of an excess of metals by higher plants and algae can initiate a variety of metabolic reactions, finally leading to global phytotoxic responses, eg, dwarf growth and chlorosis.

Metals can directly or indirectly interfere with cellular components and structures. Their effects on enzyme proteins and biomembranes were discussed extensively. Depending on the enzyme studied, metals either inhibit or stimulate its activity. They are generally considered to affect membrane permeability and to induce cell decompartmentation. In this respect, studies on the plasma membrane of root cells are highly interesting: this membrane is the first functional structural barrier encountered by metals assimilated by the root. It contains several vital metal-sensitive components responsible for the selectivity of its permeability, like ATPases and other proteins.

Beyond a threshold value the biological responses described are functions of the metal concentration in the plant tissue. Since the metal action is at the subcellular level, data on the metal content of cell compartments (mitochondria, chloroplasts, vacuoles, cytoplasm) could contribute to a better understanding of its mechanism of action.

To some extent, every metal investigated shows the general biological responses reported. The intensity of its impact is a function of its chemical characteristics, especially its redox properties on one hand and of metal speciation at the cellular level on the other hand. Further research is required to specify the biological effect, typical for every metal studied.

As an indirect effect of increased metal uptake, the physiological state of the cell can alter and 'defence mechanisms' can be induced. Phytochelatin (metal binding proteins) synthesis and induction of free radical quenching enzymes and metabolites were frequently observed. Especially the latter can protect membranes against oxidative breakdown.

Similar 'defence responses' are provoked by other forms of stress (eg, air pollutants, herbicides, cold or drought stress, etc.). Analogous responses were also observed during natural plant senescence. It is a challenging hypothesis that several kinds of stress factors demonstrate a common site of action at the subcellular level (Clijsters *et al.*, 1990) and might accelerate the process of plant senescence (Lee *et al.*, 1976a; Van Assche *et al.*, 1990). Its further evaluation, however, needs the continuation of research on senescence processes, stress effects and metal action at several levels of organisation.

References

Austenfeld, F. A. (1979), *Photosynthetica 13*, 434–438.

Bariaud, A., Bury, M., Mestre, J. C. (1985), *Physiol. Plant. 63*, 382–386.

Bazinsky, T., Wajda, L., Krol, M., Wolinska, D., Krupa, Z., Tukendorf, A. (1980), *Physiol. Plant. 48*, 365–370.

Bazzaz, F. A., Carlson, R. W., Rolfe, G. L. (1974). *Environ. Pollut. 7*, 241–246.

Beffegna, N., Marré, E., Cocucci, S. M. (1979), *Planta 146*, 387–391.

Cadenas, E. (1989), *Ann. Rev. Biochem. 58*, 79–110.

Cardinaels, C., Put, C., Van Assche, F., Clijsters, H. (1984), *Arch. Intern. Physiol. Bioch. 92*, 27–28.

Castillo, F., Greppin, H. (1986), *Physiol. Plant 68*, 201–208.

Chan, P. C., Peller, O. G., Kesner, L. (1982), *Lipids 17*, 331–337.

Cheh, A., Neilands, J. B. (1973), *Biochem. Biophys. Res. Comm. 55*, 1060–1063.

Chen, Y.-M., Lucas, P. W., Wellburn, A. R. (1991), *Environ. Pollut. 69*, 1–15.
Cheniae, G. M. (1980), in: *Methods in Enzymology,* San Pietro, A. (ed.), New York: Academic Press Vol. 69C, pp. 349–363.
Chia, L. S., Mayfield, C. I., Thompson, J. E. (1984), *Plant. Cell Environment 7*, 333–338.
Chiment, F. F., Alscher, R., Hughes, P. R. (1986), *Environ. Exp. Botany 26*, 147–152.
Christie, N. T., Costa, M. (1984), *Biol. Trace Elem. Res. 6*, 139–158.
Clarkson, D. T., Hanson, J. B. (1980), *Ann. Rev. Plant Physiol. 31*, 329–398.
Clijsters, H., Van Assche, F. (1985), *Photosynthesis Research 7*, 31–40.
Clijsters, H., Van Assche, F., Carinaels, C. (1984), *Wissenschaftliche Zeitschrift der Humboldt Universität zu Berlin, Math.-Nat. XXXIII*, 326–327.
Clijsters, H., Van Assche, F., Vangronsveld, J., Gora, L. (1990), *International Conference Environmental Contamination*, Barcelo, J. (ed.), *Barcelona*, Edinburgh: CEP Consultants; pp. 41–44.
Coombes, A. J., Lepp, N. W., Phipps, D. A. (1976), *Zeitschr. Pflanzenphysiol. 80*, 236–242.
Coppens, L., Gillis, E. (1987), *J. Plant Physiol. 127*, 153–158.
Cox, R. M., Thurman, D. A. (1978), *New Phytol. 80*, 17–22.
Decleire, M., de Cat, W., de Temmerman, L., Baeten, H. (1984), *J. Plant Physiol. 116*, 147–152.
De Filippis, L. F. (1979), *Zeitschr. Pflanzenphysiol. 92*, 39–49.
De Filippis, L. F., Pallaghy, C. K. (1976), *Zeitschr. Pflanzenphysiol. 78*, 314–322.
De Filippis, L. F., Hampp, R., Ziegler, H. (1981a), *Zeitschr. Pflanzenphysiol. 101*, 37–47.
De Filippis, L. F., Hampp, R., Ziegler, H. (1981b), *Arch. Microbiol. 128*, 407–411.
De Vos, R. (1991), Doctoral Thesis, Vrije Universiteit, Amsterdam.
De Vos, C. H. R., Schat, H., Ernst, W. H. O. (1988), *J. Cell. Biochem, Suppl. 12A*, 49.
De Vos, C. H. R., Vooijs, H., Schat, H., Ernst, W. H. O. (1989), *J. Plant Physiol. 135*, 165–169.
De Vos, C. H. R., De Waal, M. A. M., Vooijs, R., Schat, H., Ernst, W. H. O. (1991), *Physiol. Plant. 82*, 523–528.
Dhinsa, R. S. (1987), *Plant Physiol. 83*, 816–819.
Ding, A. H., Chan, P. C. (1984), *Lipids 19*, 278–284.
Elstner, E. F. (1982), *Ann. Rev. Plant Physiol. 33*, 73–96.
Elstner, E. F. (1984). in: *Pflanzentoxicologie*: Hock, B., Elstner, E. F. (eds.) Mannheim: Bibliographisches Institut; pp. 67–94.
Elstner, E. F., Osswald, W. (1984), *Naturwissenschaftliche Rundschau 37*, 52–61.
Elstner, E. F., Wagner, G. A., Schutz, W. (1988), *Current Topics in Plant Biochemistry and Physiology 7*, 159–187.
Endress, A., Suarez, S., Taylor, C. (1980), *Environ. Poll. (Series A) 22*, 47–58.
Ernst, W. (1969), *Berichte der Deutschen Botanischen Gesellschaft 82*, 161–164.
Ernst, W. (1972), *Schwermetallresistenz und Mineralstoffhaushalt*: Opladen: Westdeutscher Verlag.
Ernst, W. (1974), *Schwermetallvegetation der Erde*: Stuttgart: Gustav Fischer Verlag.
Ernst, W. H. O. (1980), in: *Cadmium in the Environment, Part I, Ecological Cycling*, Nriagu, J. O. (ed.), New York: John Wiley and Sons; pp. 639–653.
Ernst, W. H. O., Schat, H., Verkleij, J. A. C. (1990), *Evolutionary Trends in Plants, 4 (1)*, 45–51.
Farago, M. E., Mullen, W. A., Cole, M. M., Smith, R. F. (1980), *Environ. Pollution Series A 21*, 225–244.
Flückiger, W., Flückiger-Keller, H., Oertli, J. J. (1978), *J. Plant Dis. Protect. 85*, 41–47.
Foy, C. D., Chaney, R. L., White, M. C. (1978), *Ann. Rev. Plant Physiol. 29*, 511–566.
Freedman, B., Hutchinson, T. C. (1981), in: *Effect of Heavy Metal Pollution on Plants. II. Metals in the Environment*: Lepp, N. W. (ed.) London: Applied Science Publishers; pp. 35–94.

Freedman, J. F., Ciriolo, M. R., Peisach, J. (1989), *J. Biol. Chem. 264*, 5598–5605.

Gabrielli, R., Pandolfini, T., Vergnano, O. (1987), *Giornale Botanico Italiano 21*, 200–201.

Galliard, T., Chan, H. W. S. (1980), in: *Lipids: Structure and Function*: Stumpf, P. K., (ed.) New York: Academic Press; pp. 85–116.

Gaspar, T., Penel, C., Thorpe, T., Greppin, H. (eds.) (1982), *Peroxidases 1970–1980. A Survey of their Biochemical and Physiological Roles in Higher Plants.* University of Geneva, Switzerland.

Gebhart, E., Rossman, T. G. (1991), in: *Metals and their Compounds in the Environment. Occurrence, Analysis and Biological Relevance*: Merian, E. (ed.) Weinheim: VCH Verlagsgesellschaft; pp. 617–640.

Gora, L., Clijsters, H. (1989), in: *Biochemical and Physiological Aspects of Ethylene Production in Lower and Higher Plants.* Clijsters, H., De Proft, M., Marcelle, R., Van Poucke, M. (eds.) Dordrecht: Kluwer Academic Publishers; pp. 219–228.

Griffiths, W. T. (1975), *Biochem. J. 152,* 623–635.

Grill, D., Esterbauer, H., Klosch, U. (1979), *Environ. Pollut. 19*, 187–194.

Grill, E., Löffler, S., Winnacker, E.-L., Zenk, M. H. (1989), *Proc. Natl. Acad. Sci. USA 86*, 6838–6842.

Halliwell, B., Gutteridge, J. M. C. (1984), *Biochem. J. 219*, 1–14.

Hampp, R., Ziegler, H., Ziegler, I. (1973a), *Biochem. Physiol. Pflanzen 164*, 588–595.

Hampp, R., Ziegler, H., Ziegler, I. (1973b), *Biochem. Physiol. Pflanzen 164*, 126–134.

Hampp, R., Beulich, K., Ziegler, H. (1976), *Zeitschr. Pflanzenphysiol. 77*, 336–344.

Hart, B. A., Scaife, B. D. (1977), *Env. Res. 14*, 401–413.

Hazell, P., Murray, D. R. (1982) *Zeitschr. Pflanzenphysiol. 108*, 87–92.

Hildebrand, D. F., Hamilton-Kemp, T. R., Legg, C. S., Bookjans, G. (1988), *Current Topics in Plant Biochem. Physiol. 7*, 201–219.

Honeycutt, R. C., Korgman, D. W. (1972), *Plant Physiol. 49*, 376–380.

Horsman, D. C., Wellburn, A. R. (1976), in: *Effects of Air Pollutants on Plants*: Mansfield (ed.) Cambridge: Cambridge University Press; pp. 185–199.

Hoxha, Y., Jablonovic, M., Abdullai, K., Filipovic, R. (1985), *Acta Biologica et Medica Experimentalis 10*, 21–24.

Huang, C. Y., Bazzaz, F. A., Vanderhoef, L. N. (1974), *Plant Physiol. 54*, 122–124.

Hurwitz, J., Weissbach, A., Horecker, B. L., Smyniotis, P. Z. (1956), *J. Biol. Chem. 218*, 769–783.

Kacperska-Palacz, A., Uliasz, M. (1974), *Physiol. Vég. 12 (4)*, 561–570.

Kappus, H. (1985), in: *Oxidative Stress*: Sies, H. (ed.), London: Academic Press; pp. 273–310.

Karataglis, S., Symenoides, L., Moustakas, M. (1988), *J. Agron. Crop. Sci., 160*, 106–112.

Keller, T. (1974), *European J. Forest Pathol. 4*, 11–19.

Kennedy, C. D., Gonsalves, F. A. N. (1987), *J. Exp. Bot. 38*, 800–817.

Kennedy, C. D., Gonsalves, F. A. N. (1989), *Plant and Soil 117*, 167–175.

Khan, A. A. Malhotra, S. S. (1982), *Biochem. und Physiol. der Pflanzen 177*, 643–650.

Kleiner, D. (1974), *Arch. Biochem. Biophys. 165*, 121–125.

Koeppe, D. E. (1977), *Sci. Total Environ. 7*, 197–206.

Kremer, B., Markham, J. W. (1982), *Zeitschr. Pflanzenphysiol. 108,* 125–130.

Kumar, K. S., Rowse, C., Hochstein, P. (1978), *Biochem. Biophys. Res. Commun. 83*, 587–592.

Lamoreaux, R. J., Chaney, W. R. (1978), *Physiol. Plant 43*, 231–236.

Lee, F. J., Bennett, J. H. (1985), *Plant Physiol. 69*, 1444–1449.

Lee, F. J., Hassan, H. M. (1985), *J. Free Rad. Biol. Med. 1*, 319–325.

Lee, K. C., Cunningham, B. A., Paulsen, G. M., Liang, G. H., Moore, R. B. (1976a), *Physiol. Plant. 36*, 4–6.

Lee, K. C., Cunningham, B. A., Chung, K. H., Paulsen, G. M., Liang, G. H. (1976b), *J. Env. Qual. 5*, 357–359.
Levitt, J. (ed.) (1972), *Responses of Plants to Environmental Stresses*: New York: Academic Press.
Lindberg, S., Wingstrand, G. (1985), *Physiol. Plant. 63*, 181–186.
Ljutakova, S. G., Russanov, E. M., Liochev, S. I. (1984), *Arch. Biochem. Biophys. 235*, 636–643.
Lorimer, G. H. (1981), *Ann. Rev. Plant Physiol. 32*, 349–383.
Lucero, H., Andreo, C., Vallejos, R. H. (1976), *Plant Sci. Lett. 6*, 309–313.
Maier, R. (1978a), *Phyton 19*, 83–96.
Maier, R. (1978b), *Zeitschr. Pflanzenphysiol. 87*, 347–354.
Marschner, H. (1983), in: *Encyclopaedia of Plant Physiology, Vol. 15A: Inorganic Plant Nutrition*: Lauchli, A., Bieleski, R. L. (eds.) Berlin: Springer-Verlag; pp. 5–60.
Mathys, W. (1975), *Physiol. Plant. 33*, 161–165.
Mathys, W. (1977), *Physiol. Plant. 40*, , 130–136.
McBrien, D. C. H., Hassall, K. A. (1965), *Physiol. Plant. 18*, 1059–1065.
Mehlhorn, H., Seufert, B., Schmidt, A., Kunert, K. J. (1986), *Plant Physiol. 82*, 336–338.
Miszalski, Z., Ziegler, I. (1979), *Planta 145*, 383–387.
Mukherji, S., Das Gupta, B. (1972), *Physiol. Plant 27*, 126–129.
Mullins, M., Hardwick, K., Thurman, D. A. (1985), in: *International Conference on Heavy Metals in the Environment, Athens Proceedings*: Edinburgh: CEP Consultants; pp. 43–46.
Nag, P., Paul, A. K., Mukherji, S. (1981), *Indian J. Exp. Bot. 19*, 702–706.
Naiki, N. (1980), *Plant Cell Physiol. 21*, 775–783.
Nieboer, E., Richardson, D. H. S. (1980), *Environ. Poll. (Series B) 1*, 3–26.
Ohsumi, Y., Kitamoto, K., Anraku, Y. (1988), *J. Bacteriol. 170*, 2676–2682.
Okhi, K. (1976), *Physiol. Plant. 38*, 300–304.
Overnell, J. (1975), *Mar. Biol. 29*, 99–103.
Palma, J. M., Gomez, M., Yanez, J., Del Rio, L. A. (1987), *Plant Physiol. 85*, 570–574.
Pandolfini, T., Maier, R. (1985), *Giornale Botanico Italiano 119*, 55–56.
Passow, H., Rothstein, A. (1960), *J. Gen. Physiol. 43*, 621–633.
Passow, H., Rothstein, A., Clarkson, T. W. (1961), *Pharmacological Rev. 13*, 185–224.
Pauls, K. P., Thompson, J. E. (1981), *Physiol. Plant. 53*, 255–262.
Poschenrieder, C., Gunse, B., Barcelo, J. (1989), *Plant Physiol. 90*, 1365–1371.
Prassad, D. D. K., Prassad, A. R. K. (1987), *J. Plant Physiol. 127*, 241–249.
Rabe, R., Kreeb, K. H. (1979), *Environ. Poll. 19*, 119–137.
Rauser, W. E. (1984), *Plant Sci. Lett. 39*, 239–247.
Rauser, W. E. (1987), *Plant Sci. Lett. 51,* 171–175.
Rauser, W. E. (1990), *Ann. Rev. Biochem. 59*, 61–86.
Rauser, W. E., Ackerley, C. A. (1987), *Can. J. Bot. 65*, 643–646.
Rehman, S. (1984), *Bull. Env. Contamin. Toxicol. 33*, 92–98.
Robinson, P. D., Jackson, P. J. (1986), *Physiol. Plant. 67*, 499–506.
Robinson, P. D., Martin, M. N., Tabita, F. R. (1979), *Biochemistry 18 (21)*, 4453–4458.
Ros, R., Cooke, D. T., Burden, R. S., James, C. S. (1990), *J. Experim. Bot. 41*, 457–462.
Rüegsegger, A., Schmutz, D., Brunold, C. (1990), *Plant Physiol. 93*, 1579–1584.
Salhany, J. M., Swanson, J. C., Cordes, K. A., Gaines, S. B., Gaines, K. C. (1978), *Biochem. Biophys. Res. Commun. 82*, 1294–1299.
Salin, M. L. (1988), *Physiol. Plant. 72*, 681–689.
Sandmann, G., Böger, P. (1980a), *Plant Physiol. 66*, 797–800.
Sandmann, G., Böger, P. (1980b), *Zeitschr. Pflanzenphysiol. 98*, 53–59.
Sandmann, G., Böger, P. (1983), in: *Encyclopaedia of Plant Physiology, Vol. 15B: Inorganic Plant Nutrition*: Lauchli, A., Bieleski, R. L. (eds.), Berlin: Springer-Verlag; pp. 563–596.

Scandalios, J. G. (1974), *Ann. Rev. Plant Physiol. 25*, 225–258.
Schaaper, R. M., Koplitz, R. M., Tkeshelashvili, L. K., Loeb, L. A. (1987), *Mutat. Res. 177*, 179–188.
Scheller, H. V., Huang, B., Hatch, E. , Goldsbrough, P. B. (1987) *Plant Physiol. 85*, 1031–1035.
Schloß, P., Walter, C., Mäder, M. (1987), *Planta 170*, 225–227.
Schmidt, A., Kunert, K. J. (1986), *Plant Physiol. 82*, 700–702.
Shatzman, A. R., Kosman, D. J. (1978), *Biochim. Biophys. Acta 544*, 163–179.
Shioi, Y., Tamai, H., Sasa T. (1978), *Plant Cell Physiol. 19*, 203–209.
Sies, H. (1985), in: *Oxidative Stress*, Sies, H. (ed.) London: Academic Press; pp. 273–310.
Steffens, J. C. (1990), *Ann. Rev. Plant Mol. Biol. 41*, 553–575.
Stiborova, M. (1988), *Biochem. Physiol. Pflanzen 183*, 371–378.
Stiborova, M., Doubravova, M., Leblova, S. (1986), *Biochem. Physiol. Pflanzen 181*, 373–379.
Stiborova, M. Ditrichova, M., Benzinova, A (1988), *Photosynthetica 22,* 161–167.
Stobart, A. K., Griffiths, W. T., Ameen-Bukhari, I., Sherwood, R. P. (1985), *Physiol. Plant. 63*, 293–298.
Subhadra, A. V., Nanda, A. K., Behera, P. K., Panda, B. B. (1991), *Environ. Poll. 69*, 169–179.
Tanaka, K., Sugahara, K. (1980), *Plant Cell Physiol. 21*, 601–611.
Todd, J. F., Paliyath, G., Thompson, J. E. (1990), *Plant Physiol. 94*, 1225–1232.
Tomsett, A. B., Salt, D. E., De Miranda, J., Thurman, D. A. (1989), *Aspects Appl. Biol. 22*, 365–372.
Triest, L. (ed.) (1991), *Isozymes in Water Plants*, Meise: National Botanic Garden of Belgium.
Uribe, E. G., Stark, B. (1982), *Plant Physiol. 69*, 1040–1045.
Vallee, B. L. (1976), *Trends in Biochemical Science, April*, 88–91.
Vallee, B. L., Ulmer, D. D. (1972), *Ann. Rev. Biochem. 41*, 91–128.
Van Assche, F., Clijsters, H. (1983), in: *Effects of Stress on Photosynthesis*; Marcelle, R., Clijsters H., and Van Poucke, M. (eds.) The Hague: Nijhoff/Junk; pp. 371–382.
Van Assche, F., Clijsters, H. (1986a), *Physiol. Plant. 66*, 717–721.
Van Assche, F., Clijsters, H. (1986b), *J. Plant Physiol. 125*, 355–360.
Van Assche, F., Clijsters, H. (1990a), *Environ. Poll. 66*, 157–172.
Van Assche, F., Clijsters, H. (1990b), *Plant, Cell Environ. 13*, 195–206.
Van Assche, F., Clijsters, H., Marcelle, R. (1979), in: *Photosynthesis and Plant Development:* Marcelle, R., Clijsters, H., Van Poucke, M. (eds.) The Hague: Junk Publishers; pp. 175–184.
Van Assche, F., Put, C., Clijsters, H. (1986), *Arch. Intern. Physiol. Bioch. 94*, 60.
Van Assche, F., Cardinaels, C., Clijsters, H. (1988), *Environ. Poll. 52*, 103–115.
Van Assche, F., Vangronsveld, J., Clijsters, H. (1990), *Proceedings: International Conference on Environmental Contamination*; Barcelo, J. (ed.) Barcelona, Edinburgh: CEP Consultants; pp. 246–248.
Vangronsveld, J., Clijsters, H. (1992), in: *Metal Compounds in Environment and Life. Interrelation between Chemistry and Biology*: Merian, E., Haerdi, W. (eds.) Northwood: Science and Technology Letters; pp. 117–125.
Van Loon, L. C. (1986), in: *Molecular and Physiological Aspects of Plant Peroxidases*: Greppin, H., Penel, C., Gaspar, T. (eds.) Geneva: University of Geneva; pp. 405–418.
Vaughan, D., De Kock, P. C., Ord, B. G. (1982), *Physiol. Plant. 54*, 253–257.
Veltrup, W. (1982), *Zeitschr. Pflanzenphysiol. 108*, 457–469.
Verkleij, J. A. C., De Boer, A. M., Lugteborg, T. F. (1980), *Oecologia, 46*, 354–359.
Wainwright, S. J., Woolhouse, H. W. (1977), *J. Exp. Bot. 28*, 1029–1038.
Wainwright, S. J., Woolhouse, H. W. (1978), *J. Exp. Bot. 29*, 525–531.
Wallnöfer, P. R., Engelhardt, G. (1984), in: *Pflanzentoxicologie*: Hock, B., Elstner, E. F. (eds.) Mannheim: Bibliographisches Institut; pp. 95–117.

Weigel, H. J. (1985a), *J. Plant Physiol. 119*, 179–189.
Weigel, H. J. (1985b), *Physiol. Plant. 63*, 192–200.
Weigel, H. J. Jäger, H. J. (1980a), *Angew. Botanik 54*, 195–205.
Weigel, H. J., Jäger, H. J. (1980b), *Zeitschr. Pflanzenphysiol. 97*, 103–113.
Wildner, G. F., Henkel, J. (1979), *Planta 146*, 223–228.
Wilson, E. L., Burger, P. E., Dowdle, E. D. (1972), *Environ. J. Biochem. 29*, 563–571.
Woolhouse, H. W. (1983), in: *Encyclopedia of Plant Physiology, Vol. 12C: Physiological Plant Ecology III*: Lange, O. L., Nobel, P. S., Osmond, C. B., Ziegler, H. (eds.) Berlin: Springer-Verlag; pp. 245–300.
Wu, L., Thurman, D. A., Bradshaw, A. D. (1975), *New Phytol. 75*, 225–229.
Youngman, R. J. Elstner, E. F. (1984), in: *Pflanzentoxicologie*: Hock, B., Elstner, E. F. (eds.) Mannheim: Bibliographisches Institut; pp. 132–151.

7 Plants and Radionuclides

G. Shaw[1] *and J. N. B. Bell*[2]

[1]Centre for Analytical Research in the Environment, [2]Department of Biology
Imperial College at Silwood Park
Ascot, Berkshire SL5 7TE, UK

Contents

Introduction

Both higher and lower plants are of considerable strategic importance when considering the transmission of potentially harmful materials into food chains. In most cases active or passive uptake by plants represents the primary means of incorporation of contaminants into living organisms from sources within the atmosphere, hydrosphere or geosphere and as such is a process which is worthy of considerable attention. This is the case for many natural and artificially-introduced substances, but it is the latter which give particular cause for concern in terms of ecotoxicology.

Radioactive substances have existed as natural components of the Earth's ecosystems since their origins and all organisms have evolved in an environment in which they are constantly exposed to naturally-occurring 'background' radiation. It was not until the beginnings of man's exploitation of nuclear energy, however, that the existence of enhanced levels of specific radionuclides within the environment was seen as a threat to human health. The beginnings of the science of radioecology are inextricably linked to those of the nuclear programme itself in the early 1940s and since this period the interactions of radioactive materials with plants have been considered as being centrally important to our understanding of radionuclide behaviour within ecosystems as a whole. This reflects the importance of plants as primary harvesters of both energy and chemical substances, and therein lies the problem. Due to their considerable efficiency as scavengers of materials from air, soil and water, plants most often represent the principal vectors of radionuclide transfer to many animal species, including man. This transfer process, involving assimilation and concentration of radionuclides at a primary trophic level within the food chain is, arguably, the most significant in the entire sequence of radionuclide transfer within food chains; accordingly, it has received considerable attention from radioecologists, especially when considering the potential radiological dose to human populations likely from a given nuclear contamination event or ongoing discharge.

What follows is a consideration of the pathways and mechanisms responsible for the incorporation of man-made radionuclides into plants. Historically this problem has been considered with respect to its ultimate effect on human health and this chapter will, to some extent, deal with this aspect. Some pure aspects of the inter-relationships between plants and radionuclides will also be covered as it is becoming increasingly evident that the successful prediction of the behaviour of radioactive materials in the biosphere is reliant on a thorough mechanistic understanding of their interaction with plants.

Radionuclides as Environmental Pollutants

Sources, Ubiquity and Diversity of Environmentally Significant Radionuclides

Details of the sources and individual behaviour of radionuclides within the environment are beyond the scope of this chapter and worthy accounts of these topics have previously been given by Bowen (1979) and Whicker and Schultz (1982). It is worth stressing the point that was alluded to in the previous section, however, that the two major groups of radionuclides which exist are those from natural and man-made sources. Radioecologists have primarily been concerned with the behaviour of the latter category and, as a result, the bulk of radioecological literature concerns radionuclides which have been released to the environment as a result of man's activities. The production of all these radionuclides is either a direct result of the nuclear fission process, or indirectly the result of activation of elements by neutron bombardment within reactors or decay of both fission and activation products.

Environmentally significant radionuclides from artificial sources include the exclusively radioactive elements which are considered 'artificial' (ie, the actinides and technetium, which did not exist on Earth in any significant quantities prior to man's nuclear programme) as well as radioisotopes of elements which are commonly or less commonly encountered in the natural environment. The latter category includes elements with major biological significance, such as potassium, hydrogen, carbon and sulphur, and others which are not plant nutrients, for example ruthenium, cerium, strontium and caesium. There are also radioactive isotopes of transition elements (such as iron, cobalt, nickel and manganese) which are known to be important biological micronutrients. Figure 7-1 indicates the ubiquitous distribution throughout the periodic table of elements with nuclides of interest to the radio-

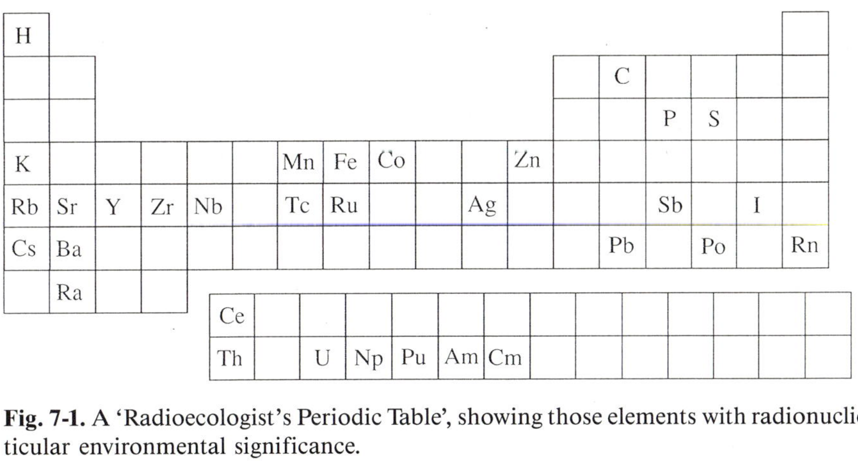

Fig. 7-1. A 'Radioecologist's Periodic Table', showing those elements with radionuclides of particular environmental significance.

ecologist and gives some indication of the scale of the problem involved in understanding the biological behaviour of a group of materials of considerable chemical diversity. As a brief summary of the more important radionuclides some broad categories can be drawn from the periodic classification of the elements.

(i) The non-metals hydrogen, carbon, phosphorus, iodine and sulphur are all elements with both highly significant biological functions and environmentally-significant radionuclides. These should behave in a similar fashion to their more abundant non-radioactive counterparts; moreover, their essentiality (in the biological sense) may result in their concentration within food chains.

(ii) The alkali metals include several environmentally significant radionuclides of which ^{86}Rb, ^{134}Cs and ^{137}Cs are important man-made examples. The group also contains an important primordial radionuclide – ^{40}K ($t_{1/2} = 1.3 \times 10^9$ years) – which is a natural radioactive isotope of the plant macronutrient, potassium (^{40}K accounts for ~0.0118% of all potassium within the environment). While Rb and Cs both possess naturally occurring stable isotopes, these are only found in the environment at trace concentrations, although due to their presence in the same group as K they may act to some extent as chemical 'analogues' for K. Indeed, Finston and Kinsley (1960) have stated that 'the ions NH_4^+, K^+, Rb^+ and Cs^+ form a homologous series which show a greater degree of physico-chemical similarity than any other series of ions, with the possible exception of the halogens', and from this it is often assumed that the ecological behaviour of Cs and Rb can be inferred from that of K. One of the major, and as yet unresolved, questions in radioecology since the early 1950s has been that of the relative interchange which is possible between these three elements, particularly in biological systems.

(iii) The alkaline earth metals calcium (Ca), barium (Ba), strontium (Sr), and radium (Ra) all form divalent cations, though the elements are chemically reactive and seldom occur in the free state (cf. alkali metals). Some of these elements have important radionuclides (^{90}Sr, ^{140}Ba and ^{226}Ra) while Ca has major radioecological significance with respect to its effect on the other three – in general Sr, Ba and Ra are considered to be approximate chemical and physiological analogues of Ca.

(iv) The noble gases are inert and because of their non-reactivity exist as monoatomic molecules. ^{222}Rn is important as an inhalation hazard in confined areas (buildings, mines and caves) but does not take part in any chemical or physiological interactions with biological tissues. The noble gases have no nutrient analogues.

(v) The heavy metals comprise a group which contains a mixture of elements with a complex and varied chemistry. Elements with important radionuclides include manganese (^{54}Mn), iron (^{55}Fe and ^{59}Fe), cobalt (^{57}Co and ^{60}Co), technetium (^{99}Tc and ^{99m}Tc), ruthenium (^{103}Ru and ^{106}Ru), lead (^{210}Pb) polonium (^{210}Po), niobium (^{95}Nb) and zirconium (^{95}Zr). Some of these are essential micronutrients (Mn, Fe and Co), some are highly toxic to organisms (eg, Pb), some are naturally-occurring but extremely scarce within the environment (Ru and Po) while some are purely products of the fission and activation process (eg, Tc). Due to their chemical complexity few general statements may be made about their ecological behaviour.

(vi) The rare earths include the lanthanide series (atomic number 57 to 71) and, because of their chemical similarity, scandium (Sc) and yttrium (Y). Perhaps the most radioecologically important nuclides within this group are ^{90}Y (the decay product of ^{90}Sr) and ^{144}Ce, a fission product.

(vii) The actinides include the naturally-occurring elements actinium (Ac), thorium (Th), protactinium (Pa) and uranium (U), as well as the transuranium elements (commonly called the 'transuranics') which are largely produced by the artificial neutron bombardment of natural U. Important transuranics include plutonium (Pu), curium (Cm), neptunium (Np) and americium (Am). These elements have complex environmental chemistries as a result of their many oxidation states (Friedman, 1976). They are generally poorly absorbed by organisms, though the naturally-occurring U is better absorbed than the artificial Pu (for this reason inhalation of Pu may be a particularly important exposure pathway for animals, including man).

Physico-Chemical Behaviour of Radionuclides in the Environment

Radionuclides occur in the environment in such minute physical quantities that they are commonly referred to as being present at sub- or ultra-trace quantities. As an illustration of this Livens and Rimmer (1988) presented values relating the masses and specific activities of several artificial radionuclides and their typical concentrations in surface soils of the UK (see Table 7-1). This indicates the very small mass usually associated with one Becquerel of radioactivity, although as the half life of a radionuclide increases so does the mass per unit of radioactivity. The molarity concept is

Table 7-1. Relationship of mass and specific activity for some artificial isotopes and typical UK concentrations (from Livens and Rimmer, 1988).

Isotope	Half-life (year)	Mass of 1 Bq (g)	Typical UK surface soil concentration range	
^{134}Cs	2.06	2.1×10^{-14}	1.1×10^{-5}	<500; exceptionally 1–2000
^{137}Cs	30.1	3.1×10^{-13}	3.1×10^{-4}	<1000; exceptionally 5–10000
^{239}Pu	2.4×10^{4}	1.8×10^{-12}	3.6×10^{-4}	<20; exceptionally up to 10000
^{237}Np	2.1×10^{6}	1.6×10^{-10}	1.6×10^{-2}	$\ll$1; exceptionally 3–5

often used in plant physiological studies and in order to put the environmental concentrations of radionuclides into a physiological perspective, Shaw and Bell (1991) calculated from the data of Cambray *et al.* (1987) that the concentrations of the two caesium isotopes, ^{134}Cs and ^{137}Cs, in the British soils worst affected by the Chernobyl accident were of the order of 10^{-13} mol dm^{-3}. These concentrations are so low as to suggest that such radionuclides do not exhibit a discrete environmental chemistry to the extent that their behaviour within environmental or biological media will not result from mass action effects. Rather, it seems likely that their physical and physiological behaviour will be strongly influenced by stable isotopes of the radionuclide in question or analogues which, though chemically distinct, may be sufficiently similar in ionic radius, charge, etc., to exert a significant competitive influence on the reaction kinetics of the radioion.

A pure sample of a radionuclide undiluted with any other isotope is referred to as being 'carrier-free' and, as we have already seen, this usually represents a very small mass of material. In radioanalytical work the adddition of a 'carrier' facilitates the handling of such small amounts of material; this usually takes the form of a relatively large quantity of the common stable isotope of the radionuclide being studied. Elements such as Tc and the actinides, for which there are no stable isotopes, inevitably exist in the 'carrier-free' state, but in the case of elements like caesium and strontium the stable nuclides ^{133}Cs and ^{88}Sr, respectively, are naturally present in environmental media. Any radiocaesium or radiostrontium released to the environment will therefore join a pool of natural carriers which will effectively influence its chemical and biological behaviour. This situation will apply to any radionuclide of an element with naturally-occurring stable isotopes, including cobalt, iron, carbon, sulphur, hydrogen, iodine, ruthenium, cerium. An important concept in this respect is that of 'specific activity'. This is the activity of a radionuclide per unit mass of its stable (and more abundant) isotope, usually expressed in Bq mol^{-1} or Bq kg^{-1}. This, however, must not be confused with the more common usage of the term in the radioecological context in which 'specific activity' is used to denote the 'activity concentration' of a radionuclide within a biological or environmental compartment – in this case the units most usually referred to are Bq kg^{-1} dry weight.

From the above it can be concluded that in many instances the introduction of an artificial radionuclide into the environment provides us with a natural tracer experiment. Indeed, this is the basis for the application of deterministic compartmental models, based on tracer kinetics, to radioecology (Whicker and Schultz, 1982). This approach is largely based on the assumption that radionuclide movements will exhibit first order kinetics although the existence of naturally-occurring tracees (stable isotopes) at relatively high abundance may result in more complex concentration-dependent kinetics. Furthermore, nutrient analogues may exert even more complex effects on processes such as radioion absorption across root plasma membranes; this will become evident later in the chapter.

Chemical Speciation of Radionuclides

The environmental impact of a toxic element is not determined solely by its gross concentration. Rather it is its chemical form, or speciation, which will control its movement through the biosphere and its interactions with living matter. This has long been recognised for metallic pollutants but, because of the analytical difficulties involved in studying carrier-free quantities of radio-contaminants in environmental media, it is really only in the last decade that this problem has been addressed in radioecological circles. In the limited space available here it is impossible to give full coverage to this subject but it will suffice to say that, in the light of the broad spectrum of elements of importance to the radioecologist (Fig. 7-1), the question of radionuclide speciation in the environment is complex. Of particular importance is the identification of chemical species of radionuclides which are mobile and therefore more likely to be available for bio-incorporation. The interested reader is directed to a recent paper by Desmet *et al.* (1991) in which this question is addressed with special reference to plant uptake of radionuclides. At a more general level, however, the collection of papers edited by Bulman and Cooper (1986) is probably the most coherent account to date of this difficult topic.

Plant Contamination by Radionuclides

Figure 7-2 summarises the major pathways of radionuclide movement within the envionment with regard to their uptake by higher plants. Following the earliest releases of artificial radionuclides to the environment in the form of fallout from atmospheric weapons tests, it was the direct deposition of radioactivity to above-ground plant organs which gave most cause for concern in the short term and this remains the case with contemporary accidents involving the release to atmosphere of nuclear contaminants. It was quickly recognised, however, that over the longer term root uptake might represent the most significant means of entry into food chains of radionuclides, particularly those such as caesium and strontium which mimic essential elements and which are therefore readily incorporated into biological tissues. Both of these sets of pathways are described below with respect to the contamination of higher plant tissues. The special role of lower plants as vectors of radionuclide incorporation into food chains is then considered.

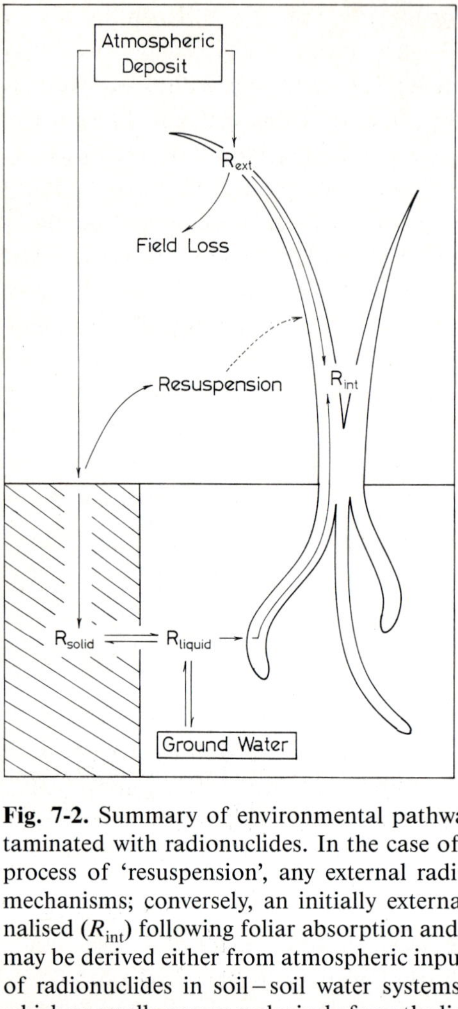

Fig. 7-2. Summary of environmental pathways by which terrestrial plants may become contaminated with radionuclides. In the case of an input from atmosphere, or as a result of the process of 'resuspension', any external radionuclide burden may be reduced by 'field loss' mechanisms; conversely, an initially external radionuclide deposit (R_{ext}) may become internalised (R_{int}) following foliar absorption and translocation. Radioactive contaminants of soils may be derived either from atmospheric inputs or from seepage in ground waters. Partitioning of radionuclides in soil–soil water systems controls their availability for root absorption, which normally occurs exclusively from the liquid phase. The chemical speciation of the nuclide in this phase, however, provides a further control on bioavailability which is highly radionuclide specific.

Foliar Contamination and Absorption of Radionuclides by Higher Plants

The direct contamination of vegetation by deposition of radioactive material from the atmosphere onto above-ground organs is important under three main circumstances. Firstly, where a release of radionuclides occurs to atmosphere during routine discharges from nuclear installations; in most cases such releases are effectively continuous, although 'spike' releases can take place, such as during 'blow-down' when coolant is released from gas-cooled reactors prior to maintenance. Secondly, when a single one-off release occurs as a result of an accident; in this case the growth stage of any crop receiving contamination will be the critical factor controlling dose to man via edible parts at harvest. Finally, there is a secondary route for contamination of plant parts near to the ground surface, arising from the deposition of contaminated soil which has been resuspended as a result of wind action or rain splash. A thorough quantitative understanding of the processes by which vegetation can be contaminated from the atmosphere and the subsequent fate of the radionuclides concerned, both on and within the plant, including loss mechanisms, is essential for the accurate prediction of the consequences of releases to air from nuclear installations, particularly under accident conditions.

Deposition Processes

Radionuclides may be deposited to the Earth's surface in a number of different physico-chemical forms, *viz.* as solid particles of various sizes, as liquids in the form of rain, snow or occult precipitation (cloud, fog and mist droplets), and as gases. Particles and gases may be deposited under dry conditions, a number of physical processes being involved, and may also be washed out from the atmosphere by various forms of precipitation.

An important parameter used to quantify dry deposition processes is the velocity of deposition (V_g) (Chamberlain, 1953). V_g is defined as the downward flux of aerosol or gas to a vegetation or soil surface, normalised to the ambient atmospheric gas or aerosol concentration above that surface. In the case of radionuclide deposition processes flux and concentration are, respectively, measured in units of radioactivity per unit area and volume, hence

$$V_g = \frac{\text{Bq cm}^{-2}\ \text{s}^{-1}}{\text{Bq cm}^{-3}} \tag{1}$$

This expression yields an empirical parameter with units of velocity (cm s^{-1}) which are independent of atmospheric concentration. V_g may be calculated on the basis of exposed leaf area, in the case of spaced plants, or in terms of the horizontal surface area for continuous swards such as cereal or grass crops. Both soil and plant are potentially available for contamination and the total velocity of deposition to a canopy can then be defined as

$$V_{g\ canopy} = V_{g\ plant} + V_{g\ soil} \tag{2}$$

In the case of gases, chemical form will play a major role in controlling the V_g to vegetation. The reciprocal of V_g is termed *r*, the total resistance to uptake of the gas, with units of s cm^{-1}. This can be broken down into a number of component resistances, each of which is controlled by different processes. For a gas which penetrates into the leaf, there are three main resistances in series.

$$r_{total} = r_a + r_s + r_{int} \tag{3}$$

r_a is the aerodynamic resistance influencing passage of the gas through the boundary layer close to the leaf surface, where there is a transition from turbulent to molecular diffusion; this is controlled partly by the molecular diffusion coefficient of the individual gas and partly by the wind speed. Penetration of the gas into the plant is controlled by the degree of opening of the stomata, which imposes a further resistance, r_s. In addition a limited amount of penetration may occur through the cuticle, which represents a high resistance, r_{ci}, and some gases may also be adsorbed onto the surface of the cuticle, this representing another resistance, r_{ce}. Within the leaf, there is a final resistance in the form of r_{int}, the internal resistance, which is controlled by the chemical sink for the gas. In the case of parallel resistances, such as r_s and r_{ci}, the total resistance equals the sum of the reciprocals of the individual values. A scheme for these resistances is shown in Fig. 7-3. Clearly any factors which alter these resistances will affect the uptake of the radioactive gas by vegetation; r_{int} will be influenced by the plant's physiological state, r_a is controlled by meteorological conditions, and r_s by both of these factors.

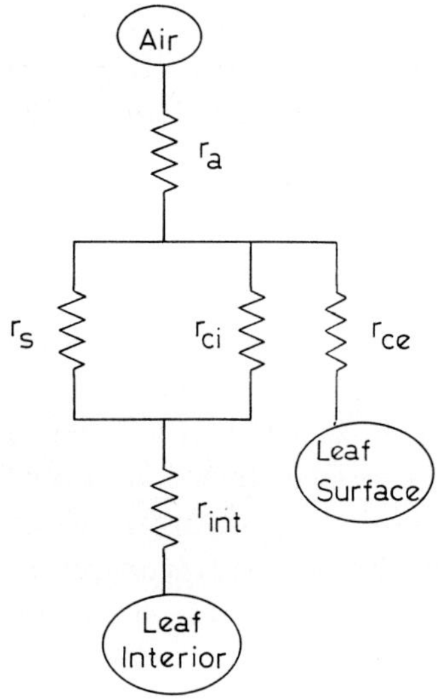

Fig. 7-3. Resistance analogue for transfer of a radioactive gas to vegetation.

In the case of particles, it is physical form rather than chemical characteristics which controls the velocity of deposition to vegetation, although the latter is important with respect to the subsequent processes of foliar absorption, translocation or wash-off. In the case of a nuclear release to atmosphere the particles are likely to exhibit a polydisperse size distribution and composition, resulting from factors such as condensation of vapours, followed by agglomeration. Larger particles are deposited close to the source or even contained at the site of initial release, with smaller particles becoming increasingly important over larger distances. After the Chernobyl accident, measurements suggested that by the time the radioactive plume had reached Sweden, most particles were around 1 μm in diameter (Chamberlain, 1991).

Particle size controls the processes by which deposition occurs and these differ in their relative efficiency. Larger particles (>20 μm) are mainly deposited under gravity by sedimentation, this being independent of wind speed. A further process is impaction, where the momentum of the particles causes collision with an obstacle, such as a leaf, due to inertia moving them forward as the air flow divides around the obstacle. The efficiency of this process increases as the obstacle diameter decreases and as the particle diameter rises. It is an important process for particles >10 μm depositing on obstacles <1 cm size when the wind velocity is >2 m s^{-1}. A further process is interception, which constitutes the filtering action of hairs and other projections on the leaf surface. In the case of particles <0.1 μm, these behave effectively as a gas and are deposited onto surfaces by Brownian diffusion. Particles within the 5–30 μm range are not readily retained by leaves as they tend to bounce off, whereas those <1 μm adhere strongly. The particle size distribution in a plume of radioactive material strongly influences the amount which will be deposited onto vegetation due to the different V_g values resulting from the different deposition processes. This is illustrated in Fig. 7-4 where a characteristic U-shaped curve is shown with a minimum V_g between 0.1 and 1.0 μm particle size. For larger particle sizes V_g is high due to the dominance of sedimentation, whereas sub-micron particles are deposited efficiently by diffusion. In the intermediate size range neither of these processes is very efficient and impaction and interception dominate. Unfortunately, when attempting to quan-

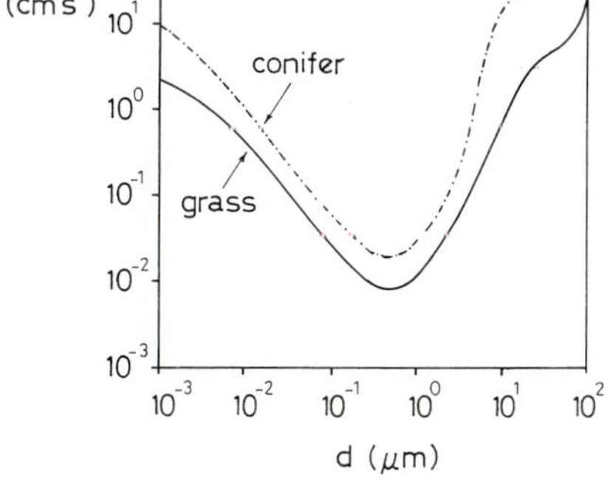

Fig. 7-4. Effect of aerosol particle size *(d)* on deposition velocity (V_g) to conifer and grass canopies (redrawn from Bonka and Horn, 1983).

tify V_g the greatest uncertainty resides within the 0.1 to 1.0 μm range, which is particularly important when considering material which has been transported long distances following a nuclear accident. Broad categories of vegetation have different efficiencies in removing radioactive particles from the air, conifers having particularly high values, as shown in Fig. 7-4. In the case of wet deposition, there is a somewhat similar pattern of scavenging efficiency, with washout of particles from the atmosphere by rain showing a minimum in the size range 0.1 to 1.0 μm (Fig. 7-5).

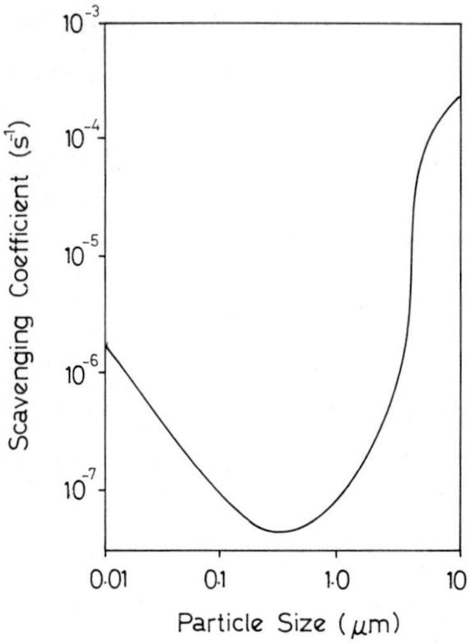

Fig. 7-5. Effect of aerosol particle size on scavenging efficiency of raindrops, falling at an intensity of 1 mm h^{-1} (redrawn from Röbig, 1979).

An understanding of the relative efficiency of different parts of a plant in intercepting radioactive particles is clearly of considerable importance in determining the hazard to humans consuming the edible portions. The uneven distribution between different plant organs for three different particle sizes is shown in Table 7-2, where a wind-tunnel study indicated a similar pattern for beech and stinging nettle in terms of the percentage catch on laminas, petioles and stems (Chamberlain and Little, 1980). Wet foliage is more efficient at capturing particles than dry leaves and thus contamination of vegetation by dry deposition may be expected to increase if this follows rain, although such an effect is less pronounced for smaller particles.

Wind-speed is an important factor controlling particle deposition. As the wind-speed rises, the deposition of smaller particles will increase due to a greater probability of penetration of the laminar boundary layer above the vegetation surface. However, for larger particles, the converse occurs with increasing wind-speed increasing losses by bounce-off, as illustrated in Fig. 7-6 for 30 μm *Lycopodium* spores; here the efficiency of particle collection declined as the wind speed was increased from 1 to 10 ms^{-1}.

Table 7-2. Percentage of total catch of particles of different sizes intercepted by different plant parts (from Chamberlain and Little, 1980).

Particle size (μm)	Plant part	Species	
		Fagus sylvatica (beech)	*Urtica dioica* (stinging nettle)
0.2	Laminas	88	86
	Petioles	7	6
	Stems	5	8
2.75	Laminas	63	68
	Petioles	30	16
	Stems	7	16
8.5	Laminas	63	68
	Petioles	17	11
	Stems	20	20

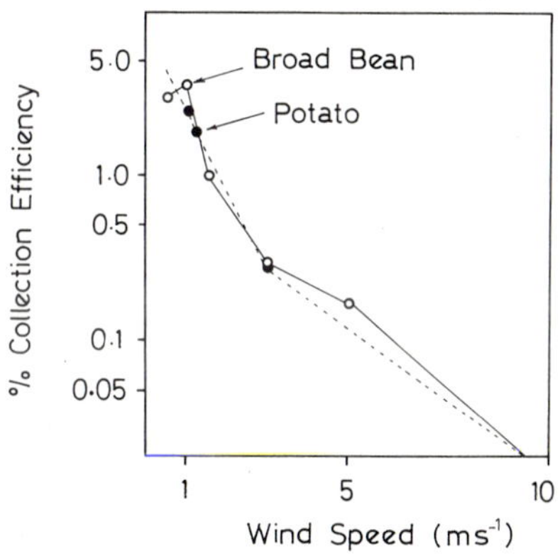

Fig. 7-6. Efficiency of collection of 30 μm *Lycopodium* spores on broad bean (*Vicia faba*) and potato (*Solanum tuberosum*) leaves (redrawn from Chamberlain, 1975).

An important parameter used in studies on the degree of contamination of a crop by aerial deposition is the 'interception fraction' (p), a dimensionless measure of the proportion of the total ground deposit initially intercepted by plants. Chamberlain (1970) determined that p is critically dependent both on the density of the intercepting herbage and an 'absorption coefficient' which he formalised as an empirical equation, as follows:

$$p = 1 - e^{-\mu W} \tag{4}$$

where μ = absorption coefficient ($m^2\ kg^{-1}$), presumably dependent on the nature of the plant canopy under consideration and W = herbage density ($kg\ m^{-2}$). Similar relationships of increasing interception fraction with increasing herbage density are found for gases, droplets and particles (Fig. 7-7).

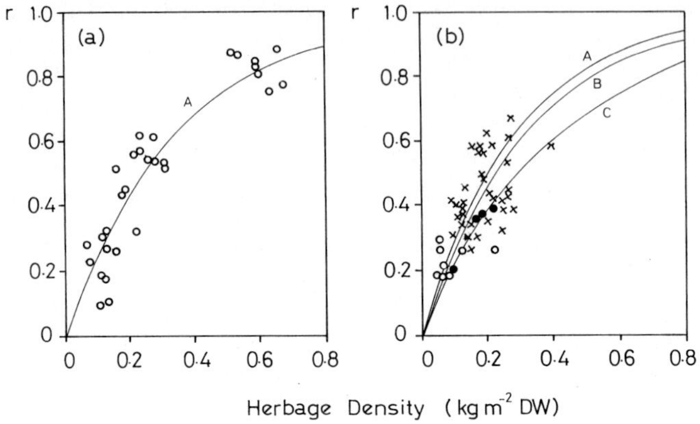

Fig. 7-7. Interception fraction (*r*) of radioactive gases, droplets and particles on grassland in relation to herbage density. (a) Radioiodine vapour (μ = 2.78 $m^2\ kg^{-1}$); (b) *Lycopodium* spores (crosses, μ = 3.33 $m^2\ kg^{-1}$); droplets (open circles, μ = 3.08 $m^2\ kg^{-1}$); droplets (closed circles, μ = 2.30 $m^2\ kg^{-1}$) (redrawn from Chamberlain, 1970).

In the case of prolonged deposition, due either to fallout from weapons detonations or operational releases from nuclear plant, Chamberlain (1970) developed a further function, the normalised specific activity (NSA) defined as:

$$\frac{\text{Radionuclide concentration in foliage (Bq kg}^{-1})}{\text{Rate of deposition to the ground (Bq m}^{-2}\text{ day}^{-1})} = \text{m}^2\text{ day}^{-1}\text{ kg}^{-1} \quad (5)$$

This parameter takes into account the growth of the plant over the period concerned as well as the loss of activity with time due to the process of 'field loss' (see below). Some examples of calculated NSA values for fall-out ^{90}Sr and ^{137}Cs on different types of slow growing vegetation are shown in Table 7-3. The highest values are clearly seen in the extremely slow-growing vegetation of the Arctic, with 1000 $m^2\ day^{-1}\ kg^{-1}$, compared with around 40 $m^2\ day^{-1}\ kg^{-1}$ for herbage under growing conditions typical of temperature latitudes.

In many ways the uptake of radioactive gases by vegetation is more complex than for particles or droplets, due to the possibility of immediate penetration to the interior of the foliage and the extra resistances involved. The behaviour of gases is of particular interest in the United Kingdom where both Magnox and Advanced Cooled Reactors routinely discharge $^{14}CO_2$ and $CO^{35}S$ with, additionally, $H_2{}^{35}S$ when blow-down occurs prior to maintenance. A number of studies have been performed which have elucidated and quantified the pathways of the radionuclides concerned. For

Table 7-3. Normalised specific activity (NSA) for vegetation in poor growing conditions (data of Chamberlain in Coughtrey *et al.*, 1990).

Radionuclide	Location	NSA (m^2 day^{-1} kg^{-1})	Vegetation
	Chilton (S. England)	45 – 113	Herbage (winter)
	Sonning (S. England)	123	Alfalfa
	Letcombe (S. England)	35 – 190	Herbage (winter)
^{90}Sr	Cwmystwyth (W. Wales)	89 – 211	Herbage
	Alaskan tundra	1000	*Calluna vulgaris*
	Finland, Alaska	> 1000	Lichens
	Finnish Lapland	60 – 250	*Betula, Salix* and *Vaccinium vitis-idaea* dried leaves
^{137}Cs	Sweden	56	Northern farms
	Finnish Lapland	200 – 760	*Betula, Salix* and *Vaccinium vitis-idaea* dried leaves

some time there has been interest in the fate of ^{35}S in COS, for which the critical group is infants drinking milk contaminated *via* the atmosphere → grass → cow pathway. In a controlled laboratory fumigation of young *Lolium perenne* swards, it was found that approximately four times more $CO^{35}S$ was taken up in the light than in the dark, indicating the importance of the stomatal resistance in controlling uptake (stomata close in the dark) (Fig. 7-8).

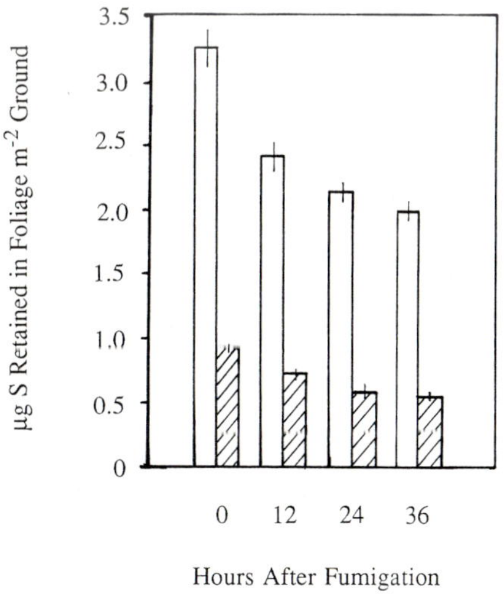

Fig. 7-8. Quantity of sulphur retained in foliage of young swards of ryegrass (*Lolium perenne*) following 20 min fumigation with $CO^{35}S$ under light (open columns) or dark (shaded columns) conditions. Vertical bars = SEM (redrawn from Kluczewski *et al.*, 1983).

In contrast, further work by the same group showed that under similar conditions the velocity of deposition of $CO^{35}S$ to a range of soils was two to three orders of magnitude lower than to vegetation in the light, indicating that this could not be a significant route for the radionuclide to enter the plant *via* the roots (Kluczewski *et al.*, 1985 a).

The Chernobyl accident provided the ideal opportunity to study the deposition of radionuclides to vegetation over a wide geographical area, after a single 'spike' release. Table 7-4 shows a selection of data on deposition velocities for ^{137}Cs and ^{103}Ru

Table 7-4. Deposition velocities reported for ^{137}Cs and ^{103}Ru after the Chernobyl accident.

Country	Deposition velocity (m s^{-1})		Comments
	^{137}Cs	^{103}Ru	
Austria	0.0016	0.0018	'Effective' V_g
Denmark	0.045	0.050	Dry, to grass
	0.04	0.05	Dry
	0.0042	0.046	Dry, to smooth surface
	0.0042	0.046	Dry only, to smooth surface
Finland	0.0027	0.0045	Dry only
France	0.5	0.5	During rainfall
	0.0005	0.0005	Dry only
	0.086	0.086	Total deposit
Germany	0.0026	0.0009	Dry only
	0.018	0.02	Total deposit
	0.0003–0.0015	–	Dry only
Greece	0.00037	–	Dry only 3–4 May 1986
	0.041	–	Dry only 9–11 May 1986
Italy	–	0.0009	Dry only
Netherlands	0.0017	0.0022	Dry only
Sweden	0.0015	0.0015	Dry only
Switzerland	0.00024	0.00024	Without fog or dew, dry only
	0.0013	0.001	With fog or dew, dry only
UK	0.0004–<0.0015	–	Dry only
	0.00053	–	Countrywide average
	0.015	0.013	Dry only
	0.0001–0.0004		

aerosols in eleven European countries after the accident, under different meteorological conditions. In general V_g values were within the range predicted from controlled experiments and earlier bomb fall-out studies for the aerosol size distribution concerned, with enhanced deposition occurring when foliage was wet due to the presence of fog and dew.

Initial Behaviour of Radionuclides after Deposition on Foliage

The fate of radionuclides after deposition onto vegetation is determined by many factors, including the chemical and physical forms of the radionuclide, climatic parameters, and the physiological state of the plant. There remain many uncertainties over the precise chemical form in which different radionuclides are deposited onto vegetation, and post-Chernobyl studies have by no means resolved all these. Clearly, material deposited in a soluble form is more likely to be absorbed into internal tissues or washed off by precipitation. However, it has been shown that the solubility of some radionuclides can change with time. For example, Mewhinney *et al.* (1987) studied the effect of continuous immersion of relatively insoluble PuO_2 particles compared with alternate 4/3 day wet/dry cycles (Table 7-5). The alternate wetting/drying treatment increased the amount of Pu dissolved from the particles by between 2 and 10 times more than the continuous immersion treatment. There was also evidence in this study of the possible importance of the interaction with a non-radioactive pollutant in that simulated acid rain at pH 3.0 substantially increased such leaching over a 7 week period compared with a distilled water treatment.

Table 7-5. Percentage of initial α radioactivity dissolved from PuO_2 particles each week after continuous immersion or subjection to 4 day/3 day wet/dry cycles (Mewhinney *et al.*, 1987).

Week	Bathing liquid	Continuous wetting		Alternate wet/dry		
		1	2	1	2	3
1	0.1 M HCl	4.0	2.9	9.1	7.8	15.0
2	0.1 M HCl	0.99	0.94	2.9	2.2	0.71
3	0.1 M HCl		0.34	3.8	3.1	0.60
4	0.1 M HCl		0.26	2.4	1.2	0.53
7	Distilled water			1.7	0.72	1.3
7	Acid 'rain' (pH 3.0)			4.9	5.9	7.7

Some radionuclides in a soluble form are readily absorbed into foliage, but the degree to which this occurs may be highly dependent on species, as shown in Table 7-6 where the amount of ^{89}Sr taken up by different crop species 72 h after application ranged over an order of magnitude. However, the subsequent fate of the radioac-

Table 7-6. Percentage of ^{89}Sr absorbed into foliage of different species 72 h after application (from Bukovac *et al.*, 1965).

Phaseolus vulgaris (French bean)	24.2[a]
Brassica oleracea var. *botrytis* (cauliflower)	15.4[b]
Raphanus sativus (radish)	12.1[ba]
Lactuca sativa (lettuce)	10.3[c]
Zea mays (maize)	1.9[d]

Means with different letters are significantly different at $P < 0.05$

tivity within the plant is highly dependent on the mobility of the radionuclide concerned. This is illustrated in Fig. 7-9 where, following a surficial application to bean leaves, around 30% of absorbed ^{137}Cs was rapidly translocated into other organs including the roots; in contrast, after a similar treatment with ^{89}Sr, scarcely any of the radionuclides was translocated, indicating the lack of mobility of this element in the phloem and reflecting the behaviour of its physiological analogue, calcium.

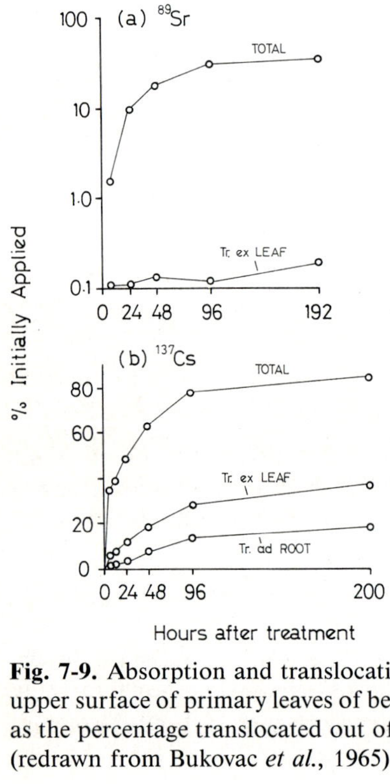

Fig. 7-9. Absorption and translocation of (a) ^{89}Sr and (b) ^{137}Cs following application to the upper surface of primary leaves of bean plants. The total quantity absorbed is indicated as well as the percentage translocated out of the leaf (Tr. ex LEAF) and into the root (Tr. ad ROOT) (redrawn from Bukovac *et al.*, 1965).

As mentioned earlier, the growth stage at which contamination *via* atmospheric deposition takes place is critically important in controlling the level of radioactivity in the edible organs at harvest. Thus an accident at different growth stages may have potentially different consequences with respect to human health and it is important to understand this phenomenon clearly in order to predict the ultimate dose to humans consuming contaminated crops. This is illustrated in Table 7-7 which shows the percentage of ^{137}Cs and ^{89}Sr in wheat grain in terms of the amount initially retained by foliage sprayed on three different occasions over the crop's growth cycle. In the case of ^{89}Sr a late application resulted in approximately 10 times more contamination of the grain harvest, compared with earlier applications, while for ^{137}Cs, which was absorbed in all cases to a much greater extent, this pattern was not observed. The importance of translocation in the case of ^{137}Cs compared with ^{89}Sr is indicated by the relatively small reduction in grain contamination at harvest when plants were sprayed at 107 days old with the newly emerging ears protected from contamination. A similar result has also been demonstrated by Åarkrog (1983).

Table 7-7. Radionuclide content of wheat grain as % of initial retention by plants sprayed at different growth stages.

Time from planting to contamination (days)	Growth stage	Content of grain as % of initial retention by plant	
		^{137}Cs	^{89}Sr
86	Just before ear emergence	5.2^{a}	0.12^{a}
107	Just after ear emergence	10.2^{b}	1.26^{b}
107 (ears protected)		5.4^{a}	0.15^{a}
136	Just before maturity	6.3^{a}	1.79^{b}

Means followed by different letters are significantly different at $P < 0.05$ (Middleton, 1959).

The Chernobyl accident revealed totally unexpected behaviour of radiocaesium in upland rough grazing areas of the British Isles, which may be partly explained by foliar absorption processes. Thus, Brown *et al.* (1991) demonstrated that *Calluna vulgaris* shoots dipped into ^{134}Cs solution showed an extremely high degree of translocation out of the source leaf into other parts of the shoots. In contrast, two other members of the Ericaceae which often grow in close association with *C. vulgaris* – *Erica tetralix* and *Vaccinium myrtillus* – showed scarcely any such translocation. Thus enhanced uptake and mobility of Cs into *C. vulgaris* foliage may have contributed to the exceptionally high levels recorded in this species after Chernobyl, although uptake *via* the roots was also shown to be elevated considerably (Bell *et al.*, 1988).

Field Loss

After initial contamination, the aerial parts of plants lose radioactivity by a process known as 'field loss'. The mechanisms by which this takes place are not fully understood but may involve the following:

- removal of particles by wind, rain or other disturbance
- die-back or weathering of leaves of their surface layers
- leaching from within foliage by the action of rain
- translocation to the subterranean plant organs
- volatilisation.

In addition, contamination will always be reduced as a result of radioactive decay, but this component can be predicted and quantified with absolute certainty compared with the other potential processes.

Field loss time courses exhibit an approximately exponential shape; consequently an important parameter used in quantifying field loss is the retention half-life, defined as the time for half of the initial deposit to be lost by processes other than radioactive decay. This can be expressed on either a mass or ground area basis. In the case of actively growing vegetation retention half-lives are remarkably similar for different species and different radionuclides, irrespective of whether these are in gaseous, soluble or insoluble forms (Table 7-8). However, it has been found consistently that under slow-growing conditions the retention half-life is greatly increased. An example of this is shown in Fig. 7-10, where a grass sward sprayed with ^{85}Sr outdoors had a retention half-life of 19 and 49 days in summer and winter, respectively. One hypothesis to explain this phenomenon was proposed by Moorby and Squire (1963). They suggested that the main mechanism for field loss under dry conditions was the shedding of epicuticular wax to which the radionuclides had become attached and that this process was much greater in summer when plants were growing actively and thus subjecting the cuticle to greater mechanical stress. The hypothesis was tested by these authors who contaminated the foliage of cabbage, potato and ryegrass with ^{85}Sr and

Table 7-8. Retention half-lives of radionuclides deposited onto actively growing vegetation (Miller and Hoffman, 1983).

	Range (days)	Geometric mean (days)
All values for herbaceous vegetation	2.8 – 34	10
I_2 vapour on herbaceous vegetation	4.5 – 14	7.2
I particulates on herbaceous vegetation	2.8 – 16	8.8
Other particulates on herbaceous vegetation	9 – 34	17
I particulates and vapour on woody vegetation	7.9 – 13	10
Other particulates on woody vegetation	12 – 23	20

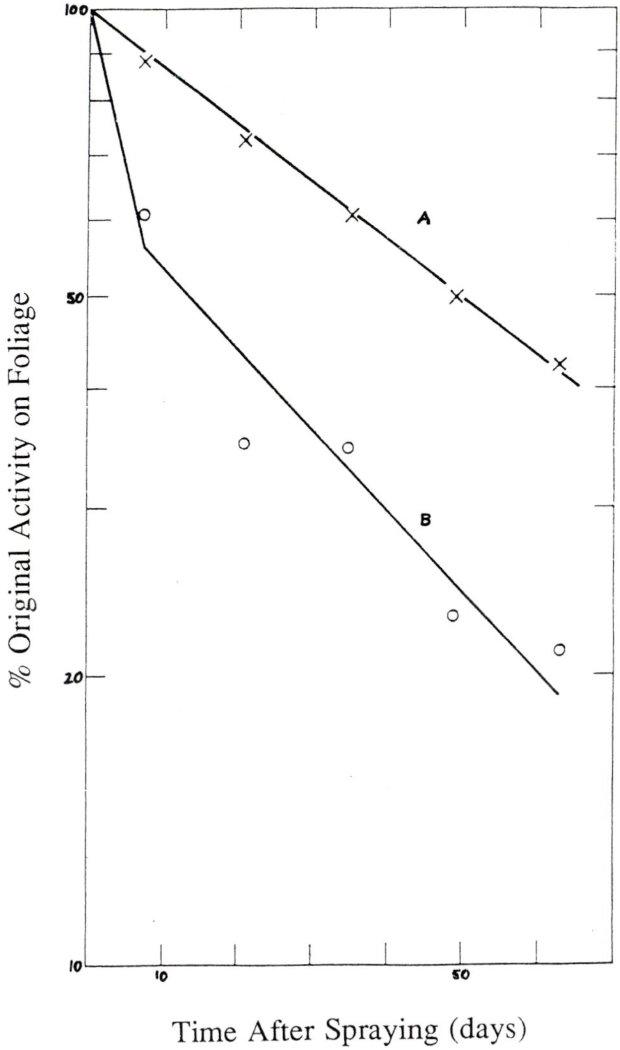

Fig. 7-10. Loss curves for ^{85}Sr sprayed onto grass swards during (a) winter and (b) summer. The 'field loss' half lives during these periods were 19 and 49 days, respectively (reproduced from Chadwick and Chamberlain, 1970).

placed the plants alongside uncontaminated ryegrass plants in a polythene tunnel through which air was drawn. For all three crops, measurable transfer of ^{85}Sr occurred onto the uncontaminated grass swards, indicating that a dry process of physical removal of the radionuclide was taking place back into the atmosphere. More recently an attempt has been made by Watterson (1989) to test Moorby and Squire's proposed mechanism more precisely. Pea plants, which were either normal, with an intricate epicuticular wax structure, or mutant, lacking such wax, were sprayed with ^{134}Cs and the retention half-lives measured. The results indicated that the presence of epicuticular wax had no significant effect on this parameter and thus the mechanism by which field-loss occurs under dry conditions still remains a matter for elucidation. As in the case of other facets of radionuclide behaviour in the environment, the Chernobyl accident provided the opportunity to verify data on retention half-lives of several radionuclides which had been generated from controlled wind-tunnel and field experiments. Table 7-9 shows that in ten European countries retention half-lives for both ^{131}I and ^{137}Cs on pastures were generally similar and within the range of previously measured values shown in Table 7-8.

Table 7-9. Ranges of retention half-lives (days) for initial decline of ^{131}I and ^{137}Cs concentration of grass and pasture vegetation after the Chernobyl accident (from Coughtrey *et al.*, 1990).

Country	^{131}I	^{137}Cs
Netherlands	7.4–11	7.4–9.2
Italy	9.4–11	7.2–13.5
Spain	–	17.4
Greece	8.3–29	16
FRG	4.3–30	7–16
Denmark	–	10–18
UK	1–11	1–23
Switzerland	6.2–17	5.3–27
France	–	26
Belgium	–	2.6–33

While field-loss measurably takes place under dry conditions, there is little doubt that precipitation may exert an effect under some circumstances. Indeed, it has been demonstrated that both physical and chemical characteristics of rain can affect the amount of radionuclide removed from foliage. Thus Cataldo *et al.* (1981) examined the removal of sub-micron (<0.1 μm) plutonium nitrate or plutonium citrate particles by simulated mist or rain from *Phaseolus vulgaris* foliage. The mist was more effective at removal than was the rain (Fig. 7-11). In the same experiment the effect of acidifying the mist was also studied. Surprisingly, acidity decreased the loss of the soluble Pu (nitrate) but increased it in the case of the insoluble form (citrate). It was suggested that this might be the result of the acid causing the citrate to dissociate into soluble hydrolytic products, while the nitrate hydrolysed to form insoluble colloidal hydroxides. An interesting point to note from this experiment is that the retention half-lives for these <0.1 μm particles were considerably larger than those recorded for larger particles. Indeed, they ranged between 164 and 1 000 days – a matter of somewhat hypothetical interest for a short-lived crop such as *P. vulgaris*!

Rain can also leach radionuclides out of the internal portions of leaves. A summary of the relative leachability of a range of radionuclides from the young leaves of *Cucur-*

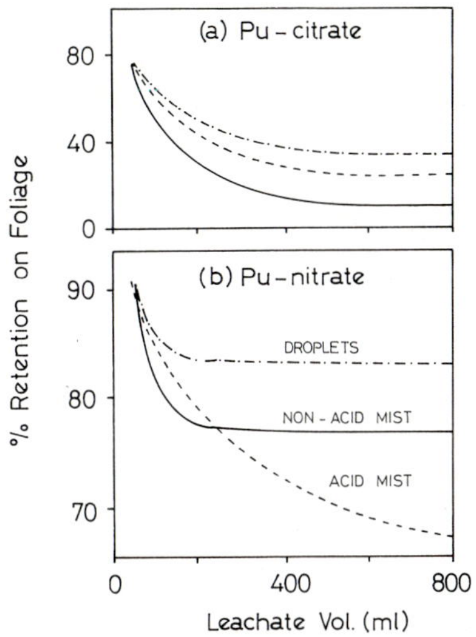

Fig. 7-11. Effect of raindrop size and acidity on the retention of (a) plutonium-citrate and (b) plutonium-nitrate by *Phaseolus vulgaris* foliage (redrawn from Cataldo *et al.*, 1981).

bita pepo is shown in Table 7-10. As a rider to these observations, however, it may be noted that Shaw *et al.* (1992) showed that cumulative rainfall did not appear to affect systematically radionuclide loss from foliar surfaces of wheat.

Table 7-10. Relative leachability of isotopes from young leaves of squash (*Cucurbita pepo*) (from Tukey *et al.*, 1958).

Readily leached	Moderately leached	Leached with difficulty
^{22}Na	^{45}Ca	^{55}Fe
^{54}Mn	^{28}Mg	^{65}Zn
^{35}S	^{32}P	
^{42}K	^{36}Cl	
$^{90}Sr/^{90}Y$		

Contamination by Rainsplash (Resuspension)

Contamination of aerial plant parts by rainsplash of soil particles to which radionuclides are attached may be of particular importance when uptake *via* the roots is low, due to properties of the soil, the plant and the element concerned. This is demonstrated in Table 7-11, which shows the results of washing the shoots of wheat which had been grown outdoors in soil contaminated with ^{137}Cs, ^{144}Ce and ^{99}Tc. In the case of ^{137}Cs a small portion was removable by washing, indicating that resuspen-

Table 7-11. Effect of washing on radionuclide contamination (Bq g^{-1} dry weight ±SEM) of spring wheat growing on a clay soil (Grogan, 1984).

Time from sowing (days)	Plant part	^{137}Cs		^{144}Ce		^{99}Tc	
		Unwashed	Washed	Unwashed	Washed	Unwashed	Washed
74	Straw	58.7 ± 5.1	52.9 ± 3.0	<0.37	<0.37	2930 ± 140	3050 ± 380
103	Straw	26.4 ± 3.3	17.7 ± 6.0	4.59 ± 1.15	<0.37	1410 ± 260	1730 ± 310
	Ears	1.8 ± 1.1	1.1 ± 0.6	<0.37	<0.37	<33	<33
132	Straw	26.7 ± 7.7	11.6 ± 2.1	2.32 ± 1.16	<0.37	844 ± 63	992 ± 29
	Ears	2.36 ± 0.2	2.3 ± 0.5	<0.37	<0.37	<33	<33

sion was relatively unimportant compared with the root uptake and translocation pathway. ^{99}Tc, in contrast, could not be removed at all, because this highly mobile radionuclide had been rapidly taken up by the roots, thereby exhausting the pool in the soil. However, ^{144}Ce, a radionuclide which is scarcely absorbed by the roots, was completely removed by washing, showing that the resuspension route was of prime importance for contamination of aerial plant parts.

A more detailed study of the role of rain in resuspension onto shoots was carried out by Dreicer *et al.* (1984). Tomato plants were grown outdoors and the quantity of soil particles <53 μm and 53–105 μm measured between 0–40 cm and >40 cm above ground level after four successive storm events. Table 7-12 shows that most deposition onto the foliage occurred within 40 cm of the ground and that there was generally an increase in this deposition with a successive number of storms. There are few studies of this type and there is a requirement for controlled experiments on the influence of different types of rain events (duration, intensity, droplet size) on both resuspension onto leaves and also removal of such contamination.

Table 7-12. Rainsplash of soil onto tomato plants after 4 consecutive storms as a function of soil particle size fraction and height above ground surface (Dreicer *et al.*, 1984).

Number of storms	Soil size fraction (μm)	Concentration of soil (mg g^{-1} wet plant tissue)	
		0–40 cm	>40 cm
1	<53	0.1 ± 0.03	0.2 ± 0.07
	53–105	0.8 ± 0.60	1.2 ± 0.37
1 + 2	<53	1.2 ± 0.18	0.1 ± 0.03
	53–105	2.5 ± 0.37	0.5 ± 0.20
1 + 2 + 3	<53	2.2 ± 0.49	0.7 ± 0.49
	53–105	2.6 ± 0.72	−0.3 ± 0.18
1 + 2 + 3 + 4	<53	2.4 ± 0.70	0.2 ± 0.04
	53–105	2.2 ± 0.49	0.5 ± 0.36

Uptake of Radionuclides by Higher Plant Roots

Following the steady accumulation of radionuclide inventories in soils around the world during the period of atmospheric nuclear weapons testing prior to 1963 interest in the contribution of root uptake processes to crop contamination burgeoned. Of particular concern were the relatively long-lived isotopes ^{90}Sr and ^{137}Cs (half-lives 28 years and 30 years, respectively) which, as chemical analogues of Ca and K, are particularly prone to biological incorporation. The principal interest during this classical period of radioecology was the equilibrium situation where an approximately constant input of radioactivity to the ground surface as a result of fallout from the atmosphere led to a steady rate of transfer into the plant from surface layers of soils. More recently, however, and particularly since the accident at the Chernobyl nuclear power plant in 1986, the effect of 'spike' releases on plant contamination *via* root uptake has become of great importance, especially when the ecosystems affected show a reduced capacity to immobilise rapidly the deposited radioactivity in a form unavailable for plant uptake. It is with respect to this latter problem that recent research has been steered towards a thorough mechanistic understanding of root absorption of radioactive materials in an attempt to appreciate the salient aspects of soil-plant interaction which are responsible for enhancing or decreasing soil-to-plant transfer of radionuclides.

The Soil-to-Plant Transfer Factor

Traditionally the overall process of incorporation of radionuclides into plant tissues from contaminated soils *via* the root uptake process has been quantified using the soil-to-plant transfer factor, or *TF*. This is defined simply as the ratio of the specific radionuclide activities in the plant tissue of concern (usually some edible organ) and the soil, with the dry masses of each material usually being taken into account:

$$TF = \frac{\text{plant specific activity (Bq kg}^{-1}\text{)}}{\text{soil specific activity (Bq kg}^{-1}\text{)}} \tag{6}$$

Calculations of *TF* values generally take into account the specific radionuclide activity in the surface 10 to 20 cm of the soil profile. This appears to be the case primarily because the bulk of root density is found in these surface layers, particularly in agricultural species such as wheat and pasture grasses, but also because, historically, radionuclide inputs from atmospheric fallout have contaminated soil surface layers with minimal downwards movement through the profile taking place even over protracted periods.

Other definitions of 'transfer factors' have been determined, usually directed to specific applications by individual authors. Two common variations on the theme are the 'inventory ratio' (*IR*) (Gilbert and Simpson, 1983), based on the degree of radionuclide contamination of soils and associated vegetation on an areal basis:

$$IR = \frac{\text{plant activity per unit area (Bq m}^{-2}\text{)}}{\text{soil activity per unit area (Bq m}^{-2}\text{)}} \tag{7}$$

and other definitions based on radionuclide activities on both a mass basis and an areal basis. One example of this was Mascanzoni's (1990) quantification of radiocaesium absorption from forest soils by fungi in Sweden following the Chernobyl accident:

$$TF = \frac{\text{`plant' specific activity (Bq kg}^{-1}\text{)}}{\text{soil activity per unit area (Bq m}^{-2}\text{)}} \tag{8}$$

While the ratios defined in Eqs. (6) and (7) conveniently yield dimensionless factors, the last of these equations yields rather cumbersome units of $m^2\ kg^{-1}$. These three 'transfer factor' equations give some idea of the range of ways in which individual radioecologists have tackled the problem of expressing the gross uptake of radionuclides by plants in some easily quantifiable form, but it is Eq. (6) which provides us with the most commonly used definition which has been adopted by the International Union of Radioecologists (IUR) as the working definition for their soil-to-plant transfer factor data base.

A major problem with the rapid proliferation of data on soil-to-plant transfer of radionuclides, especially during the latter part of the 1960s and the 1970s, was the lack of standardisation of experimental design between workers. This led to difficulties in interpreting a bewildering array of *TF* values derived for a variety of combinations of crop and soil type, sources of radio-contamination and tillage practices. In an attempt to circumvent this problem and focus research on the topic, the IUR established a working group on 'Soil-Plant Transfer Factors' in 1982 which initially brought together scientists from 15 European countries. This led to recommendations for the establishment of a number of standardised experimental protocols for the determination of *TF*s, which would then be acceptable for inclusion in a common data base which also includes important supplementary information on the experimental circumstances under which the data are obtained (Table 7-13) (IUR, 1982). This data base is constantly being updated and is probably the largest source of information on soil-to-plant transfer of radionuclides currently available.

Table 7-14 shows the ranges of *TF* values for a selection of radionuclides as published in the VIth Report of the IUR Soil-To-Plant Transfer Working Group (IUR, 1990). Values vary from 3.4×10^3 for ^{99}Tc to 1.5×10^{-7} for ^{241}Am, an overall range of 11 orders of magnitude which serves to illustrate the extremely variable behaviour observed between individual radionuclides. However, it can be seen that even for individual radionuclides the numbers of orders of magnitude spanned by *TF* values may

Table 7-13. Supplementary information appended to values of soil-to-plant transfer factors for radionuclides within the IUR data base.

Plant part
Harvested material unwashed or washed to remove resuspended soil particles
% dry matter in crop
Soil type (based on particle size analysis)
Concentration of radionuclide
Radionuclides applied artificially or as a result of contamination in the field
Depth of soil contamination
Container type
Soil pH
Duration of radionuclide in the soil
Amount of irrigation
Author's name
Identification reference
Any additional information

Table 7-14. Ranges of soil-to-plant transfer factors for selected radionuclides (source; IUR, 1989).

Radionuclide	Transfer factor		No. of orders of magnitude spanned
	Upper limit	Lower limit	
^{110m}Ag	3.9×10^{-3}	5.9×10^{-5}	3
^{241}Am	7.7×10^{-1}	1.5×10^{-7}	7
^{144}Ce	3.2×10^{-1}	2.7×10^{-2}	2
^{60}Co	4.3	4.0×10^{-4}	4
$^{137/134}Cs$	3.1×10^{1}	1.6×10^{-4}	6
^{237}Np	4.7×10^{-1}	4.1×10^{-5}	4
$^{239/240}Pu$	6.5×10^{-1}	4.0×10^{-7}	7
^{90}Sr	3.9×10^{1}	8.0×10^{-3}	5
$^{103/106}Ru$	1.5	5.0×10^{-3}	4
^{99}Tc	3.4×10^{3}	3.2×10^{-1}	5

vary from 2 in the case of Ce to 7 in the case of Pu and Am. To some extent the use of a simple empirical ratio like the *TF* is designed to ease the interpretation of such extremely variable data and the IUR data base is sufficiently extensive to allow attempts to correlate *TF*s with soil factors such as pH, organic content and soil moisture. On one hand, studies such as that of Sheppard and Evenden (1990) have indicated that *TF* values may be meaningfully correlated with the periodicity of the element under consideration. On the other hand, however, results such as those obtained by Frissel *et al.* (1990) illustrate the difficulty in identifying causal relationships between *TF*s and soil physico-chemical parameters. In the final analysis it must be realised that *TF* data are completely site and species specific, describing the degree of transfer of a particular radionuclide to one plant species under a given set of conditions and it is this situation which appears to limit the ultimate utility of the *TF* concept.

The origins of the transfer factor appear to lie in the 'transfer coefficient' originally defined by Ward *et al.* in 1965. This concept was intended by these authors to be used to assist in the quantification of ^{137}Cs secretion into cows' milk derived from contaminated feed. Unfortunately the derivation of the term was not fully explained by the original authors until the middle of the 1980s (Ward and Johnson, 1986) when it became clear that when applied to animal systems the transfer coefficient concept would only be expected to apply to an equilibrium situation in which the radionuclide in question was not subject to homeostatic control (ie, first order kinetics are implied). If, as these authors have stated, the 'physiological basis (of the transfer coefficient) is not always appreciated' when deriving data for animal systems, it is clear that the simple extrapolation of this concept to soil-to-plant transfer of radionuclides is fraught with potentially unseen pitfalls. Thus, in order to make some sense of the extensive data in the literature, especially in the wake of the Chernobyl accident, it is crucial that the mechanistic basis of plant absorption of radionuclides is understood.

Physiological Aspects of Radionuclide Absorption by Plant Roots

Radioactive pollutants of soils can be classified into two groups with respect to their potential for uptake by higher plant roots. First, there are those which are radioactive isotopes of elements which are known to be either biologically essential or chemically (and presumably physiologically) analogous to essential elements. Members of this group of significance to the root uptake pathway include nuclides of Co, Zn, Mn, Fe, K, Rb, Cs, Ra and Sr. The second group comprises nuclides of elements for which organisms exhibit no known requirement. Particularly prominent members of this group are those 'artificial' radionuclides belonging to the transuranium series, which include Pu, Am, Np and Cm. As a broad rule the degree of plant uptake of the latter elements (which are all exclusively radioactive, possessing no stable isotopes) is many times less than that of the elements in the first group, which exhibit suites of stable and radioactive nuclear species. The outstanding exception to this rule, however, is the element technetium (Tc). This was the first element to be created artificially by man, in 1937, and exhibits two nuclides, one of which (^{99}Tc) is of considerable environmental significance due to its 200000 year half-life and its remarkably high degree of bioaccumulation (see Table 7-14).

While the actinides Pu, Am, Np and Cm are critically important to assessments of the exposure of humans to environmental sources of radioactivity due to their highly hazardous α emissions, the role which the plant root absorption pathway plays in this exposure is relatively small. A conservative estimate of the degree to which plants will incorporate Pu from soil, for instance, is 10% (ie, a soil-to-plant transfer factor of

0.1) (Dahlman *et al.*, 1976) although it can be seen from Table 7-14 that *TF* values for this element are commonly six orders of magnitude lower than this. In practice the predominant pathway of plant contamination by actinide elements is usually direct contamination due to the deposition and resuspension processes covered in the previous section. In a summary of data collected on plant root uptake from contaminated sites up to 30 years after nuclear weapons testing Dahlman *et al.* (1976) concluded that the vast majority of these elements remained within the soil, associated at least in part with humic substances. Concentration ratios (soil-to-plant transfer factors) were invariably <1 in unamended soils and showed a very high degree of variability (up to eight orders of magnitude – see Fig. 7-12). Interestingly, the addition of artificial chelating agents such as DTPA increased the root uptake of Pu in these studies, probably by increasing the fraction of monomeric Pu in the soils and thereby enhancing its physico-chemical mobility. The chelation potential of a number of natural and artificial organic compounds for the actinides was investigated in a study by Bulman (1983) who concluded that both plant root and animal gut absorption of actinide elements may be enhanced by a variety of these compounds. However, the absolute absorption of the transuranics by biological tissues is still relatively low as compared with that of several other radionuclides for which the root uptake pathway is considerably more significant.

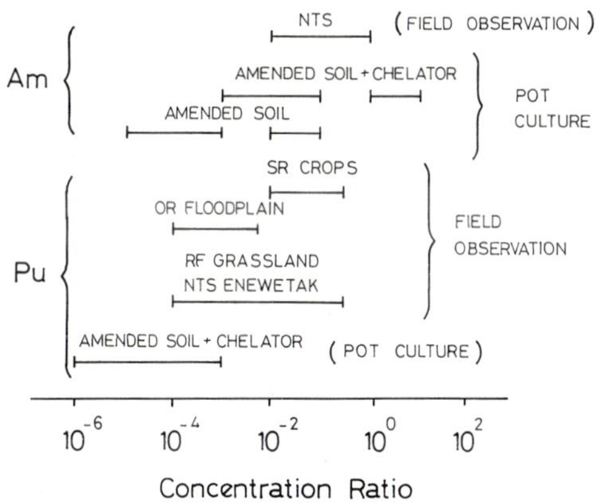

Fig. 7-12. Concentration ratios (ie, soil-to-plant transfer factors) for plutonium and americium at various sites associated with the US nuclear programme. Key to sites: NTS, Nevada test site; SR, Savannah River, South Carolina; OR, Oak Ridge, Tennessee; RF, Rocky Flats, Colorado; NTS, Enewetak Atoll, South Pacific (from Dahlman *et al.*, 1976).

The most extensively studied radionuclides with respect to biological incorporation and the contamination of food chains are ^{137}Cs and ^{90}Sr. The elements Cs and Sr, respectively, belong to groups I and II of the periodic table and as such share common physico-chemical properties with K and Ca. As analogues of these elements ^{137}Cs and ^{90}Sr are readily incorporated by biological tissues and so, historically, the existence of these nuclides within the environment as a result of weapons testing and accidental releases from the nuclear industry has resulted in a very large section of the radioecological literature being devoted to them.

^{137}Cs and ^{90}Sr are fundamentally different in their behaviour in soils. ^{137}Cs shows a very high degree of attachment and 'fixation' to soil mineral materials (particularly within the clay fraction – see Cremers *et al.*, 1988) whereas ^{90}Sr remains relatively mobile and hence available for uptake by plant roots. In the case of ^{137}Cs the relative proportion of clayey material within a contaminated soil, and particularly its mineralogy, will strongly determine the degree of soil-to-plant transfer of the nuclide (Bell *et al.*, 1988). Despite the theoretical possibility of its existence as non-exchangeable carbonates, sulphates, phosphates, silicates and sesquioxides, however, most studies have determined that considerably more than 90% of radiostrontium is in exchangeable forms (Francis, 1978). The common factor which strongly affects the plant uptake of each of these nuclides is the similarity, in physiological terms, between the radioions which exist in the environment at 'carrier-free' concentrations and the relatively highly abundant nutrient analogues, which may be present in the soil environment at concentrations many orders of magnitude greater than the radioions themselves. The potential for the control of ^{137}Cs and ^{90}Sr uptake by K and Ca is therefore great although it must not be forgotten in the case of elements such as Cs and Sr that stable isotopes (^{133}Cs and ^{88}Sr) also exist in the natural environment at concentrations considerably greater than those of their radioactive counterparts.

In radioecology the occurrence of different radioisotopes of the same element is commonly encountered. A case in point was the simultaneous deposition of ^{137}Cs (a fission product) and ^{134}Cs (an activation product) from the Chernobyl plume during

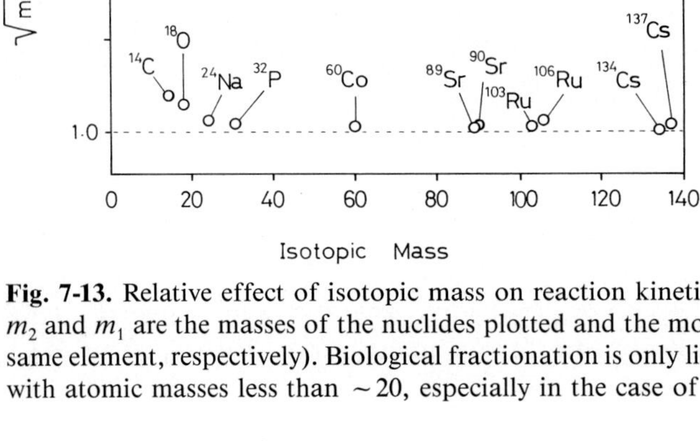

Fig. 7-13. Relative effect of isotopic mass on reaction kinetics, expressed as $\sqrt{m_2/m_1}$ (where m_2 and m_1 are the masses of the nuclides plotted and the most common stable nuclide of the same element, respectively). Biological fractionation is only likely to be important for elements with atomic masses less than ~20, especially in the case of deuterium and tritium.

its passage over western Europe in 1986. Initial observations indicated that soil-to-plant transfer factors for deposited ^{137}Cs were lower than for ^{134}Cs and gave rise to speculations that some differential plant uptake of the two nuclides was occurring.

Partial fractionation of isotopes of the same element is known to occur in kinetic systems due to differences in mass. The extent to which this is likely to hold true for a particular pair of isotopes can be estimated using the relationship developed by Bigeleisen (1949) which states that the relative effect of the mass difference of two isotopes on the overall reaction kinetics is equal to the square root of the ratio of the two masses ($\sqrt{m_2/m_1}$ – see Bowen, 1979). This relationship is plotted in Fig. 7-13 which indicates that for elements of mass number >20 the kinetic discrimination between isotopes during reactions, such as those involved during the passage of ions across plant root membranes, would be expected to be negligible. This was tested by Mills (previously unpublished) who examined the uptake of ^{137}Cs and ^{134}Cs by two upland plant species (*Calluna vulgaris* and *Festuca ovina*) and found no evidence of discrimination between the two isotopes (Fig. 7-14). By implication a similar lack of discrimination between radioisotopes of Cs and its naturally occurring non-radioactive isotope (^{133}Cs) would be expected.

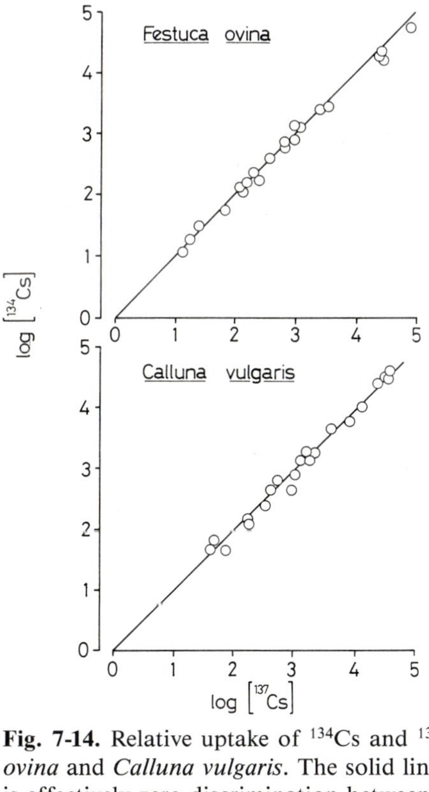

Fig. 7-14. Relative uptake of ^{134}Cs and ^{137}Cs by two common upland plant species, *Festuca ovina* and *Calluna vulgaris*. The solid line in each case is a plot of unity, showing that there is effectively zero discrimination between these two radionuclides during plant uptake (from Mills, previously unpublished).

Of more relevance to the uptake of radionuclides by plants is the question of discrimination between radionuclides and their nutrient analogues. Indeed, the question of whether the physiological mechanisms of ion uptake within the root can discriminate between the radioion and its analogue has been central to the elucidation of the environmental behaviour of ^{137}Cs and ^{90}Sr since the late 1950s. Comar *et al.* (1957) devised a measure of the degree of discrimination by plants between strontium and calcium which they termed the 'observed ratio' (*OR*), defined as:

$$OR_{Sr/Ca} = \frac{[Sr]_{plant}/[Ca]_{plant}}{[Sr]_{soil}/[Ca]_{soil}}$$

A value <1 implies that calcium is taken up preferentially whereas a value >1 implies that strontium is taken up preferentially; a value of unity indicates zero discrimination between the two elements. The biggest source of error in applying such ratios to the field situation is the difficulty in determining an accurate $[Sr]_{soil}/[Ca]_{soil}$ ratio, which reflects the extreme difficulty of relating data on the chemical extractability of a radionuclide within a soil to its actual bioavailability. As a result much of the literature evidence on this topic is based on solution culture experiments in which it has been consistently shown that $OR_{Sr/Ca}$ values tend to be greater than unity in roots whereas in stems, leaves and seeds the reverse is true. This seems to reflect the preferential translocation of Ca to the above-ground organs of plants, although as the rate of transpiration of a plant decreases Ca transport appears to be diminished in favour of Sr movement. In relation to the highly mobile K and its analogue Cs, however, the absolute degree of internal translocation of both Sr and Ca is low with both elements existing largely as immobile complexes with glucuronic acids (Mortensen and Marcusiu, 1963) and pectates (Myttenaere and Masset, 1965).

In solution culture experiments both strontium and caesium show hyperbolic absorption isotherms with respect to the external concentration of the element. Figure 7-15 (a) shows an example of a typical uptake isotherm for Sr while Shaw and Bell (1989) have demonstrated a similar isotherm for Cs. Baker (1981) has referred to such plant uptake responses as 'accumulator' functions and has identified these as being typical of the absorption of elements over which plants can exert some degree of physiological control. Typically, the nutrient elements, including K and Ca, exhibit such isotherms and it can be postulated from the similarity in the uptake patterns of K and Cs on the one hand and Ca and Sr on the other that the radioions share, to some extent, the same uptake mechanisms as the nutrient ions. This has several important implications. Firstly, the direct competition for uptake sites between radioions and nutrient ions means that the external (soil) concentration of one is increased at the expense of the uptake of the other; as the nutrient ions in question are vastly more abundant in soils than radioions it is K and Ca which will be effective in competitively excluding Cs and Sr, respectively. Secondly, the kinetics of this competition are concentration-dependent, so the assumption of first order kinetic movements of

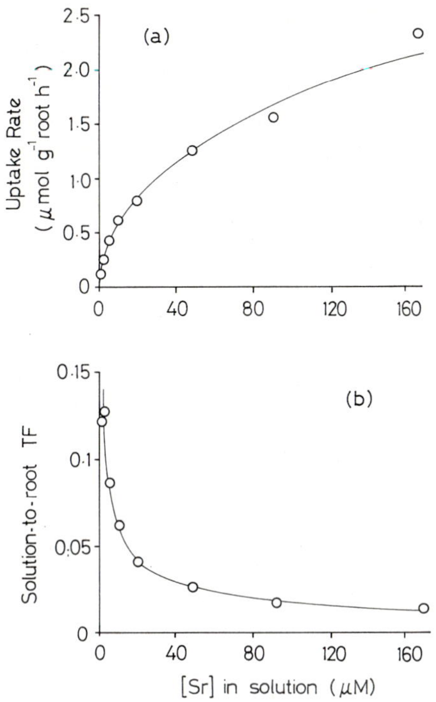

Fig. 7-15. (a) Hyperbolic absorption isotherm for strontium uptake in wheat (*Triticum aestivum*). This gives rise to the negative and non-linear relationship between the transfer factor and external concentration of strontium shown in (b) (from Hewamanna, previously unpublished).

radioions into plant tissues, as embodied in the transfer factor concept, are invalid. The result of this is that any increase in a competing ion's concentration in the medium bathing the root will result in a non-linear depression in radioion uptake, whether the competing ion is a stable nuclide of the radioisotope (see, for example, Fig. 7-15(b)) or an analogue with its own chemical identity (Whicker, 1983). Indeed, Jones *et al.* (1991) determined that, aside from the external ionic environment, the nutritional status of the plant *per se* may have the same effect.

Shaw and Bell (1991) examined this effect in the case of competition between radiocaesium and the K^+ and NH_4^+ ions during root uptake by wheat (*Triticum aestivum*). These authors formalised the observed relationships in terms of classical Michaelis – Menten kinetics which necessitates the assumption that each of these ions is taken up by identical sites associated with the root plasmalemma. Lembrechts *et al.* (1990) found a similar negative and non-linear relationship between the concentration of Ca either in soil or in solution culture and the degree of radiostrontium uptake by lettuce (*Lactuca sativa*). The principle of competitive exclusion of a radionuclide by an ion analogue may be exploited, with varying degrees of success, as a post-contamination countermeasure designed to reduce the transmission of radionuclides to either human or animal populations *via* the soil → plant pathway (Shaw, 1993). Applications of fertilisers containing either K or Ca may be more or less effective in reducing Cs and Sr uptake by plants, depending largely on the buffering characteristics of the soil in question (Nisbet, 1993). Generally these treatments are aimed at reducing the fractional molar ratio of the radionuclide with respect to the nutrient analogue so that competition between the two ions will favour the exclusion of the radioion.

Cline and Hungate (1960) applied the concept of the observed ratio to the question of discrimination between the Cs^+ and K^+ ions. These authors noticed that in bean plants (*Vicia faba*) the *OR* was non-constant when the K^+ concentration in a nutrient solution was increased, implying that some concentration-dependent effect was altering the degree of selectivity which the plant roots were able exert over these two ions. More recently, Shaw *et al.* (1992) have identified a critical threshold concentration of K^+ below which classical competition kinetics appear to describe adequately the antagonism between the two ions, but above which selectivity of the uptake mechanism is apparently 'switched' in favour of the K^+ ion. The exact identity of the uptake machanism is still a matter of debate but may be a pore in the plasmalemma which is able to accommodate several similar ions synchronously in single file (Tester, 1988); any selectivity between Cs and K uptake must be exerted at this level. At present it is clear that the long running question of whether and how plants discriminate between these ions still awaits a final answer, though the implications of the operation of such a mechanism are evidently important. Seasonal variation in the activities of the K^+ and Ca^{2+} ions in soil solutions will cause a variable degree of inhibition of radiocaesium and radiostrontium uptake and this may to some extent explain the degree of variability seen in the soil-to-plant transfer factors for these groups of nuclides.

While Cs and Sr are important because of their global distributions and because of their ability to mimic biologically essential elements, ^{99}Tc is of special radioecological interest because of its extraordinary mobility and ability to concentrate in plants despite being an 'artificial' element (Wildung *et al.*, 1977) (an excellent review of the environmental behaviour of this element is given by the collection of papers edited by Desmet and Myttenaere, 1986). The element Tc is considered to be 'extinct' on earth; its introduction into our environment over the last 50 years has provided a particularly interesting problem for radioecologists because, as well as being a novel element in the sense that biological systems have had no previous contact with it during their evolution, it has no stable isotopes. Two radionuclides of Tc exist: ^{99m}Tc, an activation product, is relatively unimportant as an environmental contaminant because of its short (~6 h) half life, whereas ^{99}Tc, a fission product of ^{238}U, is long lived (~200000 years). The environmental behaviour of this element is complex as a result of its existence in several oxidation states.

Under oxic conditions Tc^{7+} predominates and the most stable chemical species of ^{99}Tc is TcO_4^-; in well aerated soils, therefore, the pertechnetate anion is considered to be the most important chemical form of Tc (Sparkes and Long, 1988). In a similar manner to the NO_3^- ion, TcO_4^- is highly mobile in the mainly negatively charged soil environment and incorporation of this anion by plant roots can be very considerable (Wildung *et al.*, 1977; see also Table 7-14). Over the longer term, however, reduction of ^{99}Tc may occur, especially in association with microorganisms in anaerobic soil microaggregates, but also as a result of complexation with organic matter. Unfortunately, the reaction kinetics involved are so slow that it has been estimated (Van Loon, 1986) that 90% immobilisation of ^{99}Tc in a contaminated soil may take as long as 30 to 40 years (this compares with an immobilisation time for radiocaesium in most agricultural soils of less than one year). It has been shown that even over one cropping season the vast majority of ^{99}Tc within a soil will be incorporated by wheat and other crops (Grogan *et al.*, 1987), so the principal means of loss of this radionuclide from soils in agricultural ecosystems is likely to be the removal of harvested crop tissues.

Despite the extensive use of ^{99m}Tc as a tracer in medical studies the plant physiological behaviour of Tc has only recently been elucidated. Lembrechts *et al.* (1985) determined that free TcO_4^- occurs within the cells of spinach leaves, implying that Tc is transported across the plasmalemma and into the cytosol in this form (as a result the root uptake of Tc is subject to a considerable degree of antagonism from NO_3^- and other anions – Van Loon, 1986). Reduction of Tc^{7+}, however, appears to take place within the cell and Lembrechts *et al.* (1985) also determined that up to ten discrete Tc bio-organic complexes were present in spinach leaf homogenates, 80 to 90% of which had molecular weights of <6 kD. In a further study Lembrechts and Desmet (1986) argued that the reduction of TcO_4^- to Tc^{5+} is mediated by ferredoxin within chloroplasts and that the incorporation of Tc within plant tissues was therefore effected as the result of photosynthetic reduction. Stable Tc^{5+} complexes appear to be formed *in vivo* by ligand exchange with complexing agents such as thiol compounds (Lembrechts and Desmet, 1989); having been incorporated by plant tissues, however, the bioavailability of Tc to animals *via* absorption in the gastrointestinal tract appears to be reduced (Garten *et al.*, 1986).

While present knowledge of the ion absorption and metabolism of individual radionuclides such as ^{137}Cs, ^{90}Sr and ^{99}Tc is considerable there is still a major requirement for a reliable generic approach to the quantification of plant uptake of radionuclides based on mechanistic principles which can be applied equally to nuclides of widely differing chemical and physiological characteristics. Models such as SPADE (Thorne and Coughtrey, 1983), ECOSYS (Pröhl *et al.*, 1988) and RUINS (Crout *et al.*, 1990) have made important headway on this problem in the last decade, recognising some of the complex features of the soil-to-plant transfer of radionuclides which are omitted or forgotten in a simplistic transfer factor approach. A mechanistic model for soil-to-plant transfer of radiocaesium in grassland, based on the soil chemical behaviour of the nuclide, was produced by Kirk and Staunton (1989) in the wake of the Chernobyl accident but this does not account for plant growth or seasonal physiological effects and apparently remains untested. Desmet *et al.* (1991) have recently highlighted the importance of both speciation of radionuclides within the soil to their bioavailability to the plant and the effects of growth rate on the build up of radioactivity in plant tissues across a growing season. Both of these factors need to be addressed in future if a 'definitive' model for radionuclide accumulation is to be produced.

Lower Plants and Radionuclides

While much emphasis is placed by radioecologists on the contamination of agricultural produce with radionuclides there are large areas of the Earth's surface which are not exploited for the intensive culture of crops and livestock but which, nevertheless, represent important sources of foodstuffs for many animals and some human populations. An example of the importance of such areas in terms of contamination by radioactive materials was given by Meyerhof and Marshall (1990) who pointed out that in Canada approximately 90% of the land mass is non-agricultural yet is highly important as a nutritive source for both indigenous and non-indigenous peoples. Such high latitude areas are often dominated by bryophytes and lichens which, for various reasons, appear to be particularly susceptible to radio-contamination. These lower plants have been found after the Chernobyl accident, for instance, to accumulate radiocaesium to a greater degree than most higher plants (Livens *et al.*, 1991).

Bryophytes (mosses and liverworts) have evolved particularly efficient mechanisms for absorbing metals from environmental sources, an adaptation which facilitates their existence in nutrient limiting situations. The bulk of metallic pollutants collected by mosses tend to be concentrated extracellullarly by both ion exchange and particulate trapping (Richardson, 1981). An example is that of *Dicranum scoparium*, a moss species with minutely hairy leaves, which has been found to be particularly effective in trapping radioactively-tagged fly ash and uranium-containing particulates around a milling complex in Ontario, Canada (Richardson *et al.*, 1979). Similarly, Whitehead and Brookes (1969) observed that the liverwort, *Marchantia berteroana*, exhibited a relatively high enrichment coefficient for alpha activity in a uraniferous area of New Zealand and that, as a general rule, this enrichment coefficient was inversely proportional to the evolutionary status of the plant species examined (Table 7-15).

Lichens are also well known to be highly efficient natural collectors of atmospheric pollutants, including metal ions (Nieboer *et al.*, 1978) and radionuclides (Gorham, 1959). Smith and Ellis (1990) calculated V_g values for radionuclide deposition to lichens in Canada from the Chernobyl plume of 1.1 cm s^{-1} which appear to be no higher than those determined for deposition to other surfaces including grass (see Table 7-4). Once contaminated, however, lichens appear to lose their burden of contamination rather slowly. It has previously been mentioned in this chapter that the normalised specific activity of fallout ^{90}Sr and ^{137}Cs is observed to be up to 25 times

Table 7-15. Accumulation of alpha activity by various classes of plants in a uraniferous area of New Zealand, after Whitehead and Brooks (1969).

Class	Species	Alpha counts min^{-1} g^{-1} sample		
		Plant	Soil	Plant/soil
Bryophytes	*Marchantia berteroana*	1421	663	2.1
Ferns	*Blechnum capense*	939	231	4.1
	Dicksonia lanata	921	382	2.4
Monocotyledons	*Cordyline banksii*	95	87	1.1
	Uncinia teptostachya	3817	1732	2.2
Dicotyledons	*Carpodetus serratus*	149	585	0.2
	Coprosoma arborea	24	3001	0.1

higher in the Arctic than in temperate latitudes, a fact which can be ascribed to slow loss rates of activity from the cryptogams which are common in these regions (Table 7-3). Meyerhof and Marshall (1990) cite the specific example of the lichen *Alectoria nigricans* which in 1980 exhibited a remarkably high peak in ^{137}Cs activity at a latitude of 65° N. The origin of this nuclide was fallout from the pre-1963 atmospheric weapon tests which in more southerly agricultural ecosystems in Canada was quickly rendered unavailable to crops, even though the original levels of contamination were heavier further south. ^{137}Cs activity associated with lichens at all latitudes appears to fall slowly. A study around New Brunswick in Canada by Ellis and Smith (1987) indicated that the biological residence times within lichens were of the order of 1 to 2 years for ^{210}Pb and ^{210}Po (two naturally-occurring members of the ^{238}U decay series) and 5 to 8 years for ^{137}Cs. These values compare with half lives of field loss for agricultural plant species which are commonly of the order of a few days (see Table 7-8). Interestingly, Ellis and Smith found an inverse ratio between ^{40}K and ^{137}Cs levels in lichen tissues, suggesting that biological incorporation of Cs occurs and is subject to the same ion antagonism effects evident in higher plant species. Mattsson (1975) noted that ^{137}Cs and ^{144}Ce were both strongly retained at the surface of the lichen thallus, whereas the heavier and more particle reactive ^{210}Pb and Pu isotopes penetrated the thallus more completely.

Following the Chernobyl accident semi-natural ecosystems have been particularly badly affected with one of the worst cases in western Europe being the contamination of large areas of northern Sweden and Norway inhabited by the Sami people. The traditional diet of these people includes reindeer meat derived from animals whose own diet consists in large measure of lichen species (including reindeer 'moss', *Cladonia rangiferina*) which, since the summer of 1986, have been severely contaminated with ^{137}Cs. Due to the low rates of biological loss of this nuclide it is estimated that the after-effects of Chernobyl-derived fallout in the non-cultivated areas occupied by these indigenous peoples are likely to persist for many years to come (Hove and Strand, 1990).

Fungi too have been observed to be particularly efficient in concentrating radionuclides following the Chernobyl accident (see, for instance, Mascanzoni, 1990) with some Basidomycete sporocarps in Byelorussia accumulating up to 500 kBq ^{137}Cs kg^{-1} fresh matter (Guillitte and Willdrodt, 1993). Radiostrontium does not appear to be concentrated by fungi to the same degree as radiocaesium (Horyna and Randa, 1988) but considerable activities of radioactive silver (^{110m}Ag) have also been observed in fungal tissues following the Chernobyl accident (Battiston *et al.*, 1989). Transfer of radioactivity to man from contaminated mushrooms can either be direct (the practice of mycophagy is very widespread in many countries) or *via* the consumption of the tissues or milk of grazing animals, many of which selectively graze the fruiting bodies of soil-dwelling fungi (Hove *et al.*, 1990). It has been estimated that the ratio of radiocaesium activity in fungi to that in green plants is so high that consumption of a very small amount of fungal matter in the diet of ruminants may result in the introduction of a considerable radioactive body burden (Bakken and Olsen, 1990). It has also been estimated that for humans the consumption of a similarly small amount of fungal matter may result in a highly significant contribution to radiological dose due to the extremely high specific activities of fungal tissues in some areas of both eastern and western Europe (Guillitte and Willdrodt, 1993).

Many of the Basidiomycete fungi found to be efficient accumulators of radiocaesium are able to form mycorrhizas with higher plant species (Dighton and Horrill, 1988; Olsen *et al.*, 1990). Indeed, it has been determined that mycorrhizal fungi are much more prone to radiocaesium accumulation than are saprophytic fungi (Römmelt *et al.*, 1990) and this gives rise to the question of whether transmission of the high levels of radioactivity within the fungal tissue may occur into the host plant. Rogers and Williams (1986) determined that the uptake of ^{137}Cs and ^{60}Co by yellow sweet clover and Sudan grass was enhanced by vesicular-arbuscular mycorrhizal (VAM) infection and McGraw *et al.* (1979) derived a similar result for ^{137}Cs uptake by a VAM-infected tropical grass species. More recently, however, studies by Clint *et al.* (1991) and Clint and Dighton (1992) have shown that the likely patterns of uptake and translocation of radiocaesium as affected by mycorrhizal infections of various types are considerably more complex than the latter two studies would suggest. A particularly important question in this respect is the relative discrimination by individual fungal species between Cs and K compared with that by the host plant; Olsen *et al.* (1990) have reported preliminary findings which suggest that at least one species (*Cenococcum graniforme*) has the ability to discriminate against the K^+ ion in favour of the Cs^+ ion to the extent that the $OR_{Cs/K} \approx 20$. This, in part, may explain the extraordinary ability of some fungal species to concentrate radionuclides from contaminated soils, yet it is clear that much is still to be learned about this aspect of radioecology.

References

Åarkrog, A. (1983). in: *Ecological Aspects of Radionuclide Release*, Special publication series of the British Ecological Society, No 3: Coughtrey, P. J., Bell, J. N. B., Roberts, T. M. (eds.) Oxford: Blackwell Scientific Publications.

Baker, A. J. M. (1981). *J Plant Nutrition, 3*, 643–654.

Bakken, L. R., Olsen, R. A. (1990). in: *Transfer of Radionuclides in Natural and Semi-natural Environments*: Desmet, G. M., Nassimbeni, P., Belli, M. (eds.) London: Elsevier Applied Science; pp. 664–668.

Battiston, G. A., Degetto, S., Gerbasi, T., Sbrignadello, G. (1989). *J Environ Radioactivity, 9(1)*, 53–60.

Bell, J. N. B., Minski, M. J., Grogan, H. A. (1988). *Soil Use and Management, 4(3)*, 76–84.

Bigeleisen, J. (1949). *J Chem Phys, 17*, 675–678.

Bonka, H., Horn, M. (1983). in: *Transfer of Radioactive Materials in the Terrestrial Environment Subsequent to an Accidental Release to Atmosphere, CEC, Dublin, 11–15 April 1983*, pp. 41–77.

Bowen, H. J. M. (1979). *Environmental Chemistry of the Elements*, London: Academic Press.

Brooks, R. R. (1972). *Geobotany and Biogeochemistry in Mineral Exploitation*, New York: Harper and Row.

Brown, S. L., Bell, J. N. B., Minski, M. J., McNeill, S. (1991). Summary report for NERC Special Topic on Environmental Radioactivity, February 1991. Imperial College of Science, Technology and Medicine, UK.

Bukovac, M. J., Wittwer, S. H., Tukey, H. B. (1965). in: *Radioactive Fallout, Soil, Plants, Foods, Man*: Fowler, E. (Ed.) pp. 82–109.

Bulman, R. A. (1983). in: *Ecological Aspects of Radionuclide Release*, Special publication series of the British Ecological Society, N°. 3: Coughtrey, P. J., Bell, J. N. B., Roberts, T. M., (eds.) Oxford: Blackwell Scientific Publications; pp. 105–113.

Bulman, R. A., Cooper, J. R. (1986). *Speciation of Fission and Activation Products in the Environment*, London: Elsevier Applied Science Publishers.

Cambray, R. S., Cawse, P. A., Garland, J. A., Gibson, J. A. B., Johnson, P., Lewis, G. N. J., Newton, D., Salmon, L., Wade, B. O. (1987). *Nuclear Energy, 26(6)*, 77–101.

Cataldo, D. A., Garland, T. R., Wildung, R. E. (1981). *J Environmental Quality, 10(1)*, 31–37.

Chadwick, L., Chamberlain, A. C. (1970). *Atmospheric Environment, 4*, 51–56.

Chamberlain, A. C. (1953). Report HP/R 1261, AERE, Harwell, UK.

Chamberlain, A. C. (1970) *Atmospheric Environment, 4*, 57–78.

Chamberlain, A. C. (1975). in: *Vegetation and the Atmosphere, Volume I*: Monteith, J. L., (ed.) London: Academic Press.

Chamberlain, A. C. (1991). *Radioactive Aerosols*, Cambridge: Cambridge University Press.

Chamberlain, A. C., Little, P. (1980). in: *Plants and their Atmospheric Environment*: Grace, J., Ford, E. D., Jarvis, P. (eds.) Oxford: Blackwell; pp. 147–173.

Cline, J. F., Hungate, F. P. (1960). *Plant Physiology, 35*, 605–608.

Clint, G. M., Dighton, J. (1992). *New Phytologist 121*, 551–561.

Clint, G. M., Dighton, J., Rees, S. (1991). *Mycological Research, 95*, 1047–1051.

Comar, C. L., Russell, R. S., Wasserman, R. H. (1957). *Science, 126*, 428–435.

Coughtrey, P. J., Kirton, J. A., Mitchell, N. G. (1990). in: *Proceeding of Seminar on Methods and Codes for Assessing Off-site Consequences of Nuclear Accidents: Brussels*; CEC: 7–11 May 1990, Athens: EUR 13013.

Cremers, A., Elsen, A., De Preter, P., Maes, A. (1988). *Nature, 335*, 247–249.

Crout, N. M. J., Beresford, N. A., Howard, B. J., Unsworth, M. H. (1990). in: *Transfer of Radionuclides in Natural and Semi-natural Environments*: Desmet, G. M., Nassimbeni, P., Belli, M. (eds.) London: Elsevier Applied Science; p. 519.

Dahlman, R. C., Bondietti, E. A., Eyman, L. D. (1976). in: *Actinides in the Environment*, ACS Symposium Series No 35: Friedman, A. M. (ed.) Washington, DC: American Chemical Society.

Desmet, G. M., Myttenaere, C. (1986). *Technetium in the Environment,* London: Elsevier Applied Sciences.

Desmet, G. M., Van Loon, L. R., Howard, B. J. (1991). *Science of the Total Environment, 100*, 105 – 124.

Dighton, J., Horrill, A. D. (1988). *Trans Brit Mycol Soc, 91*, 335 – 357.

Dreicer, M., Hackonson, T. E., White, G. C., Whicker, F. W. (1984). *Health Physics, 46*, 177 – 187.

Ellis, K. M., Smith, J. N. (1987). *J Environ Radioactivity, 5(3)*, 185 – 208.

Finston, H. L., Kinsley, M. T. (1960). National Academy of Sciences, Nuclear Science Series 3035: Washington, DC: US Atomic Energy Commission.

Francis, C. W. (1978). *Radiostrontium movement in soils and uptake in plant.* Technical Information Center, US Dept. of Energy.

Friedman, A. M. (ed.) (1976). *Actinides in the Environment*, ACS Symposium Series No 35: Washington, DC: American Chemical Society.

Frissel, M. J., Nordijk, H., Van Bergeijk, K. E. (1990). in: *Transfer of Radionuclides in Natural and Semi-natural Environments*: Desmet, G. M., Nassimbeni, P., Belli, M. (eds.) London: Elsevier Applied Science; p. 40.

Garten, C. T., Myttenaere, C., Vandecasteele, C. M., Kirchmann, R., van Bruwaene, R. (1986). in: *Technetium in the Environment, Proceedings of a CEC Seminar*: Desmet, G., Myttenaere, C. (eds.) London: Elsevier Applied Science Publishers.

Gilbert, R. O., Simpson, J. C. (1983). *Trans-Stat, Statistics for Environmental Studies*, No 23. Battelle Memorial Institute, Pacific Northwest Laboratory, Richland, Washington, USA.

Gorham, E. (1959). *Can J Botany, 37*, 327 – 329.

Grogan, H. A. (1984). *PhD Thesis*, University of London.

Grogan, H. A., Mitchell, N. G., Minski, M. J., Bell, J. N. B. (1987). in: *Pollutant Transport and Fate in Ecosystems*: Coughtrey, P. J., Martin, M. H., Unsworth, M. H. (eds.) Oxford: Blackwell Scientific Publications; pp. 353 – 370.

Guillitte, O., Willdrodt, C. (1993). *Science of the Total Environment*, (in press).

Horyna, J., Randa, Z. (1988). *J Radional Nucl Chem Letters, 127(2)*, 107 – 120.

Hove, K., Strand, P. (1990). in: *Environmental Contamination Following a Major Nuclear Accident, Proceedings of an International Atomic Energy Agency Conference. IAEA-SM-306/39, Vol. 2, 1990c*: Flitton, S., Katz, E. W. (eds.) Vienna: IAEA; pp. 181 – 189.

Hove, K., Pederson, O., Garmo, T. H., Solheim Hansen, H., Staaland, H. (1990). *Health Physics, 59*, 189 – 192.

IUR (1982). *Report on a Workshop on the Measurement of Soil to Plant Transfer Factors for Radionuclides. Part I*, Wageningen: International Union of Radioecologists, EURATOM-ITAL.

IUR (1989). *VIth Report of the Working Group on Soil-to-Plant Transfer Factors*, 24 – 25 May 1989, Grimselpass, Switzerland: International Union of Radioecologists; p. 65.

Jones, H. E., Harrison, A. F., Poskitt, J. M., Roberts, J. D., Clint, G. (1991). *J Environmental Radioactivity, 14(4)*, 279 – 294.

Kirk, G. J. D., Staunton, S. (1989). *J Soil Sci 40(1)*, 71 – 84.

Kluczewski, S. M., Bell, J. N. B., Brown, K. A., Minski, M. J. (1983). in: *Ecological Aspects of Radionuclide Release*, Special publication series of the British Ecological Society, No 3: Coughtrey, P. J., Bell, J. N. B., Roberts, T. M., (eds.) Oxford: Blackwell Scientific Publications, pp. 91–104.

Kluczewski, S. M., Brown, K. A., Bell, J. N. B. (1985). *Atmospheric Environment, 19*, 1295–1299.

Lembrechts, J. F., Desmet, G. M. (1986). *Plant Physiology, 81*, 1003–1007.

Lembrechts, J. F., Desmet, G. M. (1989). *Health Physics, 57(2)*, 255–262.

Lembrechts, J. F., Desmet, G. M., Overbeek, H. (1985). *Environmental and Experimental Botany, 25(4)*, 355–360.

Lembrechts, J. F., Van Ginkel, J. H., Desmet, G. M. (1990). *Plant and Soil, 125*, 63–69.

Livens, F. R., Rimmer, D. L. (1988). *Soil Use and Management, 4(3)*, 63–69.

Livens, F. R., Horrill, A. D., Singleton, D. L. (1991). *Health Physics, 60(4)*, 539–545.

Mascanzoni, D. (1990). in: *Transfer of Radionuclides in Natural and Semi-natural Environment*: Desmet, G. M., Nassimbeni, P., Belli, M. (eds.) London: Elsevier Applied Science; pp. 459–467.

Mattsson, L. J. S. (1975). *Health Physics, 29*, 27–41.

McGraw, A. C., Gamble, J. F., Schenk, N. C. (1979). *Phytopathology, 69*, 1038–1041.

Mewhinney, J. A., Eidson, A. F., Wong, V. A. (1987). *Health Physics, 53(4)*, 377–384.

Meyerhof, D., Marshall, H. (1990). in: *Transfer of Radionuclides in Natural and Semi-natural Environments*: Desmet, G. M., Nassimbeni, P., Belli, M. (eds.) London: Elsevier Applied Science; p. 48–55.

Middleton, L. J. (1959). *Int J Rad Biol 1*, 387–402.

Miller, C. W., Hoffman, F. O. (1983). *Health Physics, 45(3)*, 731–744.

Moorby, J., Squire, H. M. (1963). *Radiation Botany, 3*, 163–167.

Mortensen, J. L., Marcusiu, E. C. (1963). *Soil Sci Soc Am Proc 27*, 653–656.

Myttenaere, C., Masset, M. (1965). *Physiol Veg 7(2)*, 97–104.

Nieboer, E., Richardson, D. H. S., Tomassini, F. D. (1978). *Bryologist, 81*, 226–246.

Nisbet, A. F. (1993). *Science of the Total Environment*, (in press).

Olsen, R. A., Joner, E., Bakken, L. R. (1990). in: *Transfer of Radionuclides in Natural and Semi-natural Environments*: Desmet, G. M., Nassimbeni, P., Belli, M. (eds.) London: Elsevier Applied Science; pp. 657–663.

Pröhl, G., Müller, H., Jacob, P., Paretzke, H. G. (1988). in: *Proceedings IVth International Symposium on Radioecology, the Impact of Nuclear Origin Accidents on the Environment, Cadarache, France: 14–19 March 1988.*

Richardson, D. H. S. (1981). *The Biology of Mosses*, Oxford: Blackwell Scientific Publications.

Richardson, D. H. S., Nieboer, E., Beckett, P. J., Boileau, L. Lavoie, P., Padovan, D. (1979). *The Levels of Uranium and Other Elements in Lichens and Mosses Growing in the Elliot Lake and Agnew Lake Areas, Ontario, Canada.* Report to Ontario Ministry of Environment, Government of Ontario.

Röbig, G. B. (1979). Dissertation, Justus-Liebig-Universität Gießen, Germany.

Rogers, R. D., Williams, S. E. (1986). *Soil Biology and Biochemistry, 18(4)*, 371–376.

Römmelt, R., Hiersche, L. Schaller, G., Wirth, E. (1990). in: *Transfer of Radionuclides in Natural and Semi-natural Environments*: Desmet, G. M., Massimbeni, P., Belli, M., (eds.) London: Elsevier Applied Science; p. 152–160.

Shaw, G. (1993). *Science of the Total Environment, 32(4)* 391–401.

Shaw, G., Bell, J. N. B. (1989). *J Environmental Radioactivity, 10*, 213–232.

Shaw, G., Bell, J. N. B. (1991). *J Environmental Radioactivity, 13*, 283–296.

Shaw, G., Minski, M. J., Bell, J. N. B. (1992). *Environmental and Experimental Botany, 32(4)*, 391–401.

Shaw, G. Hewamanna, R., Lillywhite, J., Bell, J. N. B. (1992). *J Environmental Radioactivity, 16*, 167–180.

Sheppard, S. C., Evenden, W. G. (1990). *J Environ Radioactivity, 11(1)*, 15–36.

Smith, J. N., Ellis, K. M. (1990). *J Environ Radioactivity, 11(2)*, 151–168.

Sparkes, S. T., Long, S. E. (1988). *The Chemical Speciation of Technetium in the Environment: a Literature Survey.* Harwell Report, AERE-R 12743 (DOE/RW/88098), Harwell, UK.

Tester, M. (1988). *J Membrane Biol, 105*, 87–94.

Thorne, M. C., Coughtrey, P. J. (1983). in: *Ecological Aspects of Radionuclide Release*, Special publication series of the British Ecological Society, No 3: Coughtrey, P. J., Bell, J. N. B., Roberts, T. M., (eds.) Oxford: Blackwell Scientific Publications.

Tukey, H. B., Wittwer, S. H., Tukey, H. B., Jr (1958). in: *Radioisotopes in Scientific Research, Volume IV – Research with Radioisotopes in Plant Biology and Some General Problems*: Oxford: Pergamon Press; pp. 304–322.

Van Loon, L. (1986). Doctoraatproefschrift Nr. 150 aan de Fakulteit der Landbouwwetenschappen van de Katholieke Universiteit te Leuven, Belgium.

Ward, G. M., Johnson, J. E. (1986). *Health Physics, 50(3)*, 411–414.

Ward, G. M., Johnson, J. E., Stewart, H. F. (1965). in: *Proceedings 2nd AEC Symposium on Fallout*: Kelement, A. Jr. (ed.) Oak Ridge, Tennessee: National Technical Information Center.

Watterson, J. W. (1989). *PhD Thesis*, University of London.

Whicker, F. W. (1983). *Radiation Research, 94*, 135–150.

Whicker, F. W., Schultz, V. (1982). *Radioecology: Nuclear Energy and the Environment Volumes I and II*, Boca Raton, Florida: CRC Press.

Whitehead, N. E., Brooks, R. R. (1969). *J Applied Ecol 6*, 301–310.

Wildung, R. E., Garland, T. R., Cataldo, D. (1977). *Health Physics, 32*, 314–317.

8 Plants as Indicators of Mineralisation and Pollution

Margaret E. Farago

Environmental Geochemistry Research
Royal School of Mines
Imperial College of Science, Technology and Medicine
Prince Consort Road, London SW7 2BP, UK

Contents

Mineral Exploration

Ecological Exploration – Geobotany

Introduction

Early mineral prospectors ecologically used the plants associated with metalliferous soils in their search for mineral deposits (Cole, 1980). This early use of plant distributions identified indicator plants, which are diagnostic of particular environmental conditions and some of which are cited in old texts (Henwood, 1857; Ramsay, 1881; Le Neve Foster, 1894). Some history of geobotanical prospecting is given by Ernst in his recent review (Ernst, 1993). Ernst has pointed out that it was Agricola, in 1556, who advised prospectors to look for abnormalities of plants or vegetation. In 1588 Thalius described a number of indicator species including *Minuartia verna*. Henwood (1857) reported the growth of *Armeria maritima* in a copper-enriched bog in Wales, recognising it as a Cu indicator plant. Figure 8-1 shows both *Minuartia verna* and *Armeria maritima* growing along a drainage stream from the bog where the soils are rich in copper.

Fig. 8-1. *Minuartia verna* and *Armeria maritima* growing along a drainage stream from the copper-rich Dolfrwynog Bog, Wales.

Cole (1980) has emphasised that geobotany for mineral prospecting is concerned with the relationships between plants and their environment in undisturbed natural terrain. It is distinct from ecological studies of disturbed contaminated land, for example, old mine workings, where metal concentrations are usually greatly in excess of those found naturally.

Vegetation has been classified into a number of reaction types (Ernst, 1993). Within the group of plants that react to their environment (reactors), those with visible (or overt) reactions can be distinguished from those with non-visible (or covert) reactions. In the case of geobotany as defined by Ernst (1993), visibility means to the unaided human eye, that is colour, morphology, presence, abundance or absence of species. Cole has pointed out (Cole and Smith, 1984) that is was not until about 1945, when rapid routine methods of analysis for large numbers of samples became available, that biogeochemistry, that is the study of the chemical composition of plants from various habitats, was used for mineral exploration in many parts of the world.

There is still some disagreement on the terms used: Ernst (1993) restricts the use of the term geobotany as defined above, whereas Brooks (1993) includes as a sub-field of geobotany, the remote sensing of vegetation cover by aerial photography or satellite imagery, which is treated by Ernst under the heading of biogeochemical prospecting. In this chapter biogeochemistry will be used as by Cole (1980) and Brooks (1993), ie, where there is chemical analysis of plant tissue. A number of books and reviews have been written on geobotanical and biogeochemical methods of mineral exploration (Cannon, 1960a; Malyuga, 1964; Brooks, 1972, 1983, 1993; Ernst, 1974, 1993; Kovalevsky, 1979, 1987; Cole, 1980).

Disclosure of Orebodies by Vegetation

Vegetation may reveal the presence of orebodies in a number of ways (Cole, 1980), these include:

1. anomalous communities of indicator plants specific to one or more elements;
2. 'out of context vegetation', that can be commonly occurring species;
3. changes in the relative abundance of characteristic species in the vegetation;
4. unusual physiognomic and/or morphological features;
5. symptoms of physiological disorders;
6. abnormal amounts of elements in the plant tissues (biogeochemistry).

Cole (1980) has suggested that anomalous plant communities are most likely to occur in areas where summer rainfall with optimum growth conditions is followed by winter drought. As the soil moisture decreases, the concentrations of mineral elements in the soil becomes critical. Anomalous plant communities can also be found in areas affected by mineralised waters, emanating from distant orebodies, occupying for example seepage zones over barren bedrock, or near springs at the margins of bogs (Fig. 8-1).

Biogeochemistry

Introduction

Biogeochemistry involves the analysis of plants for target elements, in order to locate areas of high mineralisation which can be reflected in high uptake by plants, and hence higher than average concentrations in the plant tissues. Peterson (1971, 1983) and other authors have suggested that plants growing on toxic metalliferous soils cannot prevent metal uptake but only restrict it and hence accumulate metals in their tissues to varying degrees. Therefore, the strategies of survival are tolerance and avoidance of metal toxicity. Plants differ considerably in their uptake characterisitics and for any species these may vary for different metals (Ernst, 1975; Shaw, 1989; Baker, 1981).

Concentrations of Metals in Plants as Responses to Metals in Soils

Baker (1981) has discussed three general types of plants, where in each the concentration of any particular element in the soil is reflected in the plant in a different way (Fig. 8-2). The concentration of the element in the plant part (Baker considered the aerial portions) is compared with that in the soil, and thus concentration ratios can be obtained, these ratios have been called Concentration Factors (*CF*) (Chamberlain, 1983).

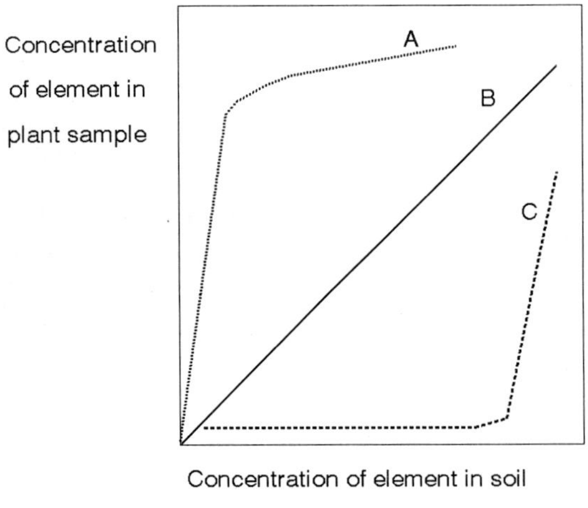

Fig. 8-2. Response of plants to increasing soil metal concentrations reflected by the metal concentrations in the aerial parts of the plants: A, Accumulator; B, Indicator; C, Excluder. (Adapted from Baker, 1981).

Accumulators – where the element is concentrated in the aerial portions of the plant, and is taken up from soils with high or low concentrations. The ratio of the concentration of the element in the plant to that in the soil is greater than 1. The concentration of the element in the plant tissues rises steeply as the soil concentration increases, until all sites within the plant are filled; it then levels off. The *CF* values reach a maximum and then decrease.

Indicators – where the uptake and transport of the element is regulated so that its concentration in the plant reflects that in the soil. Plant/soil concentration ratios are near 1.

Excluders – where the element concentrations in the aerial parts are maintained low and constant over a wide range of soil concentrations, up to a critical value, above which the exclusion mechanism breaks down. This results in unrestricted transport and toxicity. Plant soil concentration ratios are much less than 1.

The term *indicator plant* has been used somewhat differently. Qualitatively, indicator plants are diagnostic of certain environmental conditions, and these can fall into any of Baker's three categories. To avoid confusion, it has been suggested (Farago and Mehra, 1992) that the term *concentration indicator* should be used to describe plants where the plant soil concentration ratios are near 1. These three types are represented in Fig. 8-2. Some plants endemic to metalliferous soils acccumulate very high concentrations in their aerial parts and these are known as *hyperaccumulators* (Baker and Brooks, 1989; Chapters 3 and 4). Such plants are able to accumulate metals not only from soils with high metal concentrations, but also from soils with concentrations only marginally above normal.

A single plant species can act as an excluder of some elements and as an indicator or accumulator toward others. This is illustrated by *Triodia pungens* (Fig. 8-3)

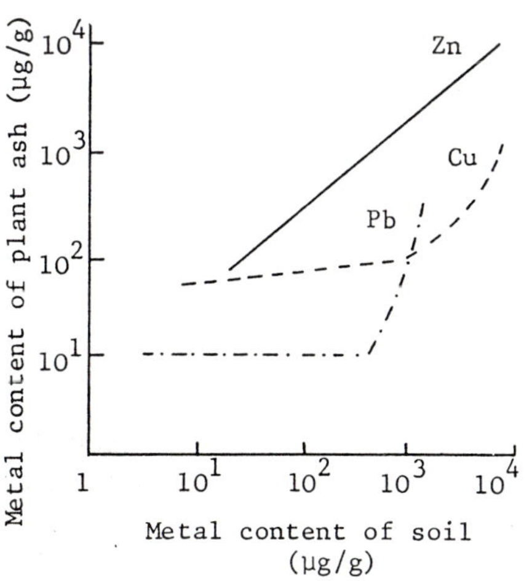

Fig. 8-3. Plant-soil relationship for Zn, Cu and Pb in the plant ash of *Triodia pungens* (from Nicolls *et al.*, 1965; Baker, 1981).

(Nicolls *et al.*, 1965), where, as the concentrations of the metal in the soil increase, Zn is taken up into the aerial parts of the plant in unrestricted quantities, whereas Pb is clearly excluded, and Cu is partially excluded.

Kovalevsky (1979, 1987) has used the term *barrier-free biosamples* for those plants that do not have an exclusion mechanism toward a particular element; whereas those plants that have a limiting value of accumulation in the aerial parts are said to have a 'physiological barrier' which operates to exclude elements from the aerial parts. The situation is illustrated in Fig. 8-4. Of these, barrier-free and high-barrier biosamples would be suitable for biogeochemical exploration. All these types are subdivisions of Baker's (1981) accumulators, and what is described as a physical barrier is more likely to be the filling of the available sites. Brooks (1993) has pointed out that Kovalevsky's nomenclature is confusing.

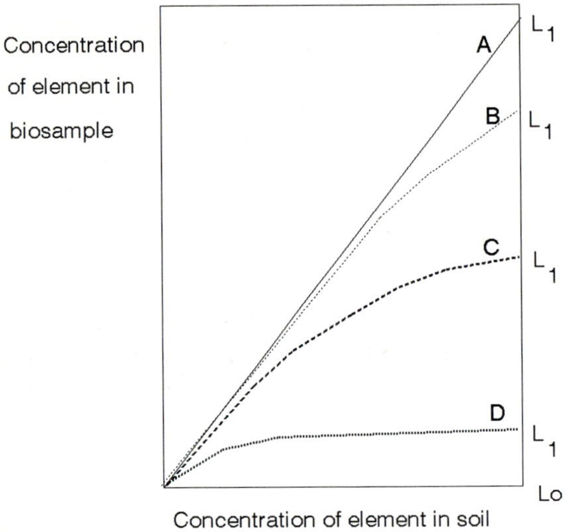

Fig. 8-4. Kovalevsky's 'barrier' concept, showing responses of plants to increasing metal concentrations in soils. L_0 is the background value of the element in the plant tissue; L_1 is the limiting value of the element in the plant tissue. A, $L_1/L_0 > 300$ (barrier-free); B, $L_1/L_0 > 30 < 300$ (high-barrier); C, $L_1/L_0 > 3 < 30$ (medium-barrier); D, $L_1/L_0 > 2 < 5$ (low-barrier) (adapted from Brooks, 1993).

Metal Concentrations in Plant Parts

Not only the aerial parts of the plant, but also the roots can be suitable organs for analysing for mineral prospecting (Malyuga, 1964). However care must be taken to ensure that root material is not contaminated by soil.

Brooks (1973) and Ernst (1993) have discussed the factors affecting the concentrations in various plant organs and their suitability for biogeochemical prospecting. For any particular target element, three sets of ratios of concentrations can be considered:

(i) *root/soil:* this concentration ratio will give information concerning the uptake of the element by the root from the soil, thus suggesting the bioavailability of that element and giving some information as to whether it is accumulated or excluded by the root;
(ii) *leaf/root:* this concentration ratio will show if there is free movement of the elements between root and aerial parts, or if the element is accumulated in either roots or leaves;
(iii) *leaf/soil:* this concentration ratio is the easiest to collect, but will depend on ratios (i) and (ii). It is the ratio most often used in biogeochemical prospecting.

Leaf age affects the metal content, older leaves having higher concentrations (Ernst, 1990a). Farago *et al.* (1980) demonstrated that old brown leaves of the indicator plant *Armeria maritima* contain much higher concentrations of Cu than young green leaves. Outer leaves of rosette plants are oldest and contain highest concentrations whereas in non-rosette plants the oldest are generally at the bottom of the plant. In trees the leaves from the outer canopy accumulate more than those from the inner canopy, because of differences in transpiration rates. The outer bark of the tree generally has higher concentrations than the inner bark and the trunkwood. Twigs have also often been used in metal prospecting.

The root is also a very useful organ (Malyuga, 1964) as it can penetrate through unmineralised overburden. Cannon (1960b) has reported a root of 200 ft (~61 m) for a live root of *Juniperus monosperma* in a mine working on the Colorado Plateau. In Sweden use has been made of organic stream bank material (stream peat) for prospecting (Brundin, 1980). This material consists of an inhomogeneous mixture of living plant roots, humus, and mineral material. Brundin *et al.* (1987) have investigated the use of the roots of two stream sedge plants, *Carex juncella* and *Carex rostrata*, since their roots constitute part of the stream peat. The authors concluded that *Carex juncella* roots are more suitable for biogeochemical prospecting than the aerial parts since when the metal concentrations were high in the growth media these were not reflected in the aerial parts. It was suggested that some exclusion mechanisms prevented very high amounts of metal reaching the leaves. An alternative explanation is that all the storage sites in the leaves become filled, so that there is a maximum loading of the leaves. The root however has many more storage sites in the root cortex (Farago, 1986).

Seasonal Variation of Metal Concentrations in Plants

There is much evidence that the concentrations of elements in most organs, but particularly in the leaves, are subject to seasonal variation. Matthews and Thornton (1982) found variations with both in the content of Cd and associated metals in plants at Shipham, and seasonal variations of lead levels have been found popular leaves in urban areas (Capelli *et al.*, 1991). Similar variations have been found with metal content of leaves from fruit tree species (Deu and Kreeb, 1993). During the fast increase in biomass in the growth spurt in the spring, there is a dilution effect and the concentration falls. Concentrations build up again until autumn. Thus in plants where all the leaves are shed, collection for biogeochemical prospecting should be carried out just before shedding. Brundin *et al.* (1987) have shown that there are seasonal variations in root concentrations, and that these can sometimes be greater than those in the aerial portions. Since the samples measured in this work were from streams banks, many factors may be playing a part, but in particular the amount of precipitation, which can effect the amount of metal available to the roots and the transpiration rates.

Remote Sensing

This more recent development of geobotany (Horler *et al.*, 1981) relies on images of the earth's vegetation from aircraft or satellites. The use of multispectral and thermal imagery allows the detection of distinctive vegetation or plant communities over orebodies. The anomalies can be detected because of the differences in reflectivities of the species of the plant communities. This largely depends on reflectance changes in the region of 0.7 μm (700 nm), the red edge, where there is a large increase in the reflectance (decrease in absorbance) of the chlorophyll in the leaves. There is some evidence that stressed vegetation has an increase in reflectance around this wavelength, the blue shift (Horler *et al.*, 1981; Carlisle *et al.*, 1986). At present there are no sensing techniques capable of the penetration of vegetation and soil, consequently the value of the technique depends on the interpretation of the data and following up with ground truth data.

The blue shift can result from other types of stressed vegetation, in addition to that on mineralised ground, these include stress from drought and from acid rain. Cole *et al.* (1987) have discussed the use of remote sensing for the identification and evaluation of stress in vegetation indicative of drought and/or mineral excesses or deficiencies. Amongst areas chosen for this study was the area of the Ceod-Y-Brenin porphyry copper deposit in Wales. The identification of this deposit followed the characterisation of the anomalous vegetation comprising of *Minuartia verna* and *Armeria maritima* (Fig. 8-1) growing in the Dolfrwynog Bog (Rice and Sharp, 1976; Cole, 1980; Farago *et al.*, 1980; Andrews and Fuge, 1986). The bog is the sump for copper that is brought in by the water that has circulated over the copper porphyry deposit under the surrounding hillsides. The copper content of the bog is very high reaching to more that 100,000 mg kg^{-1}, with pH values as low as 4.5. The natural deciduous forest consisting mainly of sessile oak, *Quercus petraea* (Mattuschka) Liebl., has been almost completely replaced by coniferous plantations, the dominant species being *Picea sitchensis* (Sitka spruce) and *Larix eurolepis* (Hybrid larch), and to a lesser extent *Picea abies* (Norway spruce) and *Pseudotsuga menziesii* (Douglas fir). Trees cannot tolerate the conditions and in the bog and *Armeria maritima* and *Minuartia verna* are the dominant species, with *Calluna vulgaris*, and *Vaccinium myrtillus* present to a lesser extent. By the analysis of data from the LANDSAT multiband scanner and LANDSAT thematic mapper, the bog and anomalous vegetation over the subcropping cupriferous veins can be clearly delineated.

Examples of plants and locations used for the geobotanical and biogeochemical prospecting for metals are given by Brooks (1972, 1983, 1993), Cole and Smith (1984), Cole (1980), Ernst (1974, 1993) and Kovalevsky (1979, 1987).

Plants as Indicators of Pollution

Introduction

Wittig (1993) has discussed the terms indicator, bioindicator, monitoring and biomonitoring and has suggested that indication seems to be 'more spontaneous and active, while monitoring seems to be continuous and passive'. The terms are also used to differentiate between qualitative and quantitative, and further, monitoring can mean continuous observation, while indication can mean on-off measurements. Perhaps it is simplest to biomonitor using bioindicators ie, reserve monitoring as a verb and indicator as a noun.

Effects of Contamination on Plants

Sources of contamination in which excessive concentrations of elements may be brought about have been enumerated by a number of authors (Freedman and Hutchinson, 1981; Ernst and Joosse, 1983; Farago and Cole, 1988; Verkleij, 1993), and may be listed as follows:

1. The exploitation of mineral resources
   Mining activities, ore tailings, spoil tips,
   Metal smelting, blast furnaces, smokes, dusts
2. Other industrial processes
   Electrolysis, cement, paint, tanning, plating, etc.
3. Domestic and industrial waste disposal practices
   Landfills, sewage sludge, incinerators
4. Energy supply
   Coal and petroleum burning, high tension lines, nuclear power plants
5. Agriculture
   Phosphate fertiliser, sewage sludge, pesticides
6. Traffic
   Leaded petrol, tyres, catalysts

Vegetation may reveal the presence of man-made contaminated sites in the same ways as it reveals orebodies:

visible symptoms of physiological disorder (eg, chlorosis);
distinctive indicator vegetation;
high concentrations of metal in tissues;
high tolerance or toxicity indexes (effect parameters);
changes in physiological parameters (biomarkers).

The most widely described effects of metal toxicity in plants are inhibited root growth, depressed shoot and leaf growth, and general chlorosis of the younger leaves (Baker and Walker, 1989). The index of metal tolerance is usually measured by the root elongation method (Wilkins, 1978). Young plants or tillers are grown both in control solution and in solutions containing known concentrations of the metal under test. The growth in the toxic solution is expressed as a percentage of that in the control solutions. Using such methods the tolerance of various species has been demonstrated. Other methods to measure toxicity and tolerance have been suggested. Wong and Bradshaw (1982) have described the ED_{50} method. The ED_{50} is defined as the effective dose of metal that produces 50% of normal root growth (in the control solution) and this method has been used to test toxicities of a number of metals, for example Wong and Bradshaw (1982) found that the order of increasing toxicities (from root growth) toward ryegrass (*Lolium perenne*) was:

$$Cu > Ni > Mn > Pb > Cd > Zn > Al > Hg > Cr > Fe$$

This method has been extended by Schat and ten Bookum (1992), who used EC_{100}, defined as the lowest effective concentration that inhibits root growth completely. These authors suggested that other parameters can also be obtained from the root elongation data: no observable effect, NOEC, and EC_{50}, equivalent to ED_{50}.

Tests on the germination of seeds in the presence of increasing metal concentrations do not appear to differentiate between tolerant and non-tolerant populations (Verkleij, 1993).

Physiological parameters, or biomarkers, might also be routinely used to estimate the stress on plants from contaminated soils. These are strong indications that phytochelatin concentration is one measure of metal-induced stress. Phytochelatins are metal-binding proteins found in vascular plants that are similar to metallothioneins (Rauser, 1987; Jackson and Alloway, 1992). Schat and Kalff (1992) have demonstrated that the phytochelatin concentration in the root apex of *Silene vulgaris* can be used quantitatively to assess the phytotoxicity from copper stress experienced by the plant. Clijsters and coworkers have developed a biological test system for assessment of phytotoxicity of contaminated land based on the changes that phytotoxic stress has on the protein patterns of iso-enzymes of peroxidase (Van Assche and Clijsters, 1990; Vangronsveld and Clijsters, 1992). These points are further discussed in Chapter 6.

Plant Communities Indicative of Metal Contamination

Wittig (1993) has compared different plant groups as bioindicators of metal pollution, and has pointed out that although mosses, lichens and fungi have been used for the monitoring of metals, higher plants have many advantages. However, the most widely used indication of polluted ground has been visible signs, bare patches, chlorotic symptoms and distinctive vegetation.

Plants indicative of mineralised habitats occur on ground contaminated by mining and industrial activities (Ratcliffe, 1974; Gemmel, 1977). Dumps and tips which are the result of mining and industrial activities often have extremely high concentrations of metals, and are usually deficient in macronutrients, so that sometimes the environment is too hostile and the ground remains bare. Thus such areas have a severe effect on the vegetation (Ernst, 1974), which is characterised by endemic species and metallophytes. The effects on vegetation from other forms of contamination may be less pronounced.

The clearest examples of indicator plants revealing old mining activities occur across Europe (Cole and Smith, 1984; Ernst, 1993). There has been mining activity in the United Kingdom since the Bronze age. Many of the methods used were inefficient, resulting in the release of large quantities of metals to the environment (Davies, 1983). In the UK areas of metal-rich soils often reflect this release from mining activities superimposed on natural inputs from bedrock sources and bedrock materials (Thornton, 1980). The most important mining areas were Devon, Cornwall, central and north Wales, the Pennines, and the southern uplands of Scotland. The once numerous mines are almost all abandoned for economic reasons, and in many of these old sites indicator plants, including *Viola lutea, Armeria maritima, Minuartia verna* and *Thlaspi alpestre*, reveal the presence of high concentrations of metals (Smith, 1979; Cole and Smith, 1984). These authors have described in particular the indicator communities in the Carperby area of the Yorkshire Pennines. Here the indicator plants not only occupy the dumps and spoil heaps but outline the extent of metal contamination along the Eller Beck stream in the meadows below the mining area. *Armeria maritima* and *Minuartia verna* in the meadow land indicate that the flood plain of the Beck is contaminated with lead and zinc and could be detrimental to grazing animals. In Derbyshire the indicator plants *Minuartia verna* and *Viola lutea* var. *calaminaria* outline wide areas contaminated by mine spoil. At High Low *Minuartia verna* delineated areas with Pb up to 100000 mg kg^{-1} and Zn up to 50000 mg kg^{-1} in the soil.

In some instances the indicator vegetation may be under threat, in the Netherlands zinc-tolerant vegetation is found in the flood plain of the River Geul (Verkleij, 1993), which is now a conservation area. The vegetation evolved when the river became contaminated from the ores which were transported along the river from the mines in Belgium. Since the closing of these mines in the 1940s the zinc concentrations in the soils have decreased, so that non-tolerant species can now become established and replace the tolerant plants (Cook *et al.*, 1972). Similar situations exist in other parts of Europe.

Plants and Atmospheric Pollution

The effects of gaseous pollutants on both crops and trees has been reviewed by Mansfield and Lucas (1990), and a number of books have appeared on the topic, for example: Unsworth and Ormrod (1982); Koziol and Whatley (1984); Treshaw (1984); Winner *et al.* (1985); Hutchinson and Meema (1987); Mathy (1988); Schulte-Hostede *et al.* (1988).

In general polluted air consists of a mixed gas – aerosol complex containing varying concentrations of sulphur dioxide, SO_2; nitrogen oxides, NO_x; ozone, O_3 and acid aerosols. Smith (1974) has suggested three categories of exposure of plants, in particular forest ecosystems, to atmospheric pollution:

(i) low exposures, which produce no visible harm;
(ii) intermediate exposure, where the obvious symptoms are reduced growth with possible increased sensitivity to disease;
(iii) gross pollution, immediately adjacent to heavily industrialised areas, notably in the vicinity of smelting operations, where plants are subjected to fumes rich in SO_2 and other pollutants together with particulate matter.

Injury to vegetation by photochemical smog in the Los Angeles basin has been a long standing problem, and the most important phytotoxic agent in the smog has been found to be ozone. Extensive ozone damage has been observed in many parts of the USA (Heggestad and Middleton, 1959; Bormann, 1982). It is now accepted that ozone is the most damaging of the air pollutants to vegetation (Ashmore *et al.*, 1987). In recent years it has become apparent that damage affecting forests in Germany has not been caused solely by acid depositions, as was once believed. Ozone concentrations have been shown to reach values that cause damage to trees in the USA, and it has been proposed that ozone and frost damage lead to a predisposition to attack by acid aerosols (Smith, 1986). Ozone toxicity appears as a brown or white flecking on the leaves. There is a wide variation in the sensitivity of species to ozone toxicity and in the sensitivities of cultivars of the same species. The high sensitivity of one variety of the tobacco plant, Bel-W3, has been exploited as biomonitor for ozone, providing a substitute for chemical methods (Ashmore *et al.,* 1987).

The absence of a particular species can also indicate pollution, for example, it has been found that the frequency of Scots Pine (*Pinus sylvestris*) decreased with increasing SO_2 concentrations in Britain and where SO_2 concentrations exceeded 0.076 ppm no trees occurred (Smith, 1986).

Metal Concentrations in Plant Bioindicators of Pollution

More recently plant analysis has been carried out to screen for the accumulation of metals indicative of pollution. Usually metal concentrations are reported in dry plant material. However, it has been suggested (Claussen, 1990) that the concentrations of pollutants are better related to the dry ash of a plant, than to the dry weight, since it eliminates the natural variations of organic matter. Heavy metal concentrations in ash of *Populus nigra* 'Italica' were reported to correlate better with values in air and soil than those in dry plant material.

Wagner (1993) has discussed the selection and standardisation of suitable bioindicator systems. A number of higher plant species have been suggested for use as passive bioindicators. These include: *Populus nigra* 'Italica', the Lombardy poplar (Wagner, 1993), *Pinus sylvestris*, Scots Pine (Laaksovirta *et al.*, 1976; Bargagli, 1993) and *Taraxacum officinale*, dandelion (Kabata-Pendias and Dudka, 1991; Djingova and Kuleff, 1993). Unlike the traditional indicator plants these have widespread availability and occur frequently in both agricultural and urbanised areas. Figure 8-5 shows comparison of screening using *Populus nigra* in different locations. The highest concentrations for zinc, lead and cadmium were found in southeastern Poland

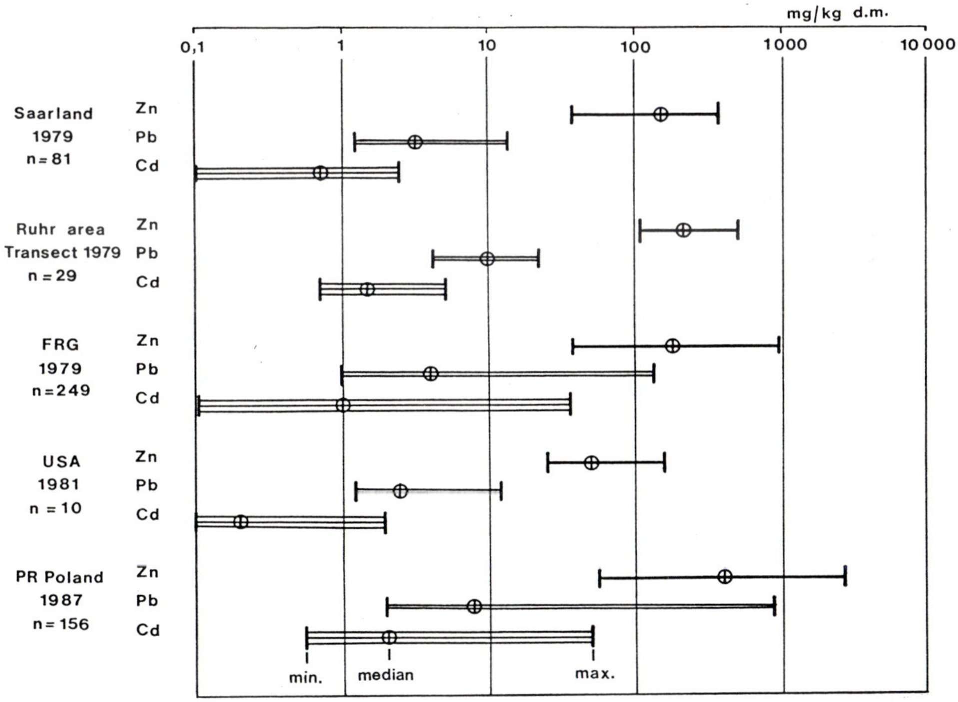

Fig. 8-5. Comparison of results for concentrations of Zn, Pb and Cd in standardised leaves of the Lombardy poplar (*Populus nigra* 'Italica') from different locations and years (from Wagner, 1993, reprinted with permission).

(Wagner, 1993). High concentrations of these elements were also found in *Taraxacum officinale* in SE Poland (Kabata-Pendias and Dudka, 1991; Djingova and Kuleff, 1993). The latter species has been used for the evaluation of pollution from various sources. Analysis of the leaves at a distance of 300 m from different types of smelter, showed that levels of pollution around a Pb–Zn smelter were high. The plant has also been used to compare urban pollution in three European cities (Djingova and Kuleff, 1993). Samples were analysed from Sofia (Bulgaria), Thessaloniki (Greece) and Rotterdam (The Netherlands) (Table 8-1). A correction for the mineral baseline was introduced by using Enrichment Factors (*EF*) (Schubert, 1985), where *EF* is the ratio between the concentration in the polluted region and that in the background (non-polluted) region. For plants collected from Rotterdam all *EF* values are positive indicating that the city is polluted with respect to background, but less so than the other two cities. The high *EF* values for Pb, Cd, Cr, As and Br in Thessaloniki were assocciated with traffic pollution. Sofia has high *EF* values for Hg and Se compared with the other two cities, the main pollution coming from coal combustion.

Table 8-1. Concentrations (mg kg^{-1}) and Enrichment Factors (*EF*) for samples of *Taraxacum officinale* collected in three European cities.

	Sofia		Thessaloniki		Rotterdam	
	mg kg^{-1}	*EF*	mg kg^{-1}	*EF*	mg kg^{-1}	*EF*
As	<0.1	<4	4.87	20	1.5	75
Br	39.3	2.4	178	11	11.6	0.8
Cd	<0.1	<0.3	0.14	4.6	0.6	1.8
Cr	2.3	7.9	3.4	12	5.25	4.7
Cu	21.7	1.5	16	1.1	13.7	1.5
Hg	0.47	2.6	<0.1	–	0.1	–
Pb	16.9	8	30	13	7	5
Sb	0.18	1.5	0.62	5.1	0.32	16
Se	0.18	1.7	<0.1	–	<0.1	–
Zn	129	2.8	51	1.1	45	1.5

from Djingova and Kuleff (1993)

Bargagli (1993) has collected data on the mercury content of pine needles from various locations (Fig. 8-6) and has suggested that Hg accumulates irreversibly in conifer leaves and that the measured concentration depends not only on exposure time but also on the environmental diffusion of the metal.

Use has also been made of the metal content of crop plants in the assessment of contaminated soils. Kabata-Pendias *et al.* (1993) suggest that legumes are promising as bioindicators of metal pollution since they have in general a relatively higher tolerance to and uptake of metal than monocotyledons. Kovács *et al.* (1993) have

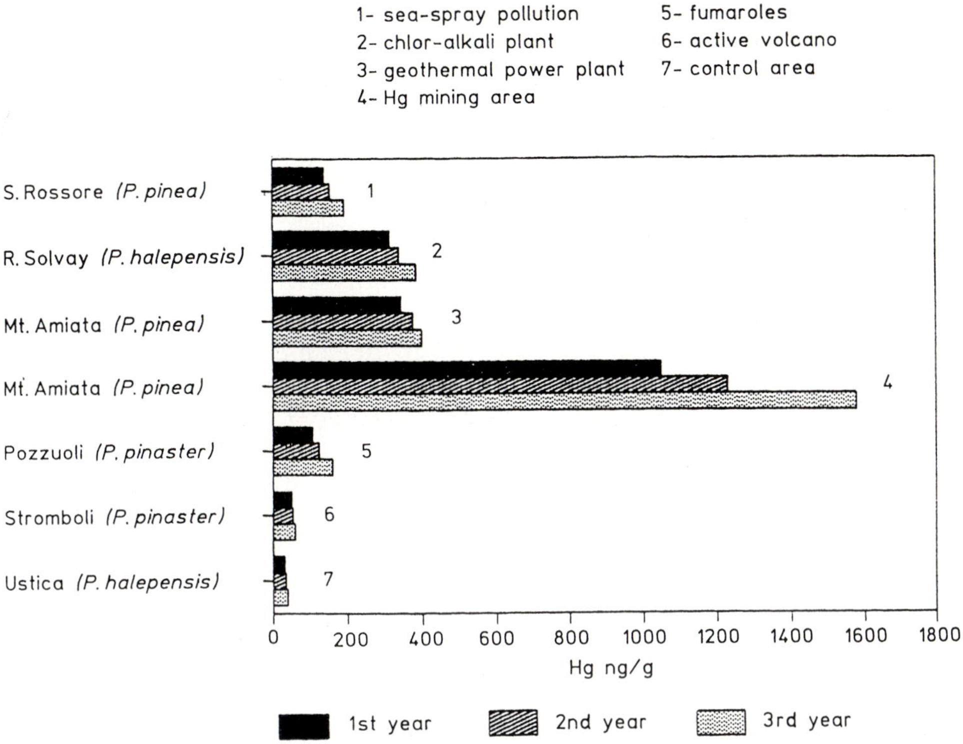

Fig. 8-6. Mercury concentrations in pine needles collected in areas with natural or anthropogenic emissions (from Bargagli, 1993, reprinted with permission).

reviewed the accumulation of metals in a number of plants including crop plants in Budapest. Xu and Thornton (1985) have examined the arsenic content of salad and vegetable crops grown in Cornwall where a history of mining and smelting since pre-Roman times has left a legacy of pollution. Soil As concentrations were between 144 and 892 mg kg^{-1} and although the As was only slightly elevated in vegetables, there were strong correlations between As in beetroot, lettuce, onions and peas and soil arsenic (total, water-soluble and acid F-extractable). Pasture herbage has been used to assess the pollution in old mining areas (Li and Thornton, 1993) and in the vicinity of smelters (Freedman and Hutchinson, 1981; Farago and O'Connell, 1983).

Some grasses growing on contaminated sites have been shown to have tolerant ecotypes (Gregory and Bradshaw, 1965), for example, *Holcus lanatus* has been shown to be Cd-tolerant (Coughtrey and Martin, 1977); *Agrostis gigantea* has been described as metal tolerant (Hogan *et al.*, 1977; Hogan and Rauser, 1979); metal co-tolerances have been described in *Deschampsia caespitosa* (Cox and Hutchinson, 1979); Kovács *et al.* (1993) have shown that the roots of *Lolium perenne* are effective accumulators of a number of heavy metals.

The use of lower plants as biomonitors has recently been detailed. Garty (1993) has extensively reviewed the use of lichens as biomonitors; the use of fungi has been reviewed by Mejstřík and Lepšová (1993) and by Wondratschek and Röder (1993); and mosses as biomonitors have been discussed by Steinnes (1993) and by Brüning and Kreeb (1993); the effects of heavy metals on bryophytes has been reviewed by Tyler (1990).

Assessment of Contamination and Ecotoxicity Using Aquatic Plants

Concentrations of trace metals have been found that are more than 100000 times greater in aquatic plants than in the water in which they grow, and they can thus be used to assess contamination. Algae have proved particularly suitable as biomonitors (Bailey and Stokes, 1985; Spenger and McIntosh, 1989; Manny *et al.*, 1991). In addition, the vulnerability of microalgae to acute short-term stress has been suggested as a practical method for the assessment of contamination, for example produced by chemical spills or sediment dredging activities (Blaise, 1993). Blaise (1993) has described laboratory studies on the effects of a 4 h exposure to a number of metals on the cellular ATP and on the 96 h recovery of *Selenastrum capricornutum*. Dose-response curves indicate a rapid effect on energy metabolism, but the author suggests that recovery offers a better insight into the long-term consequences of acute exposure.

Higher aquatic plants have also been used for contaminant assessment and toxicity testing (Manny *et al.*, 1991; Wang, 1991) and considerable interest has been shown in the use of duckweeds for the assessment of toxicity in aquatic environments (USEPA, 1985; Taraldsen and Norberg-King, 1990; Wang, 1990). Duckweeds (*Lemna* species) are small free floating plants of worldwide occurrence that can be cultured in the laboratory for toxicity testing. Huebert and Shay (1993) have pointed out that there are a number of factors to be considered when using duckweeds for toxicity testing and that previous protocols must be evaluated critically. One factor of importance in the consideration of both accumulation and toxicity and aquatic plants, is the chemical species of the metal in the aqueous medium (Borgmann, 1983).

In-situ Remediation of Pollution Using Plants

There have been many attempts to reduce the migration of toxic metals from soils, and at the same time improve the appearance of derelict sites by the introduction of metal-tolerant ecotypes (Bradshaw *et al.*, 1965; Gemmel, 1977; Bradshaw and Chadwick, 1980). Many such sites are poor in micronutrients and must first be treated with soil amendments such as lime and fertilisers. The maintenance of plant cover at such sites minimises soil erosion and reduces contamination from wind-blown dust. It has been suggested that hyperaccumulator plants have great potential for the decontamination of soils polluted with metals (Baker *et al.*, 1991). Thus cropping with metal-accumulating plants and harvesting the biomass could be a practical way of removing metals from surface-contaminated soils.

The free floating aquatic plant, *Eichhornia crassipes* (MART.) Solms, the water hyacinth, is a highly prolific weed that infests many waterways in the southern USA, Africa and Asia. This plant has the ability to take up metals from solution and to reduce BOD, and has been used successfully to treat lagoon effluent and sewage waste (Wolverton and McDonald, 1979; Haider *et al.*, 1984; Farago and Parsons, 1985; Nor, 1990). It appears to be one of the most generally tolerant plants known and can also be used to assess contamination in water (Farago *et al.*, 1989).

References

Andrews, M. J., Fuge, R. (1986). *Applied Geochemistry 1*, 519.

Ashmore, M. R., Bell, J. N. B., Reily, C. L. (1987). *Nature 327*, 417–418.

Bailey, R. C., Stokes, P. M. (1985), in: *Aquatic Toxicology and Hazard Assessment: Seventh Symposium. STP 854*: Caldwell, R. D., Purdy, R., Bahner, R. C. (eds.) Philadelphia: American Society for Testing and Materials; pp. 5–26.

Baker, A. J. M. (1981). *J Plant Nutr 3*, 643–654.

Baker, A. J. M., Brooks, R. R. (1989). *Biorecovery 1*, 81–126.

Baker, A. J. M., Walker, P. L. (1989). *Chemical Speciation and Bioavailability 1*, 7–17.

Baker, A. M. J., Reeves, R. D., McGrath, S. P. (1991), in: *In situ Bioreclamation: Applications and Investigations for Hydrocarbon and Contaminated Site Remediation:* Hinchee, R. E., Olfenbuttel, R. F. (eds.) London: Butterworth-Heinemann: pp. 600–605.

Bargagli, R. (1993), in: *Plants as Biomonitors. Indicators for Heavy Metals in the Terrestrial Environment*: Markert, B. (ed.) Weinheim: VCH; pp. 461–484.

Blaise, C. R. (1993), in: *Ecotoxicology Monitoring*, Richardson, M. (ed.) Weinheim: VCH; pp. 92–97.

Borgman, U. (1983), in: *Aquatic Toxicology*: Nriagu, J. O. (ed.) New York: John Wiley; pp. 47–72.

Bormann, F. H. (1982). *Ambio 11*, 338–346.

Bradshaw, A. D., Chadwick, M. J. (1980). *The Restoration of Land. Studies in Ecology No 6*, Oxford: Blackwell.

Bradshaw, A. D., McNeilly, T. S. (1965), in: *Ecology and the Industrial Society;* Goodman, G. T., Edwards, R. W., Lambert, J. M. (eds.) Oxford: Blackwell.

Brooks, R. R. (1972). *Geobotany and Biogeochemistry in Mineral Exploration*, New York: Harper and Row.

Brooks, R. R. (1983). *Biological Methods of Prospecting for Minerals*, New York: Wiley.

Brooks, R. R. (1993), in: *Plants as Biomonitors. Indicators for Heavy Metals in the Terrestrial Environment*: Markert, B. (ed.) Weinheim: VCH; pp. 127–153.

Brundin, N. H. (1980), in: *Geomedical Aspects in Present and Future Research*: Lag, H. G. (ed.) Oslo: Universitetsforlaget; pp. 35–37.

Brundin, N. H., Ek, J. I., Selinus, O. C. (1987). *J Geochem Explor 27*, 157–188.

Brüning, F., Kreeb, K. H. (1993), in: *Plants as Biomonitors. Indicators for Heavy Metals in the Terrestrial Environment*: Markert, B. (ed.) Weinheim: VCH; pp. 395–401.

Cannon, H. L. (1960a). *Science 132*, 591–598.
Cannon, H. L. (1960b). *US Geol Surv Bull 1095-A*, 1–50.
Capelli, M. Manfredi, V. R., Moretti, G. F., Trenti, A. (1991), in: *Carcinogenic and Mutagenic Metal Compounds 3. (Interrelation between Chemistry and Biology)*: Merian, E., Bronzetti, G. L., Haerdi, W. (eds.) New York: Gordon and Breach.
Carlisle, D., Berry, W. L., Kaplan, I. R. (eds.) (1986). *Mineral Exploration. Biological Systems and Organic Matter*, Englewood Cliffs: Prentice Hall.
Chamberlain, A. C. (1983). *Atmos Environ 17*, 693–697.
Claussen, T. (1990). *Plant Soil 127*, 91–96.
Cole, M. M. (1980). *Trans Inst Min Metall (Sect B Appl Earth Sci)* B73-N91.
Cole, M. M. Smith, R. F. (1984). *Trans Inst Br Geogr NS 9*, 477–493
Cole M. M., Farago, M. E., Mehra, A., Vulcovic P. (1987), in: *Proc 20th Int Symp Remote Sensing of the Environment.* Michigan: ERIM; pp. 889–898.
Cook, S., Lefebvre, C., McNeilly, Y. (1972). *Evolution 26*, 366–372.
Coughtrey, P. J., Martin, M. H. (1977). *New Phytol 79*, 273–280.
Cox, R. M., Hutchinson, T. C. (1979). *Nature 279*, 231–233.
Davies, B. E. (1983), in: *Applied Environmental Geochemistry*: Thornton, I. (ed.) London: Academic Press; pp. 425–460.
Deu, M., Kreeb, K. H. (1993), in: *Plants as Biomonitors. indicators for Heavy Metals in the Terrestrial Environment*: Markert, B. (ed.) Weinheim: VCH; pp. 577–591.
Djingova, R., Kuleff, I. (1993), in: *Plants as Biomonitors. Indicators for Heavy Metals in the Terrestrial Environment*: Markert, B. (ed.) Weinheim: VCH; pp. 435–460.
Ernst, W. H. O. (1974). *Schwermetallvegetation der Erde*, Stuttgart: Fischer.
Ernst, W. H. O. (1975), in: *Proc. Int. Conf. on Heavy Metals in the Environment, Vol. II,* Toronto, p. 121.
Ernst, W. H. O. (1990a), in: *Element Concentrations Cadasters in Ecosystems*: Lieth, H., Markert, B. (eds.) Weinheim: VCH; pp. 17–40.
Ernst, W. H. O. (1990b), in: *Heavy Metal Tolerance in Plants: Evolutionary Aspects*: Shaw, A.. J. (ed.) Boca Raton: CRC Press; pp. 21–27.
Ernst, W. H. O. (1993), in: *Plants as Biomonitors. Indicators for Heavy Metals in the Terrestrial Environment*: Markert, B. (ed.) Weinheim: VCH; pp. 107–126.
Ernst, W. H. O., Joosse, E. N. G. (1983). *Umweltbelastung durch Mineralstoffe; Biologische Effekte*, Jena: VEB Gustaf Fischer Verlag.
Farago, M. E. (1986), in: *Frontiers in Bioinorganic Chemistry*: Xavier, A. V. (ed.) Weinheim: VCH.
Farago, M. E., Cole, M. M. (1988), in: *Metal Ions in Biological Systems Vol 23. Nickel and its Biology*: Sigel, H., Sigel, A. (eds.) New York: Marcel Dekker.
Farago, M. E., Mehra, A. (1992), in: *Metal Compounds in Environment and Life, 4, (Interrelation between Chemistry and Biology)*: Merian, E., Haerdi, W. (eds.) Northwood: Science and Technology Letters; pp. 163–169.
Farago, M. E., O'Connell, J. T. (1983). *Min Environ 5*, 71–78.
Farago, M. E., Parsons, P. J. (1985). *Environmental Technology Letters 6*, 165–174.
Farago, M. E., Mullen, W. A., Cole, M. M., Smith, R. F. (1980). *Environ Pollut (Ser A) 21*, 225–244.
Farago, M. E., Mehra, A., Banerjee, D. K. (1989). *Environ Geochem Health 11*, 149–156.
Freedman, B., Hutchinson, T. C. (1981), in: *Effect of Heavy Metal Pollution on Plants. Vol. 2. Metals in the Environment*: Lepp, N. W. (ed.) London: Applied Science; pp. 35–37.
Garty, J. (1993), in: *Plants as Biomonitors. Indicators for Heavy Metals in the Terrestrial Environment*: Markert, B. (ed.) Weinheim: VCH; pp. 195–263.
Gemmel, R. P. (1977). *Colonization of Industrial Wasteland. Institute of Biology. Studies in Biology No 80*, London: Edward Arnold.

Gregory, R. P. G., Bradshaw, A. D. (1965). *New Phytol 64*, 131–143.

Haider, Z. S., Malik, K. M. A., Rahman, M. M. (1984), in: *Proceedings of the International Conference on the Water Hyacinth, 7–11 February 1983. UNEP, Nairobi*, Thyagarajan, G. (ed.); pp. 41–57.

Heggestadd, H. E., Middleton, J. T. (1959). *Science 129*, 208–211.

Henwood, W. J. (1857). *Edinb New Phil J 5*, 61–63.

Hogan, G. D., Rauser, W. E. (1979). *New Phytol 83*, 655–670.

Hogan, G. D., Courtin, G. M., Rauser, W. E. (1977). *Can J Bot 55*, 1038–1042.

Horler, D. N. H., Barber, J., Barringer, A. R. (1981). *Remote Sensing in Geological and Terrain Studies*, London: Remote Sensing Society.

Huebert, D. B., Shay, J. M. (1993). *Environ Toxicol Chem 12*, 481–483.

Hutchinson, T. C., Meema, K. M. (1987). *Effects of Atmospheric Pollutants on Forests, Wetlands and Agricultural Ecosystems*, Berlin: Springer-Verlag.

Jackson, A. P., Alloway, B. J. (1992), in: *Biogeochemistry of Trace Metals*: Adriano, D. C. (ed.) Boca Raton: Lewis Publishers; pp. 109–158.

Kabata-Pendias, A., Dudka, S. (1991). *Environ Geochem Health 13*, 108–113.

Kabata-Pendias, A., Piotrowska, M., Dudka, W. S. (1993), in: *Plants as Biomonitors. Indicators for Heavy Metals in the Terrestrial Environment*: Markert, B. (ed.) Weinheim: VCH; pp. 485–494.

Kovaks, M., Turcsanyi, G., Kaszab, L., Szoke, P. (1993), in: *Plants as Biomonitors. Indicators for Heavy Metals in the Terrestrial Environment*: Markert, B. (ed.) Weinheim: VCH; pp. 495–505.

Kovalevsky, A. L. (1979). *Biogeochemical Exploration for Mineral Deposits*, New Delhi: Amerind Press.

Kovalevsky, A. L. (1987). *Biogeochemical Exploration for Mineral Deposits*, Utrecht: VNU.

Koziol, M. J., Whatley, F. R. (1984). *Gaseous Air Pollutants and Plant Metabolism,* London: Butterworth Scientific.

Laaksovirta, K. Olkkonen, H., Alakuijada, P. (1976). *Environ Pollut 11*, 247–255.

Le Neve Foster, Sir C. A. (1894). *A Treatise on Ore and Stone Mining*, London: Griffin.

Li, X., Thornton, I. (1993). *Environ Geochem Health 15*, 135–144.

Malyuga, D. P. (1964). *Biogeochemical Methods of Prospecting*, New York Consultants Bureau.

Manny, B. A., Nichols, S. J., Schlosser, D. W. (1991). *Hydrobiologia 219*, 333–344.

Mansfield, T. A., Lucas, P. W. (1990), in: *Pollution: Causes, Effects and Control:* Harrison, R. M. (ed.) Cambridge: Royal Society of Chemistry.

Mathy, P. (1988). *Air Pollution and Ecosystems,* Dordrecht: D. Reidel Publishing Co.

Matthews, H., Thornton, I. (1982). *Plant Soil 66*, 181–193.

Mejstřík, V., Lepšová, A. (1993), in: *Plants as Biomonitors. Indicators for Heavy Metals in the Terrestrial Envionment*: Markert, B. (ed.) Weinheim: VCH; pp. 366–378.

Nicolls, O. W., Provan, D. M. J., Cole, M. M., Tooms, J. S. (1965). *Trans Inst Mining Metall (Sect B) 74*, 696–799.

Nor, Y. M. (1990). *Chem. Spec Bioavail 2*, 85–91.

Peterson, P. J. (1971). Sci Prog Oxf 59, 505.

Peterson, P. J. (1983), in: *Metals and Micronutrients, Uptake and Utilisation by Plants:* Robb, D. A., Peirpoint, W. S. (eds.) London: Academic Press.

Ramsay, Sir A. (1881). *Mem Geol Surv Gr Brit 3*, 611.

Ratcliffe, D. A. (1974). *Proc Roy Soc Lond*, 355–372.

Rauser, W. E. (1987), in: *Metallothionein II: Proceedings of the Second International Conference on Metallothionein and Other Low Molecular Weight Metal-Binding Proteins*: Kaegi, J. H. R., Kojima, Y. (eds.) Basel: Birkhäuser Verlag; pp. 301–308.

Rice, R., Sharp, G. J. (1976). *Trans Inst Min Metall 85,* 1–13.
Schat, H., Kalff, M. (1992). Cited in Verkleij (1993).
Schat, H., ten Bookum, W. M. (1992). *Heredity, 68,* 219–229.
Schubert, R. (1985). *Bioindication in terrestrischen Okosystemen*, Stuttgart: G. Fischer.
Schulte-Hostede, S., Darrall, N. M., Black, L. W., Wellburn, A. R. (1988). *Air Pollution and Plant Metabolism,* London: Elsevier Applied Science.
Shaw, A. J. (1989). *Heavy Metal Tolerance in Plants: Evolutionary Aspects*, Boca Raton: CRC Press.
Smith, S. (1986), in: *Understanding our Environment*: Hester, R. E. (ed.) London: Royal Society of Chemistry; pp. 263–294.
Smith W. H. (1974). *Envion Polut 6,* 111–129.
Sprenger, M., McIntosh, A. (1989). *Arch Environ Contam Toxicol 18,* 225–231.
Steinnes, E. (1993), in: *Plants as Biomonitors. Indicators for Heavy Metals in the Terrestrial Environment*: Markert, B. (ed). Weinheim: VCH; pp. 381–394.
Taraldsen, J. E. Norberg-King, T. J. (1990). *Environ Toxicol Chem 9,* 761–767.
Thornton, I. (1980), in: *Inorganic Pollution and Agriculture. Proceedings of the ADAS Open Conference of Soil Scientists, London*, pp. 105–125.
Treshow, M. (1984). *Air Pollution and Plant Life*, Chichester: John Wiley.
Tyler, G. (1990). *Bot J Linn Soc, 104,* 231–253.
Unsworth, M. H., Ormrod, D. P. (1982). *Effects of Gaseous Air Pollution on Agriculture and Horticulture,* London: Butterworth Scientific.
USEPA (1985). *Lemma Acute Toxicity Test.* United States Environmental Protection Agency *Fed. Reg. 50,* 39331–39333.
Van Assche, F., Clijsters, H. (1990). *Environ Pollut 66,* 175–172.
Vangronsveld, J., Clijsters, H. (1992), in: *Metal Compounds in Environment and Life: Interrelation between Chemistry and Biology*: Merian, E., Haerdi, W. (eds.) Northwood: Science and Technolgoy Letters; pp. 163–169.
Verkleij, J. A. C. (1993), in *Plants as Biomonitors. Indicators for Heavy Metals in the Terrestrial Environment:* Markert, B. (ed.) Weinheim: VCH; pp. 415–424.
Wagner, G. (1993), in: *Plants as Biomonitors. Indicators for Heavy Metals in the Terrestrial Environment*: Markert, B. (ed.) Weinheim: VCH; pp. 425–434.
Wang, W. (1990). *Environ Research 52,* 7–22.
Wang, W. (1991). *Water Air Soil Pollut 59,* 381–400.
Wilkins, D. A. (1978). *New Phytol 80,* 623–625.
Winner, W. E., Mooney, H. A., Mansfield, T. A. (1985). *Effects of Air Pollutants on Plants,* Cambridge: Cambridge University Press.
Wittig, R. (1993), in: *Plants as Biomonitors. Indicators for Heavy Metals in the Terrestrial Environment*: Markert, B. (ed.) Weinheim: VCH; pp. 3–27.
Wolverton, B. C., McDonald, R. C. (1979). *Ambio 8,* 1–9.
Wondratschek, I., Röder, U. (1993), in: *Plants as Biomonitors, Indicators for Heavy Metals in the Terrestrial Environment:* Markert, B. (ed.) Weinheim: VCH; pp. 343–363.
Wong, M. H., Bradshaw, A. D. (1982). *New Phytol 91,* 255–261.
Xu, J., Thornton, I. (1985). *Environ Geochem Health 7,* 131–133.

9 Analytical Techniques for Plant Analysis

Margaret E. Farago and Aradhana Mehra

Environmental Geochemistry Research
Royal School of Mines
Imperial College of Science, Technology and Medicine
Prince Consort Road, London SW7 2BP, UK

Contents

Introduction

The determination of the concentrations of elements in plant tissues is very important in a number of fields. These include agriculture, ecology, botanical exploration for minerals, and environmental surveys for pollutants.

In agriculture, excesses or deficiencies of essential and beneficial elements (Epstein, 1972; Hewitt and Smith, 1975; Webber, 1981) have profound effects on the health and hence the yields of plants. Chlorosis is often the visible deficiency symptom (Kabata-Pendias and Pendias, 1984). However, disturbances of metabolic processes and consequent losses of biomass production may occur before deficiency symptoms are recognised. Thus the analysis of plants for mineral and nutrient status is particularly important in agriculture. The agricultural agencies of many countries publish recommended analytical methods (for example, HMSO, 1986).

Plant analysis, and that of soils, is very important in the understanding of the factors influencing the plant communities that thrive over certain soils and rocks (for example, Cole, 1992). Botanical methods for mineral exploration have been reviewed by Brooks (1983). These methods include geobotany, the variation in patterns of plants with changes in mineralisation, and biogeochemistry, which relies on the analysis of plants.

Experiments on the tolerance of plant species to metals and their uptake can give information about the evolution of a phylogenetic series of taxa within a genus (Baker *et al.*, 1983). Kabata-Pendias and Pendias (1984) have detailed the elemental contents of many plants and soils and have discussed the factors influencing uptake. Their review covers the literature to the early 1980s.

Plant analyses also play a large part in many environmental surveys, particularly in the determination of pollutants, for example the impact on vegetation from smelter emissions has been reviewed by Freedman and Hutchinson (1981). More recent data are discussed in Markert (1993). Analysis, particularly of environmental material including plants and soils, has gained in importance during the last few decades. However, the demands for analyses have sometimes outstripped skills in the handling of advanced techniques and the appreciation of the reliability of analytical data. A great challenge of today is to generate reliable data of high accuracy and precision. Some useful reviews of environmental analysis have appeared in the literature (ACS, 1983; Fergusson, 1990). Stoeppler (1991) and Ure (1990) have reviewed determination of heavy metals. Methods for trace element analysis and applications are described in detail up to the mid 1970s in the text by Pinta (1978). Recently, Markert (1992) has discussed multi-element analysis of plant materials.

Analytical Tasks

Analytical Difficulties

Depending upon the research facilities and expertise available to a laboratory, plant analysis can be carried out to concentrations as low as ng kg^{-1}. Stoeppler (1991) has defined different categories and analytical tasks depending on their order of difficulties:

(i) *Difficulties moderate to low.* Concentration ranges from μg g^{-1} up to mg g^{-1} can be determined with comparatively simple and inexpensive methods, including colorimetry, FAAS, emission spectrometry with plasma (mainly ICP) excitation, XRF and polarography.

(ii) *Difficulties increased.* Precise and accurate determinations at the μg g^{-1} level requires experienced staff and sensitive methods, for example GFAAS, voltammetry and ICP-MS.

(iii) *Difficulties high.* Concentrations range from μg g^{-1} down to ng g^{-1} levels. Here significant risks from contamination and losses can be expected in addition to problems of detection limits and selectivity of the methods employed. Methods will include the most sensitive modes of AAS (hydride and cold vapour, graphite furnace) frequently in combination with chromatographic techniques.

(iv) *Difficulties extreme.* Trace concentrations at the ng g^{-1} level and below require experienced staff, clean room techniques, and rigid quality control. Trace analysis has been discussed by Tölg and Tschopel (1987) and Tölg (1988).

Ranges of Elements in Plants

Markert (1992) has pointed out that the concentrations of elements in plants range from the ultra-trace (eg, Ru, 0.00001 $\mu g\ g^{-1}$) through to the major elements (eg, C, 45%). Thus the difficulties involved in elemental determinations will vary considerably with the element or groups of elements under consideration. Markert (1992) has constructed a table showing the elemental concentrations of a 'Reference Plant' (Table 9-1) using his own and other published data from the literature. No concentration data from accumulators or rejector plants were included. Markert has compared the elemental concentration data of a number of plants with those of this hypothetical Reference Plant. By presenting the results as bar chart spectra, showing the relative deviation from the Reference Plant as a percentage against the element, fingerprint spectra can be obtained for individual or related plants. These types of charts can be obtained only when multi-element determinations have been carried out on a plant sample. It is expected that this type of information will increase in the near future.

Planning

Planning for an analytical programme should be oriented not only towards the particular task and its urgency, but also towards the laboratory equipment and staff expertise available. Planning should be carried out using the following general guidelines:

- the purpose of analysis
- the most important elements to be analysed
- organising measurement programmes with respect to place and time
- quality assurance and quality control requirements
- documentation of data generated
- personnel requirements and costs.

The analytical methodology must meet realistic expectations regarding sensitivity, accuracy, reliability, precision, interferences, matrix effects, limitations, cost, and the time required for analysis (Keith *et al.*, 1983). Selection of the optimum analytical method is one of the most important factors influencing the reliability of data (Keith, 1980, 1991; ACS , 1983).

Table 9-1. Elemental concentrations in 'Reference Plant'. Modified from Markert (1992). Values in $\mu g\ g^{-1}$ except where % shown.

			E? Al	80	Ag	0.2
Essential[a]	H	6.5%	Ga	0.1	Au	0.001
marconutrients	C	44.5%	In	0.001		
	N	2.5%	TI	0.05	Cd	0.05
	O	42.5%			Hg	0.1
	P	0.2%	Ge	0.01		
	S	0.3%	E? Sn	0.2	Ce	0.5
	Mg	0.2%	Pb	1	Pr	0.05
	K	1.9%			Nd	0.2
	Ca	1%	As	0.1	Sm	0.04
			Sb	0.1	Eu	0.008
			Bi	0.01	Gd	0.04
	Na	150			Tb	0.008
	Rb	50	E? F	2	Dy	0.03
	V	0.5	E? Br	4	Ho	0.008
	Cr	1.5			Er	0.02
Essential[a]	Mn	200	Sc	0.02	Tm	0.004
micronutrients	Fe	150	Y	0.2	Yb	0.02
	Co	0.2	La	0.2	Lu	0.003
	Ni	1.5				
	Cu	10	Ti	5		
	Zn	50	Zr	0.1	Th	0.005
	Mo	0.5	Hf	0.05	U	0.01
	B	40				
	Si	0.1%				
	Cl	0.2%	Nb	0.05		
	I	3	Ru	0.00001[b]		
	Se	0.02	Rh	0.00001[b]		
			Pd	0.0001[b]		
	Li	0.2				
	Cs	0.2	Ta	0.001		
	Be	0.001	W	0.2		
	E? Sr	50	Os	0.000015[b]		
	Ba	40	Ir	0.00001[b]		
			Pt	0.00005[b]		

Concentration data from Markert (1992).

a Essentiality data from Hewitt and Smith (1975) and Farago (1986).

b Value calculated by Markert 1992.

E? represents beneficial or possibly essential elements.

Sampling and Sample Storage

Sampling Strategies

The objectives of sampling are broadly divided into exploratory (surveillance) or monitoring (assessment) goals. Preliminary information about the site of material being analysed is provided by exploratory sampling. Monitoring is usually intended to provide information on the variation in concentration of specific analytes over a particular period of time or within a specific geographic area.

There are three primary sampling approaches: random, systematic, and judgmental. Samples collected for research studies may allow use of prior knowledge in order to obtain as much useful data as possible, but sampling for local purposes often requires absolutely random samples. Often a combination of judgmental, systematic, or random sampling is the most feasible approach. The principles of environmental sampling in general have been discussed in a number of publications (Keith, 1988, 1991; ACS, 1990, Baiulescu *et al.*, 1991).

Experimental results are dependent upon the samples which are representative of a population. Samples from a chosen site usually need to be representative of that site at the time the samples are taken. Thus, sampling must be carried out so that the analytical results reflect the seasonal or local variations during the period of investigation. Both seasonal and species variation in metal content of plants have been found, for example Matthews and Thornton (1982) found variations with both in the content of Cd and associated metals in plants at Shipham, and seasonal variations of lead levels have been found in poplar leaves in urban areas (Capelli *et al.*, 1991). Usually representative samples are sought, but sometimes non-representative, or targetted, samples are required (eg, samples from a site near an industrial fallout). The sampled portion should be small and informative of the population being investigated. For multiple analyses, sufficiently large samples must be taken, these provide replicate samples but incur additional storage costs.

The number of individual samples will depend on the kind of information required (Hoffmann and Schweiger, 1981). If an average compositional value is required, a large number of randomly selected samples may be obtained, combined, and blended to provide a reasonably homogeneous composite sample from which a sufficient number of sub-samples are analysed. If composition profiles on the variability of the sample population are of interest, many samples will need to be collected and analysed individually.

The minimum number of samples necessary for specific confidence limits have been discussed (ACS, 1983). An empirical approach, the *N-N-N*-concept has been suggested, where equal numbers (*N*) of samples, blanks and spiked blanks are analysed.

Sampling operations can contribute to both systematic and random error. These components are interactive, and they are evaluated using appropriate statistical methods. Some aspects of sampling may cause large variations and yet cannot be treated strictly by statistics (Keith, 1991). Field blanks should be handled exactly the same way as test environmental samples. If field blanks are not available, blank samples that best simulate a sample that does not contain the analyte should be obtained. When environmental samples are made to investigate localised contamination, measurements of concentrations at sites recognised as uncontaminated (control sites) are required. Such samples demonstrate whether the site under study is contaminated or truly different from the norm. Local control sites are usually adjacent or very near the test sample sites. In contrast to a local control site, an area control site is in the same area, eg, a city or county, as the sampling site but not adjacent to it.

Variations in the elemental compositions of plants and the use of computer aided sampling have been discussed by Markert and Klausmeyer (1990). The natural variations in the element composition can be divided into 5 systematic levels as shown in Table 9-2.

Table 9-2. Systematic classification of different levels of elemental variations within plants.

Biological level	Possible reasons for biological variations in concentrations
1. Species	Genetic variabilities
2. Population	Different climatic and edaphic conditions; genetic variabilities
3. Stand, within an ecosystem	Different edaphic and microclimatic conditions; genetic variabilities
4. Individual	Different microclimatic and microedaphic conditions; stage of development; seasonal changes, exposure; genetic variabilities
5. Plant compartments	Transportation and delocalisation of elements within plant compartments

Storage

Care must be taken to avoid loss or gain of elements from samples during storage and preservation. Samples should be safeguarded from loss, tampering, or misidentification. This requires use of a unique sample identification system, safe storage facilities, and a sample management system in which all steps are fully documented. Sample storage is often necessary in cases where immediate analysis is not possible, where reference samples have to be kept, or if surplus sample material is required for calibration or reference during a longer period. Changes in the composition of the original materials are possible in samples stored for long periods. Samples that could be involved in potential litigation should be protected with a chain-of-custody documentation system (Keith *et al.*, 1983).

Sample Preparation and Decomposition

Preparation

Sample preparation comprises the treatments the plant material must undergo between sampling and the definitive determination of the element, it serves to: bring the elements of interest into a chemical and physical form suitable for the method of determination; eliminate interfering substances; and in some cases enrich the elements of interest. Sample preparation may involve physical operations such as drying and grinding and/or chemical operations such as digestion. These procedures are potential sources of contamination and losses (Fergusson, 1990).

Decomposition and Ashing

For many methods of elemental analysis, the organic constituents of plant material are eliminated either by dry ashing or by wet digestion (Sulcek and Povondra, 1989). Stoeppler (1991) has summarised the most frequently used decomposition methods for trace and ultratrace analysis of biological and environmental materials.

Dry Ashing

Dry ashing is simple and inexpensive. Large numbers of samples can be treated at the same time and the quantity of the sample is not limited. Usually the sample is heated in a crucible of platinum, silica or pyrex to a relatively high temperature, between 400 °C and 700 °C, in a muffle furnace. Dry ashing is not recommended for samples with trace metal contents unless special furnaces, and very accurate temperature control methods are employed. The more volatile metals such as lead, mercury, and cadmium may be partially lost at even low temperatures.

Wet Ashing

Wet digestions can be carried out in open vessels or in closed systems. In open systems, increased consumption of reagents requires the use of especially pure reagents. Considerable progress in the purification of reagent chemicals has been made over the years. Mitchell (1982) has described some of the methods for producing analytical reagents of highest purity. Losses of volatile elements during wet digestion can be minimised by reflux systems or by open systems coupled with volatile element trapping (Knapp, 1984).

Closed systems minimise consumption of reagents and avoid losses of volatile elements. Decomposition vessels should have low trace metal contents and be of extremely pure quartz and PTFE and should be carefully purified prior to use in ultrapure acids or acid vapour (Tschopel *et al.*, 1980) and elemental blanks should be determined for each decomposition vessel. Matusiewicz (1991) has described the special advantages and problems associated with acid vapour-phase digestion procedures.

Microwave assisted wet digestion has attracted considerable attention and has been successfully applied to plant material. Both open and closed vessels have been used, but the most popular approach is the sealed bomb method (Kingston and Jassie, 1988; Sulcek and Povondra, 1989; Matusiewicz, 1991). Karanassios *et al.* (1991) describe microwave stopped flow digestion systems that can give rapid (ie, less than 5 min) reproducible extractions of elements of environmental concern from plant samples.

A comparison of different mineral acids for wet digestion of plant material for AAS analysis was carried out by Dokiya *et al.* (1975), who concluded that nitric acid was best for the determination of Cu, Mn and Pb by flameless AAS, whereas a nitric sulphuric mixture was best for Fe, Mn, Cu and Zn by FAAS. Most literature reports dissolve plant material in nitric acid (Middleton and Stuckey, 1954) or in nitric and perchloric acids (Isaac and Johnson, 1975; Williams, 1978, Thompson and Wood, 1982). This method will not dissolve aluminium containing particles from soil or other contaminations (Pierson and Evenson, 1988; Ramsey *et al.*, 1991) (see also *Standard Reference Materials*).

Sensitivity and Detection Limits of Analytical Techniques

The concentration of analytes that can be measured in various materials has been decreasing over the years as sensitivity and detection limits of analytical techniques have improved. The method detection limit (MDL) is the order of magnitude of the smallest quantity or concentration of substance which can be detected in principle; the limit of detection (LOD), on the contrary, is a precisely calculable statistical value for a particular, defined analytical procedure. The instrument detection limit (IDL) is the smallest signal above background noise that an instrument can detect reliably. It is expressed either as an absolute limit (in units of mass, eg, ng), or as a relative limit (in terms of concentration, eg, g mL^{-1}).

Some absolute detection limits are shown in Table 9-3.

Table 9-3. Absolute detection limits (ng).

Method	Detection limits
X-ray fluorescence	1
Colorimetry	$1-10^{-2}$
Anodic stripping voltammetry	$1-10^{-2}$
Fluorimetry	$0.1-10^{-2}$
Flame AAS	$0.1-10^{-3}$
ICP-AES	$0.1-10^{-4}$
Electrothermal AAS	$10^{-3}-10^{-5}$
Mass spectrometry	$10^{-3}-10^{-6}$
Neutron activation analysis	$10^{-6}-10^{-8}$

Data from Tölg (1974), Liteanu and Rica (1980) and Walsh *et al.* (1986)

Standard Reference Materials

Most of the instrumental direct methods, which are employed in routine analyses, are highly matrix-dependent, therefore methods must be validated with certified standard reference materials. Standard plant reference materials (SRMs), with certified elemental contents can be obtained from NIST, National Institute of Standards and Technology of the United States Department of Commerce or from BCR, Community Bureau of Reference, Commission of the European Community (Colinet *et al.*, 1982; Criepink, 1990; Horwitz and Albert, 1991).

A recent paper (Ramsey *et al.*, 1991) has demonstrated that low recoveries of Al measured using NIST tomato leaves are not indicative of a negative bias of an analytical method for the determination of Al in plant samples. A major problem in Al determinations in vegetation is contamination from soil particles. In this study it was found that normal 'total' attack by nitric/perchloric acids gave only 40% recovery of Al in NIST tomato leaves, and 69% in citrus leaves. Similar low results had been reported previously (Pierson and Evenson, 1986). When HF was added to the digestion mixture, the Al concentrations, measured by ICP-AES were close to the certified values. The accuracy of the ICP-AES determination was thus validated. Scanning electron microscopy was used to identify non-leaf 'soil' particles. The contribution of the Al from these 'soil' sources was estimated by computer controlled Electron Probe Microanalysis (EPMA). There was agreement in the Al contribution from the soil particles and the missing Al, not found using nitric/perchloric acid attack. The authors concluded that soil contamination is responsible for the missing Al, which was not solubilised by 'normal' acid digestion. Thus EPMA allows the extent of soil contamination of vegetation to be estimated.

Ramsey and Coles (1988) have analysed a wide range of plant reference materials by ICP-AES and have shown that there is a negative bias of up to 30% for some elements due to matrix effects. The application of the Parameter-Related Internal Standard Method (PRISM) (Thompson and Ramsey, 1985) reduced this bias for all elements. Recently Horwitz and Albert (1991) have discussed the variability in concentration estimates of standard reference materials and have suggested that many laboratories lack a quality assurance programme (see *Quality Assurance and Quality Control*) that would enable the investigation of occurrences of gross deviation from certified values.

Analytical Methods

Flame Atomic Absorption Spectrometry (FAAS)

Atomic spectrometry is based on the generation of free atoms which can absorb or emit radiation due to defined transitions of the valence electrons of the outer shell of the atom. Comprehensive and critical reviews of atomic spectroscopy and its uses appear in *Journal of Analytical Atomic Spectrometry*. Advances in AAS and fluorescence spectrometry have been reviewed by Hill *et al.* (1991). Branch *et al.* (1991) has updated the use of AS for the analysis of clinical and biological materials, foods and beverages and discussed methods for individual elements. Cresser *et al.* (1991, 1992) have reviewed environmental analysis, including those for soils and plants and have included summary tables of methods.

Specific element identification is feasible in the wavelength range from somewhat below 200 nm to approximately 850 nm. Flame AAS has an element coverage of around 70 elements and is used for determination of most inorganic elements in biotic matrices. Under optimal conditions day-to-day precision is around, sometimes even below, 2%. In FAAS, detection limits are at a level of $1-100$ ng mL^{-1} (Welz, 1985). These can be improved by using slotted-tube atomic trap techniques (Brown and Taylor, 1985), and injection techniques for reducing the volumes required (Berndt and Jackwerth, 1975). Problems of physical (radiation) interferences have been reduced by improving the optical system of FAAS. Chemical interferences can be overcome by matrix matched calibration, background correction and addition of compounds (eg, lanthanum salts) that enhance atomisation efficiency. However the concentration range is limited, above certain levels the relationship between absorbance and concentration may not be linear. Aspects of AAS are discussed in Cantle (1982) and problems encountered in trace element determination in plant material by FAAS are discussed by Hoenig and de Boerger (1983).

Graphite Furnace Atomic Absorption Spectrometry (GFAAS) or Atomic Absorption with Electrothermal Atomisation (ETAAS)

Graphite furnace atomic absorption spectrometry (GFAAS) is the most sensitive established atomic spectrometric method for a large number of elements except those forming thermally stable oxides and carbides (B, Hf, Ti, lanthanons) and non-metals. The detection limits are in the sub-ng mL^{-1} range which are competitive with other sensitive techniques such as neutron activation analysis. The introduction of Zeeman background compensation has brought about significant progress in the reduction of systematic errors. These can be further reduced by the treatment of graphite: by coating with pyrographite or metal oxides, by using glassy carbon tubes; and by operation with controlled and constant temperatures during atomisation. The use of the stabilised temperature platform furnace (STPF) for the determination of a number of trace elements in biological samples has been discussed by Subramanian (1991).

One of the advantages of GFAAS is the direct analysis of solid samples without prior decomposition, this technique has been reviewed by Bendicho and de Loos-Vollebregt (1991). The solid samples can be introduced directly or as a slurry. Calibration can be with aqueous standards, synthetic solids, standard additions or with SRMs. SRMs have been used for calibration for solid sampling of plant material (Schmidt and Falk, 1987). Examples of the use of solid samples with ETAAS are shown in Table 9-4. In many cases a matrix modifier is used with the sample, this allows the matrix to be volatilised at the pyrolysis stage without analyte loss, particularly important with volatile analytes such as Cd and Pb. (Ure, 1990).

Table 9-4. Applicatons of solid sampling with ETAAS (GFAAS) to plant analysis.

Elements	Direct/slurry	Reference
P	Direct	Langmyhr and Dahl (1981)
Pb, Rb, Cd, Cu, Mn, Ni	Direct	Grobenski *et al.* (1982)
Al, Cd, Co, Cu, Mn, Ni, Pd, Zn	Direct	Akatsuka and Atsuya (1989)
Pb, Cd	Direct	Baxter and Frech (1990)
As, Cd, Cr, Cu, Pb, Mn	Direct	Vollkopf *et al.* (1985)
Cd	Slurry	Fagioli and Landi (1983)
Cu, Cr, Fe, Pb	Slurry	Carrion *et al.* (1988)
Mn, Fe, Cu, Cr, Al, Mg	Slurry	Miller-Ihli (1990a)

There is interest in multielement analysis using GFAAS, one approach uses a continuum source (CSAAS). The application of CSAAS to the analysis of biological materials has been discussed by Miller-Ihli (1990b).

Solvent Extraction

Preconcentration before atomic spectrometry by extraction of the analyte into a solvent, usually complexed with a coordinating agent has been discussed by Cresser (1978). Procedures include ammonium pyrrolidine dithiocarbamate (APDC) or 8-hydroxy-quinoline as complexing agent and methyl isobutyl ketone (MIBK) as solvent. The extracts can be analysed in the flame or using the graphite furnace. The use of dithizone with chlorinated hydrocarbons is discouraged with FAAS, where poisonous phosgene can be produced in the flame. However chlorinated solvents have been used with the graphite furnace for analysis of plants and soils (Ure and Hernandez-Artiga, 1977).

Vapour Generation

Hydride and cold-vapour techniques represent a special combination of chemical separation and pre-enrichment with AAS determination, resulting in higher powers of detection for elements with volatile hydrides, eg, As, Bi, Se, Sb, Hg. Recent literature on vapour generation has been reviewed by Hill *et al.* (1991). Some examples of the use of hydride generation for the analysis of plant material are given by Muse *et al.* (1989), Leuka *et al.* (1990) and Ainsworth and Cooke (1990). Hydride generation can also be used with ICP-EAS (see below) and applications have been reviewed (Nakahara, 1991).

Inductively Coupled Plasma Atomic Emission Spectrometry (ICP-AES)

Atomic emission spectrometry (AES) is also called optical emission spectrometry (OES). It is the oldest atomic spectrometric multielement method which originally involved the use of flame, electric arc or spark excitation. Recently there has been considerable innovation in new sources: plasma sources and discharges under reduced pressure. Littlejohn *et al.* (1991) have reviewed recent advances in the field of atomic emission spectrometry, including fundamental processes and instrumentation.

The usefulness of the ICP technique in elemental analysis was demonstrated by the pioneering work of Greenfield *et al.* (1964) and Wendth and Fassel (1965). The technique has been developed and refined and is the most rapidly growing method of elemental analysis. The ICP source is based on the inductive heating of a gas flowing in a tube assembly (torch) and is operated at frequencies between 1 and 100 MHz.

The ICP source has proved to be very useful as an emission source because of its high stability, low noise and freedom from interferences when operated under appropriate conditions. However, interferences have been noted in the literature (Walsh and Howie, 1986; Thompson and Ramsey, 1985). Since the interferences are largely from the elements Al, Fe and Ca, they are less important in plant analyses than in those for soils, where these elements are usually major components.

Applications of ICP-AES, which is one of the more popular methods at present, to plant analyses are reported in Branch *et al.* (1991), and Cresser *et al.* (1991). The practical detection limits of ICP-AES for plant tissue analysis are reported by Thomson and Walsh (1988) and these are compared with the typical concentrations in plants tissues (Bowen, 1966) in Table 9-5.

Table 9-5. Practical ICP detection limits in relation to typical concentrations of elements in plant tissues (μg g^{-1}).

Element	Practical detection limit[a]	Typical concentration in plant tissue[b]
As	2	0.2
Cd	0.5	0.6
Cu	0.25	14
Mo	0.5	0.9
Ni	1.5	3
Pb	2.5	2.7
Zn	0.4	100

[a] From Thompson and Walsh (1988); the values allow a dilution factor of 10 during decomposition.

[b] From Bowen (1966).

Other Sources for Atomic Emission Spectrometry

For special applications direct current plasma (DCP) (Leis *et al.*, 1989) and microwave-induced plasma (MIP) may be used. The MIP first became widely used as a spectroscopic radiation source after a stable discharge at atmospheric pressure had been obtained (Beenakker, 1977; Beenakker *et al.*, 1978). The MIP is not capable of taking up wet aerosols, but is useful for the excitation of dry aerosols, produced by electrothermal evaporation from a graphite furnace (Aziz *et al.*, 1982). Direct sample insertion has been discussed recently by Blain and Savin (1992).

Graphite furnace emission spectrometry, with the furnace operated under reduced pressure, has been explored. In FANES, furnace atomic non-thermal excitation spectrometry, an additional discharge is applied for analyte excitation (Falk *et al.*, 1984). Littlejohn *et al.* (1991) have reviewed recent advances in the field of atomic emission spectrometry, including fundamental processes and instrumentation.

Laser Spectrometry

The extreme sensitivity of the laser methods can give very reproducible result. The main atomisation methods used in laser spectrochemistry are electrothermal atomisation (eg in graphite tubes or cups), atomisation in plasmas (eg, in ICP), analytical flames or by sputtering (in a glow discharge lamp or by ion beam bombardment), or the laser itself (Piepmeier, 1986). The application of laser spectroscopy is a promising technique in elemental micro- and trace analysis. Currently, the most serious disadvantage is the very high price of the laser system which, however, may soon be replaced by far less expensive semiconductor diode lasers. Then laser spectrometry could become a routine laboratory technique for elemental analysis. Recent literature on laser spectrometry is given by Piepmeier (1986), Letokhov (1986), Niemax (1989), Lawrenz and Niemax (1989) and Niemax and Sdorra (1990).

Atomic Fluorescence Spectrometry (AFS)

The fluorescence technique combines the advantages of the large dynamic range of emission techniques with the simplicity and high selectivity of absorption techniques. Flame sources have been extensively used, however, for elements with refractory oxides, the ICP source has been found to be more satisfactory for AFS. A system for hollow cathode lamp excited ICP-AFS, as proposed by Demers and Allemand (1981), is commercially available as a modular simultaneous multielement ICP system. Although fluorescence techniques often offer two orders of magnitude sensitivity improvement over absorption, the multielement approach for AFS has not yet been commercially successful. Also promising for the future is the laser-excited furnace AFS where the detection limits for most elements are comparable to those of ICP-AES and for some elements, for eg, As, Cd, Pb, Tl, Lu, even lower (Omenetto and Human, 1984). The future for AFS techniques has been discussed by Stockwell and Corns (1992).

X-Ray Spectrometry

When atoms are subjected to radiation of a suitable energy, electrons from the inner (K, L, M, ...) atomic shells are eliminated. Following this, electrons from the outer shells drop into free positions with the emission of comparatively simple elctromagnetic radiation in the energy range from 0.6–120 keV (wavelength 2–0.01 nm) which is unique and can be exactly computed for each excited element. The frequency of the emitted lines is proportional to the square of the nuclear charge number of the respective element. Therefore, after wavelength or energy dispersion, this radiation can be used for qualitative and quantitative determination of elements. The absolute power of detection for conventional wavelength- and energy-dispersive X-ray fluorescence (XRF) is highest for elements with an atomic number around 30. The detection limits worsen towards both higher and lower atomic numbers. The main limiting factor for the power of detection is the Compton and Raleigh scattering of the primary beam. Moreover, interelement effects necessitate it to calibrate with standards having a matrix matched to the analysis samples (William, 1987) or extensive software connections for eliminating matrix effects (Kikkert, 1983). Instrumentation and applications have been discussed by Bacon *et al.* (1990). Several applications of XRF methods to the analysis of plant material have been reported by Gladney *et al.* (1989), Martin (1988), Rethfelden (1980) and Klopfenstein and Nussbaum (1976). Martin (1988) reported the main advantage of WDXRF in the analysis of plant material is that little sample preparation is required except to form the ground material into a pellet. Sixteen elements were determined in tobacco and a comparison was made with ICP-AES for analysis of SRM 1572 citrus leaves.

Mass Spectrometry (MS)

Mass spectrometry is the most universal instrumental method for multi-element determination as it permits the determination of all elements and their isotopes in both solids and liquids. Mass spectrometry of metals is based on the principle of generation of gaseous ions thermally, by a spark of high potential, or by an electron- or ion-current. The generated ions are separated by energy-mass focusing in strong magnetic fields. The qualitative detection or quantitative determination of charged particles is achieved by direct-reading electronic systems. The detection limits for virtually all elements are low. Bacon *et al.* (1990) have reviewed applications of inorganic mass spectrometry.

Inductively Coupled Plasma Mass Spectrometry (ICP-MS)

For analysis of solutions, ICP-mass spectrometry (ICP-MS) is very promising (Houk *et al.*, 1980; Houk, 1986; Bacon *et al.*, 1990). Recent advances in separation and preconcentration techniques are discussed by Horvath *et al.* (1991). Bacon *et al.* (1990) report that although ICP-MS is a multi-element technique, recent papers tend to concentrate on a small number of target elements. With isotope dilution mass spectrometry (IDMS), detection limits are further reduced (Heumann, 1988); IDMS is also suitable for accurate speciation in very low concentration levels of elements (Heumann, 1990). For the direct analysis of solid samples, glow discharge mass spectrometry (GD-MS) (Harrison *et al.*, 1986) is of interest. Tölg (1988) has suggested that a substantial improvement in the absolute detection power of GD-MS, as applied to micro analysis, can be expected, at least in comparison with the ICP as ion source.

LAMMA, LMMS and SI-MS

Laser microprobe mass analyser (LAMMA) or laser microprobe mass spectrometry (LMMS) applies laser beams for ionisation (Adams, 1983; Feigl *et al.*, 1984). A special edition for *Fresenius' Zeitschrift für Analytische Chemie* (1981) is devoted to early results using LAMMA. The techniques and applications of LMMS have been reviewed by Bacon *et al.* (1990) and Clarke (1989). Anion spectra for several species of fungi have been reported using this technique (Heinrich, 1989). The spectra are very characteristic and the concentrations are in the order $K > Mg > Na \gg Ca$. This is one of the most sensitive methods and has great potential for the study of plant tissues. Secondary ion mass spectrometry (SI-MS) applies mass spectrometers to the measurement of sputtered ions (Benninghoven, 1987, 1988). Both LAMMA and SI-MS are very sensitive means to determine trace element distributions on surfaces of solid materials, in biological structures and in extremely small samples (Benninghoven *et al.*, 1987, 1988; Adams *et al.*, 1988; Jochum *et al.*, 1988; Odom *et al.*, 1988; Bacon *et al.*, 1990).

Neutron Activation Analysis (NAA)

The use of activation methods for the analysis of biological material has been reviewed by Bowen (1980) and Cornelis (1985). In Instrumental Neutron Activation Analysis (INAA), no chemical separations or concentrations are made before placing the plant material into the core of a nuclear reactor. In the core it is bombarded with neutrons. Many of the elements present in the sample undergo nuclear reactions, of which the most common are of the (n, γ) type. The radioactive products of these reactions decay with the emission of characteristic gamma radiation which is measured using a detector. INAA is thus potentially a useful multielement method for the study of plant samples (Cornelis *et al.*, 1982). Czauderna (1984) compared INAA with NAA, where chemical separations are made initially, and found that INAA is not always quicker since it is often necessary to wait a long period of time for interfering radionuclides to decay to a sufficiently low level, and even then a long counting period may be required to give acceptable precision. Some examples of NAA methods applied to plant samples are given by Farago and Parsons (1983), Jones and Johnston (1989), Glooshenko (1989), Gladney *et al.* (1989) and Ahmad *et al.* (1979).

Electrochemical Methods

Voltammetric methods are based on the potential-dependent redox reaction of the species to be analysed at the surface of a suitable electrode. According to the terminology recommended by IUPAC (International Union of Pure and Applied

Chemistry) the original method applying a dropping mercury electrode is called polarography whereas voltammetry comprises all approaches that evaluate current – potential relations which includes polarography. As each metal has a certain redox potential at a certain working electrode, it frequently allows the sequential simultaneous determination of several metals. For voltammetric determinations, metals to be determined should be present completely dissolved in a non-interfering analyte. Even traces of residual organic matter can severely interfere, hence all traces of biological and environmental materials must be removed before carrying out voltammetric determinations. If necessary, alkali salts can be added as supporting electrolytes.

Differential pulse methods have gained considerable importance in contemporary voltammetric trace analysis (Wang *et al.*, 1985). For metal levels in the analyte solution above 50 to 500 g L^{-1}, direct determination with dropping mercury electrodes (DME) is commonly applied and this method is termed differential pulse polarography (DPP). Where element concentrations are lower, stripping voltammetric methods, anodic stripping voltammetry (ASV) and cathodic stripping voltammetry (CSV) are used. For a determination in the sub-ng g^{-1} range and upper-pg g^{-1} range only, the highest sensitive stripping technique with adsorptive accumulation, adsorptive stripping voltammetry, has to be applied (Henze, 1990). Voltammetric measurements are reproducible and can be repeated without any observable change, they are best calibrated by the standard addition procedure (Seiler, 1988). Procedures for the determination of cobalt in plant digests by adsorption stripping voltammetry have been described by Gilbert *et al.* (1988). The method was validated by neutron activation analysis. Methods for multielement determination in biological materials have been reported (Adeloju *et al.*, 1985).

Colorimetry

In spite of its limited sensitivity, colorimetry is still useful in determination of elemental concentrations in the g range or higher (Seiler, 1988). Its main advantage is that the needed instrument, a spectrophotometer, is common in every laboratory. Colorimetric trace metal determinations are based, commonly after sample decomposition, on selective separations from interfering ions (Abbasi *et al.*, 1988). Automated colorimetric procedures are described for the determination of N and P in trees (Stewart *et al.*, 1990). Modern spectrophotometers provide high stability, low noise, and the advantages of computerised background control. However, for total metal determinations in environmental samples, this method is less frequently applied and has been replaced by atomic spectroscopic and electrochemical methods (Stoeppler, 1991).

Chromatography

Chromatography is based on the principle that different compounds under distinct conditions travel at different speeds through a system consisting of a mobile and a stationary phase thus allowing for separation of numerous compounds in the same analyte solution. In gas chromatography (GC) the mobile phase is an inert gas. The compound to be determined is volatilised by heat – this can cause problems with compounds of low thermal stability. For metal determinations in liquids, atomic spectrometric detectors are directly coupled with gas chromatographs (Ebdon *et al.*, 1986). Inorganics usually have to be transformed into appropriate organometallic compounds. In liquid chromatography, the mobile phase is always a liquid. This allows for comparatively lower temperatures as compared with gas chromatographic techniques and makes the determination of thermally less stable compounds possible (Knox *et al.*, 1978). Today, liquid chromatography is mainly applied as high performance liquid chromatography (HPLC). Detection methods for HPLC include spectrophotometry and electrochemical techniques. HPLC-GFAA (graphite furnace atomic absorption) techniques have been developed (Brinckman *et al.*, 1977; Stockton and Irgolic, 1979). Using an autosampler, the period between analyses can be set to 45–60 s. A number of determinations including Hg, Pb, Sn, Se and As are reported using these systems (Irgolic, 1991). Inductively coupled plasma emission spectrometers (ICP-AES) are very useful as HPLC detectors (Irgolic, 1985). Ion chromatography uses various chelating cation exchange resins and this technique allows the simultaneous analysis of cations and anions (Jones and Tarter, 1985). The most recent advances in analytical separation methods are reviewed in a recent volume (Krull, 1991); gas and liquid chromatography interfaced with a number of sensitive detection methods are described. The methods have great potential in plant analysis, particularly in the determination of the chemical species present.

Quality Assurance and Quality Control

Quality assurance (QA) has been described as a system of activities that assures the producer or user of a product that defined standards of quality with a stated level of confidence are met. Quality control (QC) differs in that it is an overall system of activities that controls the quality of a product so that it meets the needs of the users (Taylor, 1987). In other words, QC consists of the technical activities to control and assess the quality of the measurements, while QA is the management system that ensures that an effective QC system is working as planned.

The two parameters most often used to assess measurement quality objectives are bias and precision. Bias is defined as a systematic deviation (error) in data. Precision is defined as random variation in data. One objective of any sampling quality assurance programme is to provide the type and number of quality control samples necessary to control and minimise the effects of bias and precision in the sampling effort. To control and minimise errors, careful precautions should be taken during sample collection and storage, sample preparation, and analytical determination.

The quality and utility of analytical data depend critically on the validity of the sample and the adequacy of the sampling programme and storage procedures (see *Sample Preparation and Decomposition*).

Sample preparation involves physical and chemical treatments which are potential sources of bias, variance, contamination, and mechanical loss. Sample preparation should be planned carefully and documented in sufficient detail to provide a complete record of the sample history. Furthermore, samples taken specifically to test the quality assurance system should be subjected to the same preparation steps as the test sample.

Analytical measurements should be made with properly tested and documented procedures. These procedures should utilise controls and calibration steps to minimise random and systematic errors. There are basically two types of controls: (a) those used to determine whether or not an analytical procedure is in statistical control, and (b) those used to determine whether or not an analyte of interest is present in a studied population but not in a similar control population. The purpose of calibration is to minimise bias in the measurement process. Calibration or standardisation critically depends upon the quality of the chemicals in the standard solutions and the care exercised in their preparation. Another important factor is the stability of these standards once they are prepared. Calibration check standards should be freshly prepared frequently, depending on their stability (Keith, 1991). No data should be reported beyond the range of calibration of the methodology. Appropriate quality control samples and experiments must be included to verify that interferences are not present with the analytes of interest, or, if they are, that they be removed or accommodated.

References

Abbasi, S. A., Hameed, A. S., Nipaney, P. S., Soni, R. (1988). *Analyst, 113*, 1561–1565.

ACS (1983). *Principles of Environmental Analysis*, The American Chemical Society Committee on Environmental Improvement. Washington DC: ACS; Reprinted from *Anal Chem 55*, 2210–2218.

ACS (1990). *Principles of Environmental Sampling*, The American Chemical Society Committee on Environmental Improvement. Washington DC: ACS; Reprinted from *Environ Sci Technol 24*, 610–617.

Adams, F. (1983). *Spectrochim Acta, 38B*, 1379–1393.

Adams, F., Gijbels, R., van Grieken, R. (eds.) (1988). *Inorganic Mass Spectrometry,* New York: John Wiley and Sons.

Adeloju, S. B., Bond, A. M., Briggs, M. H. (1985). *Anal Chem, 57*, 1386–1390.

Ahmad, S., Chaudrey, M. S., Quereshi, I. H. (1976). *J Radioanal Chem, 54*, 331–342.

Ainsworth, N., Cooke, J. A. (1990). *Environ Pollut 65,* , 65–77.

Akatsuka, K., Atsuya, I. (1989). *Anal Chem, 61*, 216–220.

Aziz, A., Broekaert, J. A. C., Leis, F. (1982). *Spectrochim Acta, 37B*, 381–389.

Bacon J. R., Ellis, A. T., Williams, J. G. (1990). *J Anal Atomic Spectrom, 6*, 243R–321R.

Baiulescu, G. E., Dimitrescu, P., Zugravescu, P. Gh. (1991). *Sampling*, Chichester: Ellis Horwood.

Baker, A. J. M., Brooks, R. R., Pease, A. J., Malaisse, F. (1983) *Plant Soil, 73*, 337–385.

Baxter, D. C., Frech, W. (1990). *Fresenius Z Anal Chem, 337*, 253–.263

Beenakker, C. I. M. (1977). *Spectrochim Acta, 32B*, 173–178.

Beenakker, C. I. M., Bosman, B., Boumans, P. W. J. M, (1978). *Spectrochim Acta, 33B*, 371–381.

Bendicho, C., de Loos-Vollebregt, M. T. C. (1991). *J Anal Atomic Spectrom, 6*, 353–374.

Benninghoven, A., Rudenauer, F. G., Werner, H. W. (1987). *Secondary Ion Mass Spectrometry: Basic Concepts, Instrumental Aspects, Applications and Trends*, New York: John Wiley and Sons.

Benninghoven, A., Huber, A. M., Werner, H. W. (1988). *Secondary Ion Mass Spectrometry SIMS VI.* Chichester: John Wiley and Sons.

Berndt, H., Jackwerth, E. (1975). *Spectrochim Acta, 30B*, 169–177.

Blain, L., Savin, E. D. (1992). *Spectrochim Acta, 47B*, 3994–4023.

Bowen, H. J. M. (1966). *Trace Elements in Biochemistry*, New York: Academic Press.

Bowen, H. J. M. (1980). *CRC Crit Revs, 10*, 127–174.

Branch, S., Crew, H. M., Halls, D. J., Taylor, E. (1991). *J. Anal Atomic Spectrom, 6*, 69R–136R.

Brinckman, F. E., Blair, W. R., Jewett, K. L., Ivenson, W. P. (1977). *J Chromatogr Sci, 15*, 493–503.

Brooks, R. R. (1983). *Biological Methods of Prospecting for Minerals*, New York: John Wiley and Sons.

Brown, A. A., Taylor, A. (1985). *Analyst, 110*, 579–582.

Cantle, E. J. (ed.) (1982). *Atomic Absorption Spectrometry*, London: Elsevier.

Capelli, M., Manfredi, V. R., Moretti, G. F., Trenti, A. (1991). in: *Carcinogenic and Mutagenic Metal Compounds 3. (Interrelation between Chemistry and Biology)*: Merian, E., Bronzetti, G. L., Haerdi, W. (eds.) New York: Gordon and Breach.

Carrion, N., de Benzo, Z. A., Moreno, B., Fernandez, A., Eljuri, E. J., Florez, D. (1988). *J Anal Atomic Spectrom, 3* , 479–483.

Clarke, N. S. (1989). *Chem. Brit, 25*, 484–486.

Cole, M. M. (1992). in: *The Ecology of Areas with Serpentinized Rocks. A World View*: Roberts B. A., Proctor, J. (eds.) Netherlands: Kluwer Academic Publications.

Colinet, E., Griepink, B., Muntau, H. (1982). Commission of the European Communities. Eur 8119 EN.

Cornelis, R. (1985). *J Trace Microprobe Tech, 2*, 237–265.

Cornclis, R., Hoste, J., Versieck (1982). *Talanta, 29*, 1029–1034.

Cresser, M. S. (1978). *Solvent Extraction in Flame Spectroscopic Analysis*, London: Butterworth.

Cresser, M. S., Armstrong, J., Dean, J., Ramsey, M. H., Cave, M. (1991). *J Anal Atomic Spectrom, 6*, IR–67R.

Cresser, M. S., Armstrong, J., Dean, J., Watkins, P., Cave,, M. (1992). *J Anal Atomic Spectrom, 7*, IR–66R.

Czauderna, M. (1984). *Int J Appl Radiat Isot, 35*, 681–684.

Demers, D. R., Allemand, C. H. (1981). *Anal Chem, 53*, 1915–1921.

Dokiya, Y., Yamamoto, Y., Toda, S., Fuwa, F. (1975). *Spect Lett, 8(10)*, 759–769.

Ebdon, L., Hill, S., Ward, R. W. (1986). *Analyst, 111*, 1113–1138.

Epstein, E. (1972). *Mineral Nutrition in Plants*, London: John Wiley and Sons.
Fagioli, F., Landi, S. (1983). *Anal Lett, 16 (A17 & 18)*, 1435–1477.
Falk, H., Hoffmann, E., Ludke, Ch. (1984) *Spectrochim Acta, 39B*, 283–294.
Farago, M. E. (1986). in: *Frontiers in Bioinorganic Chemistry*: Xavier, A. V. (ed.) Weinheim: VCH Publishers.
Farago, M. E., Parsons, P. J. (1983). in: *Platinum, Gold and Other Chemotherapeutic Agents. ACS Symposium Series, 209*, Lippard, S. J. (ed.) Washington DC: ACS; pp. 297–311.
Feigl, P. K. D., Krueger, F. R., Schveler, B. (1984). *Microchim Acta, II*, 85–96.
Fergusson, J. E. (1990). *The Heavy Elements, Chemistry, Environmental Impact and Health Effects*, Oxford: Pergamon Press; Chapter 4.
Freedman, B., Hutchinson, T. C. (1981). in: *Effect of Heavy Metal Pollution on Plants, Vol. 2, Metals in the Environment*: Lepp, N. W., (ed.) London: Applied Science.
Fresenius' Zeitschrift für Analytische Chemie (1981). *Sessions of the LAMMA-Symposium, Düsseldorf, Oct. 1980, 308*, pp. 193–320.
Gilbert, M. J. M., Powell, H. K. J., Fardy, J. J. (1988). *Anal Chim Acta, 207*, 103–109.
Gladney, E. S., Neifert, P. E., Bowen, N. W. (1989). *Anal Chem, 61*, 1834–1836.
Glooschenko, W. A., (1989). *Environ Pollut, 57*, 27–33.
Greenfield, S., Jones, I. L., Berry, C. T. (1964). *Analyst, 89*, 713–720.
Griepink, B. (1990). in: *Element Concentration Cadasters in Ecosystems*: Lieth, H., Markert, B. (eds.) Weinheim: VCH Publishers; pp. 181–206.
Grobenski, Z., Leymann, R., Tamm, R., Welz, B. (1982). *Mikrochim Acta, 1*, 115–125.
Harrison, W. W., Hess , K. R., Markus, R. K., King, F. L. (1986). *Anal Chem, 58*, 341A–356A.
Heinrich, G. (1989). *J Plant Physiol, 133*, 770–772.
Henze, G. (1990). in: *NATO ASI Series, Vol. G23, Metal Speciation in the Environment*: Broekaert, J. A. C., Gucer, S., Adams, F. (eds.) Berlin: Springer-Verlag.
Heumann, K. G. (1988). in: *Inorganic Mass Spectrometry*: Adams, F., Gijbels, R., van Grieken, R. (eds.) New York: Wiley; pp. 301–376.
Heumann, K. G. (1990). in: *NATO ASI Series, Vol. G23, Metal Speciation in the Environment*: Broekaert, J. A. C., Gucer, S., Adams, F. (eds.) Berlin: Springer-Verlag.
Hewitt, E. J., Smith, T. A. (1975). *Plant Mineral Nutrition*, London: English University Press.
Hill, S. J., Dawson, J. B., Price, W. J., Shuttler, I. L., Tyson, J. F. (1991). *J Anal Atomic Spectrom, 6*, 187R–228R.
HMSO, (1986). *The Analysis of Agricultural Materials*, Ministry of Agriculture Fisheries and Food Reference Book 427, 3rd edn.
Hoenig, M., de Borger, R. (1983). *Spectrochim Acta, 38B*, 873–880.
Hoffmann, G., Schweiger, P. (1981). *Staub Reinhalt, 41*, 443–444.
Horvath, Z., Lasztity, A., Barnes, R. M. (1991). *Spectrochim Acta Rev, 14*, 45–78.
Horwitz, W., Albert, R. (1991). in: *Biological Trace Element Research, Multidisciplinary Perspectives ACS Symposium Series 445*: Subramanian, K. S., Iyengar, G. V., Okamoto, K. (eds.) Washington DC: American Chemical Society.
Houk, R. S. (1986). *Anal Chem, 58*, 97A–104A.
Houk, R. S., Fassel, V. A., Flesch, G. D., Svec, H. J., Gray, A. L., Taylor, C.E. (1980). *Anal Chem, 52*, 2283–2289.
Irgolic, K. J. (1985). in: *Environmental Inorganic Chemistry*: Irgolic, K. J., Mastell, A. E. (eds.) Weinheim: VCH Publishers.
Irgolic, K. J. (1991). in: *Trace metal Analysis and Speciation. Journal of Chromatography Library Series, Vol 47*: Krull, I. S. (ed.) Amsterdam: Elsevier.
Isaac, R. A., Johnson, W. C. (1975). *J Assoc Off Anal Chem, 58*, 436–440.
Jochum, K. P., Matus, L., Seufert, H. M. (1988). *Fresenius' Z Anal Chem, 331*, 136–139.
Jones, V. K., Tarter, J. G. (1985). *Int Lab, 15/9*, 36–39.

Jones, K. C., Johnston, A. E. (1989). *Environ Pollut, 57*, 199–216.

Kabata-Pendias, A., Pendias, H. (1984). *Trace Elements in Soils and Plants*, Boca Raton: CRC Press.

Karanassios, V., Liu, F. H. L. B., Salin, E. D. (1991). *J Anal Atomic Spectrom, 6*, 457–463.

Keith, L. H. (1980). in: *Dynamics, Exposure and Hazard Assessment of Toxic Chemicals*: Hague, R., Woburn, M. A. (eds.) Ann Arbor: Ann Arbor Science Publishers; pp. 41–45.

Keith, L. H. (ed.) (1987). *Principles of Environmental Sampling*, Washington DC: American Chemical Society.

Keith, L. H. (1991). *Environmental Sampling and Analysis: A Practical Guide*, Chelsea, MI: Lewis Publishers.

Keith, L. H., Crummett, W., Deegan, J., Libby, R. A., Taylor, J. K., Wentler, G. (1983). *Anal Chem, 55*, 2210–2218.

Kingston, H. M., Jassie, L. B. (1988). *Introduction to Microwave Sample Preparation, Theory and Practice*, Washington DC: American Chemical Society.

Kikkert, J. N. (1983). *Spectrochim Acta, 38B*, 1497–1508.

Klopfenstein, D., Nussbaum, E. (1976). *Proc Indiana Acad Sci, 85*, 339–342.

Knapp, G. (1984). *Fresenius' Z Anal Chem, 317*, 213–219.

Knox, J. H., Done, J. N., Gilbert, M. T., Pryde, A., Wall, R. A. (1978). *High Performance Liquid Chromatography*, Edinburgh: University Press.

Krull, I. S. (ed.) (1991). *Trace Metal Analysis and Speciation, Journal of Chromatography Library Series, Vol 47*, Amsterdam: Elsevier.

Langmyhr, F. J., Dahl, I. M. (1981). *Anal Chim Acta, 131*, 303–306.

Lawrenz, J., Niemax, K. (1989). *Spectrochim Acta, 44B*, 155–164.

Leis, F., Broekaert, J. A. C., Waechter, H. (1989). *Fresenius' Z Anal Chem, 333*, 2–5.

Letokhov, V. S. (1986). *Laser Analytical Spectrochemistry*, Bristol: Adam Hilger.

Leuka, M. Pander, K. K., Pander, B. B. (1990). *Environ Pollut, 66*, 89–99.

Liteanu, C., Rica, I. (1980). *Statistical Theory and Methodology of Trace Analysis*, Chichester: Ellis Horwood.

Littlejohn, D., Jowitt, R., Sparkes, S. T., Thorne, A. P., Walton, S. J. (1991). *J Anal Atomic Spectrom, 6*, 137R–185R.

Markert, B. (1992). in: *Biogeochemistry of Trace Metals*: Adriano, D. C. (ed.) Boca Raton: Lewis Publishers.

Markert, B. (ed.) (1993). *Plants as Biomonitors,* Weinheim: VCH.

Markert, B., Klausmeyer, N. (1990). *Toxicol Environ Chem, 25*, 201–212.

Martin, J. M. (1988). *J Assoc Off Anal Chem, 71*, 1038.

Matthews, H., Thornton, I. (1982). *Plant Soil, 66*, 181–193.

Matusiewicz, H. (1991). *Spectroscopy International, 3(2)*, 22–27.

Middleton, G., Stuckey, R. E. (1954). *Analyst, 79*, 138–142.

Miller-Ihli, N. J. (1990a). *Fresenius' Z Anal Chem, 337*, 271–274.

Miller-Ihli, N. J. (1990b). *Talanta, 37*, 119–125.

Mitchell, J. W. (1982). *Talanta, 29*, 993–1002.

Muse, J. O., Tudino, M. B., d'Huicque, L., Troccoli, O. E., Carducci, C. N. (1989). *Environ Pollut, 58*, 303–312.

Nakahara, N. (1991). *Spectrochim Acta Revs, 14*, 95–109.

Niemax, K. (1989). in: *Handbook of Laser Science and Technology, Vol. 4*: Letokhor, V. S., Shank, C. V., Shen, Y. R., Walther, H. (eds.) Chichester: Horwood Academic Publishers.

Niemax, K., Sdorra, W. (1990). *Applied Optics, 29*, 5000–5006.

Odom, R. W., Lux, G., Fleming, R. H., Chu, P. K., Niemeyer, I. C. , Blattner, R. J. (1988). *Anal Chem, 60*, 2070–2075.

Omenetto, N., Human, H. G. C. (1984). *Spectrochim Acta, 39B*, 1333–1343.

Piepmeier, E. H. (1986). *Analytical Applications of Lasers*, New York: John Wiley and Sons.
Pierson, K. B., Evenson, M. A. (1986). *Analyt Chem, 58*, 1744–1748.
Pinta, M. (1978). *Modern Methods for Trace Element Analysis*, Ann Arbor: Ann Arbor Science Publishers.
Ramsey, M. H., Coles, B. J. (1988). *Paper presented at 4th Biennial National Atomic Spectroscopy Symposium, York.*
Ramsey, M. H., Dong, D., Thornton, I, Watt, J., Giddens, R. (1991). *Environ Geochem Health, 13*, 114–118.
Rethfelden, H. (1980). *Fresenius' Z Anal Chem, 301*, 308.
Schmidt, K. P., Falk, H. (1987). *Spectrochim Acta Part B, 42*, 431–443.
Seiler, H. G. (1988). in: *Metal Ions in Biological Systems, Vol 23, Nickel and its Role in Biology*: Sigel, H. (ed.) New York: Marcel Dekker, Inc.
Stewart, M. T. L., Hopmans, P., Flinn, D. W., Hillman, T. J. (1990). *Environ Pollut, 63*, 155–177.
Stockton, R. A., Irgolic, K. J. (1979). *Int J Environ Anal Chem, 6*, 313.
Stockwell, P. B., Corns, W. T. (1992). *Spectroscopy World, 4*, 14–18.
Stoeppler, M. (1991). in: *Metals and their Compounds in the Environment, Occurrence, Analysis and Biological Relevance*: Merian, E. (ed.) Weinheim: VCH.
Subramanian, K. S. (1991). in: *Biological Trace Element Research, Multidisciplinary Perspectives, ACS Symposium Series 445*: Subramanian, K. S., Iyengar, G. V., Okamoto, K. (eds.) Washington DC: American Chemical Society.
Sulcek, Z., Povondra, P. (1989). *Methods of Decomposition in Inorganic Analysis*, Boca Raton, FL: CRC Press.
Taylor, J. K. (1987). *Quality Assurance of Chemical Measurements*, Chelsea, M. I.: Lewis Publishers, Inc.
Thompson, M., Ramsey, M. H. (1985). *Analyst, 130*, 1413–1422.
Thompson, M., Walsh, J. N. (1988). *Handbook of Inductively Coupled Plasma Spectrometry, 2nd edn.*, London: Blackie; pp. 130–160.
Thompson, M, Wood, S. (1982). in: *Atomic Absorption Spectrometry*: Cantle E. J. (ed.) London: Elsevier.
Tölg, G. (1974). *Talanta, 21*, 327–345.
Tölg, G. (1988). *Fresenius' Z Anal Chem, 331*, 226–235.
Tölg, G., Tschopel, P. (1987). *Anal Sci, 3*, 199–208.
Tschopel, P., Kotz, L., Shulz, W., Veber, M., Tölg, G. (1980). *Fresenius' Z Anal Chim, 302*, 1–14.
Ure, A. M. (1990). in: *Heavy Metals in Soils*: Alloway, B. J. (ed.) London: Blackie.
Ure, A. M., Hernandez-Artiga, M. P. (1977). *Anal Chim Acta, 94*, 195–197.
Vollkopf, U. Grobenski, Z., Tamm, R., Welz, B. (1985). *Analyst, 110*, 573–577.
Walsh, J. N., Howie, R. A. (1986). *Applied Geochemistry, 1*, 161.
Wang, J., Farias, P. A. M., Mahmoud, J. S. (1985). *Anal Chim Acta, 172*, 57–64.
Webber, J. (1981). in: *Effect of Heavy Metal Pollution on Plants, Vol 2, Metals in the Environment*: Lepp, N. W. (ed.) London: Applied Science.
Welz, B. (1985). *Atomic Absorption Spectrometry, 2nd edn.*, Weinheim: VCH.
Wendt, R. H., Fassel, V. A. (1965). *Anal Chem, 37*, 920–922.
William, K. L. (1987). *An Introduction to X-ray Spectrometry*, Boston, MA: Allen and Unwin.
Williams, E. V. (1978). *J Food Technol, 13*, 367–384.

10 Phytochemistry: Methods for the Study of Inorganic Species in Plant Tissues

Margaret E. Farago

Environmental Geochemistry Research
Royal School of Mines
Imperial College of Science, Technology and Medicine
Prince Consort Road, London SW7 2BP, UK

Contents

Introduction

The search for the exact sites of ion and compound deposition in plants and the nature of the chemical species involved has been carried out with a number of aims: The elucidation of various mechanisms of phytochemical importance; mineral uptake and utilisation; toxicity and tolerance exhibited by many individual species; and the study of plants that accumulate or hyperaccumulate metals. The literature for a number of individual elements to the late 1970s has been reviewed (Farago, 1981) and the phytochemistry of metal hyperaccumulators has been reviewed by Baker and Brooks (1989).

Extraction Techniques

Early work on the distribution of elements in plant tissues relied on extraction techniques. In many cases radioactive isotopes have been used in an effort to locate inorganic species in various plant extracts. Some workers have used sequential extraction schemes often in combination with chromatographic separation methods. Farago and Mahmoud (1983) pointed out that there are inherent difficulties in searching for metal complexes from plant sources by extraction methods. The extraction procedures themselves release many potential ligands. The complexes isolated may be those with the greatest thermodynamic stability in the extraction mixtures, and not those that existed *in vivo*. This is particularly relevant when dealing with kinetically labile ions. Nevertheless extractions followed by other techniques such as chromatographic separations have suggested that a number of metal–ligand complexes are present in plant metal-accumulators. These range from organic acids to carbohydrate of the pectin type, to amino acids and proteins and some of the investigations will be described below.

In 1960, Hewitt and Notton estimated the phosphorus in fractions of tomato and cauliflower leaves using ^{32}P coupled with extractions. Bowen *et al.* (1962) fed tomato plants with a number of radioactive elements, the fresh tissues were then extracted with a series of extractants. The results posed a number of interesting questions, for example, the association of cobalt and iron with proteins, and the authors suggested that a high speed centrifuge should be used to separate intracellular particles of different sizes. The nature of the metal compounds extracted was also questioned. Some inconclusive paper chromatography was also reported.

The African ‘copper flower’ *Becium homblei*, was studied by Reilly (1969) and a correlation was found between the copper and nitrogen concentrations in the plant, suggesting the binding of Cu to proteins. A simple sequential extraction scheme showed a high percentage of Cu in the final residue, again suggesting strong bonding. Paper chromatography using butanol–acetic acid–water solvent indicated two pep-

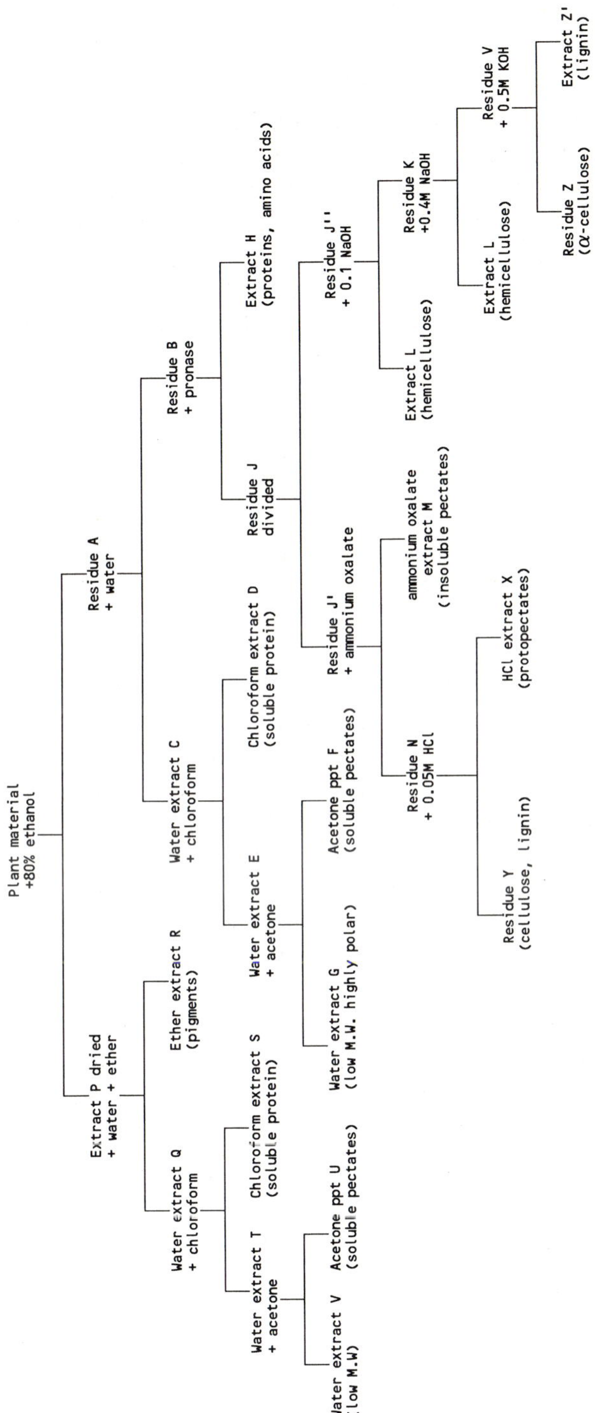

Fig. 10-1. Sequential extraction scheme (from Farago and Mahmoud, 1983).

tide or amino acid complexes in the extract. However, no specific amino acids bound to copper were identified (Reilly, 1972).

A more complex extraction scheme than those of Hewitt, Bowen, and Reilly described above, was used to show that ^{65}Zn is primarily associated with pectin in the root residue (Peterson, 1968). This was modified and used to indicate the association of copper in *Armeria maritima* (Mill) Willdenow (Farago *et al.*, 1980) and Ni in *Hybanthus floribundus* (Farago and Mahmoud, 1983). The scheme is shown Fig. 10-1, and some typical results are shown in Table 10-1. In these last two investigations radiotracers were not required, since the plants accumulated Cu and Ni, respectively, and metal determinations were carried out by flame atomic absorption spectrometry on the individual fractions.

Table 10-1. Sequential extraction of leaves of *Hybanthus floribundus* from Western Australia, showing percentages of Ni in original dried sample (13000 μg g^{-1}). Extraction scheme as in Fig. 10-1 (from Farago and Mahmoud, 1983).

Extract/ppt	Compounds extracted	% of total Ni	
		Scheme 1 *J′* residue	Scheme 2 *J″* residue
Ether extract R	Pigments	0.1	0.1
Chloroform extract S	Soluble protein	0.6	0.6
Acetone ppt U	Soluble pectates	34	36
Water extract V	Low molecular weight	0.1	0.1
Chloroform extract D	Soluble protein	1.3	1.4
Acetone ppt F	Soluble pectates	36.5	39
Water extract G	Low molecular weight, highly polar	6.9	7.4
Pronase extract H	Proteins + amino acids	3.3	3.6
Ammonium oxalate extract M	Insoluble pectates	12	–
0.05M HCl extract X	Protopectates	4.7	–
Residue Y		0.2	–
0.1M NaOH extract L	Hemicellulose	–	7.2
0.44M NaOH extract W	Polysaccharides	–	3.3
1.5M KOH extract Z′	Lignin	–	0.9
Residue Z	α-cellulose	–	0.4
	Total	99.7	100

Much work has been carried out on Ni and other metal accumulators by Brooks and co-workers (Baker and Brooks, 1989). For example, Ni was suggested to be in association with malate in the aqueous extracts of Ni accumulating plants (Kersten *et al.*, 1980). Eluent from a Sephadex G-10 column indicated three types of Ni, aquo, citrate and malate complexes. with the latter being the most abundant. The extract was methylated and malate was confirmed by GC-MS. High voltage electrophoresis has also been used on extracts to identify nickel complexes (Lee *et al.*, 1977).

Recently, Homer *et al.* (1991) have characterised the Ni-rich extract from the Philippine plant *Dichapetalum geloniodes* subsp. *tuberculatum*, leaves of which contained 25000 $\mu g\ g^{-1}$ of nickel. Using a Sephadex G-10 column the Ni-rich aqueous eluent (molar mass < 500 D) was subjected to GC-MS analysis, showing the presence of citric and malic acids, plus a trace of tartaric acid. Attempts were made further to separate the Ni-containing species by gel filtration, however, is was shown that there is a complex series of equilibria set up on the column between Ni citrate, malate and aquo species. The citrate complex is the most thermodynamically stable complex, but there was insufficient citrate present in the extract to complex all the nickel. The authors report therefore that it is meaningless to assign fractions of the plant Ni in the extract to particular complexes, thus underlining the points made by Farago and Mahmound (1983) concerning labile complexes in plant extracts, and Homer *et al.* (1991) further stress the need for more work on the characterisation of nickel complexes in such systems.

An aqueous extract from a cobalt hyperaccumulator (Morrison *et al.*, 1981), *Haumaniastrum robertii*, contained 14% Co. Since the molecular weight was about 5200 per Co, it could not be proteinaceous. Electrophoresis gave a complicated pattern, and gas-liquid chromatography following methylation gave 2 peaks which were not identified.

Paper chromatography of an aqueous extract, after separation with Sephadex gel (G25), was used by Farago and Mullen (1979) in an investigation of soluble copper in aqueous extracts of copper-tolerant *Armeria maritima*. The roots of copper-tolerant *A. maritima* contain high concentrations of the amino acid proline and copper (Farago and Mullen, 1979, 1981) and the results indicated that water soluble copper (about 20%) was in combination with proline whereas the insoluble copper was associated with carbohydrates (Farago *et al.*, 1980).

Kwan and Smith (1990a, b) have used a number of techniques in the study of the distribution and chemical forms of Cd and Tl in *Lemna minor* that had been grown in the presence of ^{204}Tl or ^{115m}Cd. In addition to sequential extractions, three enzymes were applied sequentially to extracts of *Lemna* tissue. The results of this treatment emphasised the chemical differences between the two elements; a relatively small amount of Tl was associated with the cell wall fraction, whereas a significant amount of Cd is bound to cell wall pectins and proteins. Tl is a monovalent ion Tl^{+}, with low complexing ability, like K^{+}, whereas Cd^{2+} is known to have an affinity for a number of ligands. Using Sephadex gels calibrated with a series of highly purified protein and vitamin standards, fractions were collected for radio- and protein assay. A major portion of the Cd in the soluble phase was found to be bound to three groups of proteinaceous and polypeptide fractions, the distribution between these depended on the length of exposure of the plants to Cd. Most of the Tl in the soluble phase was the free ion.

Radioactive ^{109}Cd has been used in a study of the unicellular giant alga *Acetabularia acetabulum* (Karez *et al.*, 1991). After incubation with ^{109}Cd, extracts were eluted from a Sephacryl S-200 column. After short incubations of 1 – 2 h ionic and low molecular weight compounds were eluted, thus little or no cadmium was immobilised, but for incubation times of one week or more cadmium was associated with fractions with molecular weights of more than 25000 D and between 70000 and 20000 D.

Smith *et al.* (1989, 1991) exposed *Lemna minor* to ^{51}Cr(VI) in the form of chromate and demonstrated that extracts showed Cr associated with complexes of about 2000 D and > 100000 D. Aqueous extracts were subjected to gel permeation chromatography (Sephadex G50) and both ^{51}Cr and protein activity were followed. Although the Cr fractions corresponded to protein fractions, the authors were unable to demonstrate that these were Cr-induced proteins, since the protein profiles for Cr-treated and untreated plants were the same. Clearly more work is required on the chemical identification of metal complexes in such extracts.

A recent volume on trace metal analysis and speciation (Krull, 1991) has reviewed the latest instrumentation and methodology for determination of the chemical forms present in the sample. Various separation methods, such as gas and liquid chromatography interfaced with a number of sensitive detection methods are described. These methods have potential applications in the determination of chemical species in plant extracts particularly for organometallic and kinetically non-labile species of As, Sn, Cr, Fe, Te, Se, V, Hg and Al. Such methods have been applied to Al complexes in tea. The tea plant accumulates Al and the chemical species of it in tea leaves has been investigated by chromatography interfaced with ICP-MS (Baxter *et al.,* 1989). A tea infusion was chromatographed using a Superose column and monitored by ICP-MS at 27 amu. Two Al peaks were obtained corresponding to 13000 and 6700 D and it was suggested that these are phenolic thearubugen complexes. Tea leaves have also been studied by ^{27}Al NMR (see *Nuclear Magnetic Resonance Spectroscopy*).

Differential Ultracentrifugation

Various cell organelles can be separated from one another, by first breaking cell walls then using the centrifuge to isolate particles having different densities. The homogenisation of the plant cells can cause considerable damage to organelles. Work must be carried out at as low a temperature as possible since many oxidation reactions, while not taking place within the intact cell can occur rapidly when the cell wall is disrupted. The large pH changes that occur on homogenisation can produce osmotic shock in some organelles, thus a buffer is used as the homogenising medium.

A number of studies using radiotracers, extraction and centrifugation lead to the conclusion that Zn in *Agrostis tenuis* is associated with carbohydrate in the cell wall

(Bradshaw, 1969; Turner and Marshall, 1970, 1971, 1972). The use of the differential centrifugation by Brooks *et al.* (1981) showed 72% of the nickel in the Iberian hyperaccumulator *Alyssum serpyllifolium* Desv. subsp. *malacitanium* Rivas Goday was in the supernatent, suggesting a polar complex of nickel in the vacuoles.

Microelectrodes

The use of microelectrodes, specific to particular elements has allowed the measurement of fluxes across membranes. Bowling (1976) described the implantation of ion specific microelectrodes into the epidermal and particular cortical cells of *Zea mays*. Early work on the determination of ion activity in the cytoplasm of roots using specific ion electrodes was carried out by Etherton (1968) using a K^+ electrode; and by Gerson and Poole (1972) using a Cl^- electrode. More recently, Newman *et al.* (1987) using ion-selective microelectrodes measured the fluxes of H^+ and K^+ in corn roots. The effects of Al on the Ca^{2+} fluxes in the roots of Al-tolerant and Al-sensitive wheat cultivars were examined by Huang *et al.* (1992) using a Ca^{2+} selective electrode. Both cultivars maintained similar Ca^{2+} uptake in the absence of Al. Al supplied as $AlCl_3$ inhibited Ca uptake in the Al-sensitive cultivars (the effect was reversible on the removal of Al), whereas the Al-tolerant cultivar was relatively unaffected. It was concluded that Al inhibits Ca^{2+} influx across the root plasmalemma.

Histochemistry and Stains

Plant material is killed, fixed and sectioned for microscopic investigation. The sections are stained with spot reagents which give vivid colours with particular metals. The classical methods of fixing, dehydrating and wax embedding can all produce translocation of elements. In a modified method stains were applied to sections of leaves that had been immersed in a mixture 1 : 1 v/v) of ethanol and glycerol and colour photography was used to produce a permanent record (Farago and Mahmoud, 1983). Using this method nickel was clearly located in the outer cell wall in *Hybanthus floribundus* (Lindl.) F. Muell. Many stains have been described by Gurr (1958).

The uptake if iron often requires Fe(II) as the bioavailable form, and uptake from soils containing insoluble Fe(III) is accomplished by phenolic compounds which are exuded by certain plants (Chaney and Bell, 1987). A particular use of staining has been in the study of the sites if iron reduction in plants, using Prussian blue stain (PB) (Ambler *et al.*, 1971; Brown and Ambler, 1974). The method consists of placing the roots in a nutrient solution containing Fe(III) and ferricyanide. Since PB can be produced both from Fe(III) and $Fe(CN)_6{}^{4-}$ and from Fe(II) and $Fe(CN)_6{}^{3-}$, reduction of either source of iron in the presence of the roots will produce PB at the sites of

Fe reduction. Sections of roots can then be viewed under the microscope and the stain located. Bell *et al.* (1988) have evaluated the use of ferricyanide and the use of nitro-BT, [2,2′-di-*p*-nitro-phenyl-5,5′-diphenyl]-3,3′-(3,3′-dimethoxy-4,4′-biphenylene)-di-tetrazolium chloride. With ferricyanide, iron-stressed tomato plants were mainly stained on the younger roots hairs located on the laterals or primary root tip. Nitro-BT is thought to compete with Fe(III) at the same transmembrane ferric reduction site (Sijmons and Bienfate, 1983). A purple diformazan precipitate is produced on reduction. Although the roots were stained in a similar way to those incubated with ferricyanide, Bell *et al.* (1988) point out that further research is required to determine if nitro-BT reduction completes with ferric reduction at the same site in the tomato. The iron-stress redox activity has been shown to be localised on the plasma membrane in tomato roots (Buckout *et al.*, 1989), and electron microscopic examination (see next section) of the roots stained with Prussian blue indicated that the PB had accumulated between the plasma membrane and the cell walls of the root hairs and epidermal cells (Wergin *et al.*, 1988).

Electron Microscopy and Microprobe

When a specimen is examined with an electron microscope it is bombarded with a beam of high velocity electrons. As a consequence, some electrons are absorbed by the specimen, some pass through, some are deflected and scattered. If one of the inner electrons of an element in the specimen is removed by the electron beam then an electron falls from a higher energy shell to fill the hole, and X-rays are emitted. Each element has its characteristic X-ray spectrum. Analysis can be carried out both by energy dispersive spectrometry (EDS) and wavelength dispersive spectrometry (WDS). Electron microanalysis was applied to the study of plant material by Lauchli and coworkers (Lauchli and Schwarder, 1966; Lauchli *et al.,* 1971). The minimum level of an element that may be detected is dependent on the matrix and the element itself, the limit in plant material is around 0.1%. Micronutrients and other trace amounts are therefore beyond the limits of detection. Thus most literature studies have been restricted to the observations of Na, K, Ca, Mg and P. Kunoh (1990) has reviewed the limitations of X-ray microanalytical methods and reported that a semi-quantitative comparison of different methods of preparation of plant tissues showed the highest Si concentration from freeze drying and the lowest from conventional fixation and critical point drying. Preparation of tissues for analysis also may lead to mobilisation and redistribution of chemical species.

X-ray microanalysis has been applied to the study of plants containing platinum group metals (Farago and Parsons, 1983a, b; Parsons and Farago, 1987). *Eichhornia crassipes* plants treated with Pt metals accumulate these in their roots. Pt was detected on root surfaces using a scanning electron microscope with an energy dispersive spectrometer (SEM-EDS). Further examination with the microprobe analyser, showed that Pt was localised in the epidermal region of the roots.

The location of uranium from sea water taken up by the alga *Dunaliella* was achieved by EDAX and the microscope in the STEM mode (Spry and Bochem, 1981).

Williams (1986) has described the use of the scanning electron microscope in the study of biominerals, and in particular silicon in grass plants, where it functions as a structural support and protective device. Most biominerals are composites and contain an organic polymer, often a protein or polysaccharide. The silicified hairs of *Phalaris canariensis* were shown to be concentrically cylindrical and contain polysaccharides and silica. Investigation of the silica by NMR (see *Nuclear Magnetic Resonance Spectroscopy*) has shown it to be of the type $SiO_n(OH)_{4-2n}$. Further work has been discussed by Perry (1987, 1989) who has shown varying amounts of other elements during mineralisation, and it is suggested that in the plant hairs these are associated with differing spatial requirements of the polysaccharides (Perry and Yun, 1991).

Morrison *et al.* (1981) have used the proton microprobe to map K, Ca, Mn and Co by scanning across a leaf. The instrument is similar to the electron microprobe, but has a sensitivity of the order of $1-10\ \mu g\ g^{-1}$. It can also be operated in air instead of vacuum, these two differences rendering the method very suitable for plant analyses. The results showed that in the region of potassium deficiency there were elevated levels of calcium and cobalt.

X-Ray Photoelectron Spectroscopy (XPS)

XPS involves the photoejection of electrons from core levels, leading to valuable information concerning the bonding of the chemical species involved. The ejected photoelectrons have kinetic energy E_K, then the binding energy, E_B can be calculated from the equation

$$E_K = h\nu - E_B$$

Thus the photoelectron spectrum consists of a number of peaks at discrete values of E_K corresponding to values of E_B. The position of E_B identifies the atom, and since the binding energies are sensitive to chemical bonding and oxidation state, small variations can give an indication of the atom's chemical environment.

EPS was used to show that Pt on precipitates on the surfaces of roots of *Eichhornia crassipes* grown in nutrient with added $(NH_4)_2[PtCl_6]$ was Pt(IV) and had not been reduced by the roots (Parsons and Farago, 1987). The technique was used to study Al on the surfaces or barley roots, in Al-sensitive and tolerant plants grown in the presence of Al. Al-tolerant roots contained 21% Al as the phosphate, whereas the sensitive roots contained only 1.3% Al (Millard *et al.*, 1990). The method which is useful for the examination of surfaces, suffers, like electron microscopy, from the drawbacks of using dried material under vacuum.

Synchrotron Radiation: Extended X-Ray Absorption Fine Structure (EXAFS) and X-Ray Absorption Near Edge Structure (XANES)

These techniques have potential in the study of plants *in vivo* or of dried samples as well as the study of compounds obtained by extraction methods. The techniques involve the excitation of K- and L-shell X-ray fluorescence using a high intensity synchrotron radiation source. The energy of the highly collimated X-ray beam is scanned through the absorption edge of the target elements, measuring fluorescence for trace constituents and transmission for others, the resultant spectra are compared with those from model compounds. The techniques and their applications have been discussed by Gordon and Jones (1991) and early work appears in Winick and Doniach (1980). The techniques generally require no chemical preparation and are sensitive in the concentration range of μg^{-1}. Plant samples have been shown to give spectra, for example Gordon and Jones (1991) show the XANES spectrum for iron in NIST orchard leaves SRM 1751, where the Fe concentration is 300 $\mu g\ g^{-1}$. The development of the X-ray microprobe has great potential in plant studies, not only can elements be located spatially, but the coordination of the element and its chemical species be determined *in situ*.

EXAFS techniques can also be used to study compounds extracted from plants, and recently this technique has been used to study phytochelatins. Phytochelatins are metal binding compounds found in several plant species (Grill *et al.*, 1985a, b, 1988). The phytochelatins, such as those from cell cultures of *Rauwolfia serpentina*, are functionally similar to the metallothioneins; they contain only three amino acids (Glu, Cys and Gly) with general formula $(\gamma\ EC)_nG$ where $n = 2-11$. The parent compound where $n = 1$, is glutathione. In the order Fabales, peptides were found which were similar to the phytochelatins, but where alanine replaces glycine. These were named the homo-phytochelatins (Grill *et al.*, 1986). Recently EXAFS studies have been carried out on Cd-phytochelatin complexes extracted from cell suspension cultures of *R. serpentina* treated with Cd^{2+} (Nolting *et al.*, 1991). In the native complexes the Cd is exclusively coordinated to sulphur from 4 cysteine moieties, there are no Cd–Cd interactions. On the addition of more Cd to the native Cd complexes, more Cd becomes coordinated, it is suggested to oxygen and nitrogen donors.

Nuclear Magnetic Resonance Spectroscopy (NMR)

NMR is observed with isotopes whose nuclei exhibit nuclear spin. These nuclei absorb radiation when they are in a magnetic field. This absorption is measured in various types of NMR experiment (Roberts and Jardetzky, 1985). NMR techniques have the great advantage that information can be obtained using living plant material. The use of NMR in plant biochemistry has been reviewed (Roberts, 1984, 1987; Martin 1985).

The most studied nucleus, particularly for *in-vivo* studies has been ^{31}P. Naturally abundant ^{31}P occurs in metabolites that move freely in the tissue. The most abundant of these are: inorganic phosphate; sugar phosphate; nucleotide and nucleoside di- and triphosphates. The assignments of the signals have been discussed by Roberts (1987). The chemical shift of P in phosphates is sensitive to pH over the physiological range of 5.8 to 7.8. Estimations of intracellular pH can thus be made by comparison of chemical shifts *in vivo* with those *in vitro* (Roberts *et al.,* 1981). Use of this method has shown that in oil palm cell suspension cultures, there is loss of pH control with sudden increases in the external pH (Fox and Ratcliffe, 1990).

Studies of processes related to energy metabolism in plants can also be made using ^{31}P NMR observations of tissue concentrations of ATP, of the products of ATP hydrolysis and of ATP-Mg^{2+} interactions (Roberts, 1886).

In some experiments the concentrations of metals in specific intracellular compartments can be observed indirectly by their effects on ^{31}P signals from orthophosphate ($^{31}P_i$). Paramagnetic ions, such as Mn^{2+}, have a broadening effect on the $^{31}P_i$ signals. Kime *et al.* (1982) reported that when roots of *Zea mays* were incubated with Mn^{2+}, the $^{31}P_i$ resonance from vacuolar phosphate broadened with time much more than those from the cytoplasm. However, later work by Pfeffer *et al.* (1986) demonstrated that initially on exposure to Mn^{2+}, the cytoplasmic signals broadened. The vacuolar P_i signal then broadens over a much longer period of Mn^{2+} incubation and remains broadened after removal from Mn^{2+}, in contrast to the cytoplasmic P_i signals. These results show that Mn^{2+} is effectively removed from the cytoplasm by the vacuolar transport system and that there is little leakage from the vacuole. The same authors also demonstrated that treatment of the maize roots with 50 μM $Al_2(SO_4)_3$ at pH 4.0 for 20 h reduced respiratory rates and reduced concentrations of several nucleotide phosphates. Thus effects on intercellular phosphate metabolism are probably a primary toxic lesion of Al. The uptake of phosphate in maize roots has also been studied recently by ^{31}P NMR (Tu *et al.*, 1990).

^{15}N NMR has been used to investigate nitrogen metabolism particularly using ^{15}N enrichment. ^{14}N has a high natural abundance (99.63%), but it has a high quadripole moment, thus broad peaks are obtained, except where the local field around the nitrogen is symmetrical. Several species can be determined, these include NO_3^-, NH_4^+ , urea, betaine and some amino acids (Richards and Thomas, 1974). ^{14}N NMR spectra from tissues of *Hordeum vulgare*, *Zea mays* and *Pisum sativum* have been reported (Belton *et al.,* 1985) and the method has been used to explore nitrate uptake and reduction in these species.

Both ^{27}Al and ^{31}P NMR have been used to study the uptake of Al^{3+} by cell suspensions of the alga *Chlorella saccaraphila* (Folsom *et al.,* 1986). The results were interpreted as indicating Al complexation with cytoplasmic phosphates, and inhibition of glucose metabolism.

It was demonstrated by Sillerud and Heyser (1985) that signals from intra- and extracellular $^{23}Na^{+}$ could be distinguished. Using suspensions of cultured millet cells, they added the negatively-charged paramagnetic complex, dysprosium(III) triphosphate, which interacts with the positively charged sodium ion and alters the chemical shift. The intracellular signal is unchanged since the paramagnetic complex is not membrane permeable. ^{23}Na NMR has also been used to study Na accumulation in maize root tips (Gerasimowicz *et. al.*, 1986).

Aluminium complexation in tea leaves has been studied using ^{27}Al NMR (Nagata *et al.*, 1992). Intact leaves, mature, young and buds, were studied, together with a number of Al model compounds. It was conluded that Al is bound to chatechins, with some bound to phenolic and organic acids. Signals assigned to Al–F were found only in mature leaves.

Electron Spin Resonance Spectroscopy (ESR)

Electron spin resonance (ESR) also called electron paramagnetic resonance (EPR), can be used to study paramagnetic species. The theory and applications of ESR spectroscopy have been described (Berliner and Reuben, 1985). When the electrons of the paramagnetic substance are placed in an applied magnetic field, each takes up one of two possible states, aligned with or against the applied field. These two states differ in energy and will be thermally occupied. Electrons in the lower energy level can, with an applied field H_o be excited to the upper level by absorbing microwave energy of frequency v; where v is determined by the resonance equation

$$h\,v = g\,\beta\,H_o$$

The value of g, which is characteristic of the microwave centre, is obtained by holding the microwave frequency constant and varying H_o, and is displayed as the derivative of the microwave absorption against the magnetic field. Some nuclei have spin, and if such a spin has value I, then the ESR spectrum will be split into $(2\,I+1)$ lines, the magnitude of the splitting is the A value. Many of the oxidation states of the transition elements are paramagnetic and therefore detectable by ERS sprectroscopy, those of biological interest are shown in Table 10-2.

V(IV) is paramagnetic and suitable for study by ESR spectroscopy. Vanadium accumulates in some species of *Amanita* toadstools (Bertrand, 1943; Frausto da Silva,

Table 10-2. ESR-detectable elements of interest and potential interest in plant studies.

Ion	Nucleus	% Natural abundance	Nuclear spin	Electron spin
V(IV) (VO^{2+})	^{51}V	99.76	7/2	1/2
Mn(II)	^{55}Mn	100	5/2	5/2
Fe(III)	^{57}Fe	2.19	1/2	1/2, 5/2
Co(II)	^{59}Co	100	7/2	1/2, 3/2
Ni(II)	^{61}Ni	1.13	3/2	1
Ni(III)	^{61}Ni	1.13	3/2	1/2
Cu(II)	^{63}Cu	69.09	3/2	1/2
	^{65}Cu	30.91	3/2	
Mo(V)	^{95}Mo	15.72	5/2	1/2
	^{97}Mo	9.46	5/2	

1989). Amavadine, the vanadium-containing substance, had been isolated from the toadstools by Bayer and Kneifel in 1972. The same authors (Kneifel and Bayer, 1973) had suggested that amavadine was a square pyramidal complex containing V(IV) as the VO^{2+} moiety. Gillard and Lancashire in 1983 briefly compared the ESR spectrum of frozen segments of *Amanita muscaria* with amino acid complexes of VO^{2+} and concluded that they were not good models for amavadine. Subsequent comparisons of the ESR spectra of dried and fresh *Amanita* specimens with a series of square pyramidal complexes of VO^{2+} (Krauss *et al.*, 1984) demonstrated further that this type of structure was probably not that of amavadine (Frausto da Silva, 1989). Carrondo *et al.* (1988) synthesised and determined the structure of the vanadium complex of *N*-hydroxylimino-α,α′-diacetic acid as the ammonium salt of $[V(IV)(HIDA^{3-})_2]^{2-}$. The vanadium is 8-coordinate, with a distorted dodecahedral structure, each ligand being quadrideentate. The ESR spectrum of this complex is similar to that of the *Amanita* specimens and to extracted amavadine. Recently Armstrong *et al.* (1991) have reported that the similar complex with *N*-hydroxylimino-α,α′-dipropionic acid $[V(HIDPA)_2]^{2-}$is 8-coordinate. The spectroscopic properties of the two complexes are identical with those of amavadine. In addition the *K*-edge X-ray spectra of the compounds show that the coordination is essentially the same as in specimens of *Amanita muscaria.*

Vanadium uptake by higher plants has been discussed by Morrell *et al.* (1986). ESR spectroscopy was used to demonstrate the bioreduction of V(V) to V(IV) during uptake by plants. V(V) has all electrons paired and therefore does not give an ESR signal. Barley roots grown in the presence of V(IV) supplied as vanadyl sulphate $[VO]So_4$ showed the characteristic spectrum of V(IV) (Fig. 10-2), however, roots grown in the presence of V(V) supplied as NH_4VO_3 also showed this characteristic spectrum. The results strongly support the hypothesis of the bioreduction by the plant roots. The authors suggest that the redox reactions might involve cell wall polysaccharides (see below).

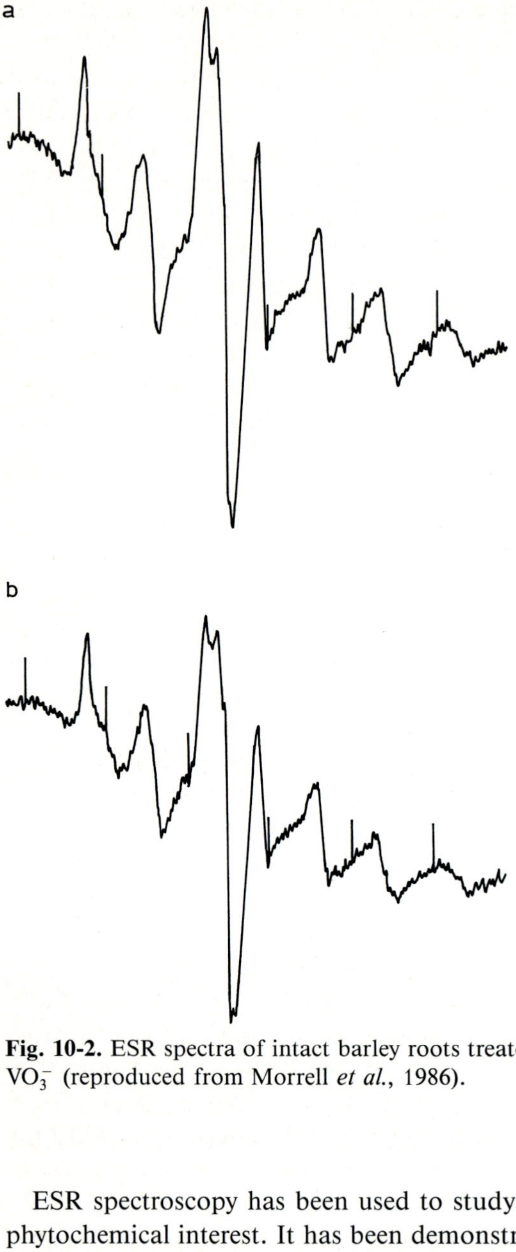

Fig. 10-2. ESR spectra of intact barley roots treated with vanadium as either (a) VO^{2+} or (b) VO_3^- (reproduced from Morrell *et al.*, 1986).

ESR spectroscopy has been used to study a number of other redox reactions of phytochemical interest. It has been demonstrated that reduction of Cr(VI) to Cr(V) and Cr(III) can be brought about in the cortex of garlic (*Allium sativum*) roots (Micera and Dessi, 1988). Plant roots equilibrated with 250 ppm Cr(VI) as dichromate showed spectra attributed to Cr(V) species (g_0 = 1.970) even after short incubation times. From the hyperfine splitting it was concluded that the Cr(V) was coordinated primarily to oxygens. With longer contact time, a second signal attributed to Cr(III) was detected. The Cr(V) signals were also detected from the cloves and the leaves, indicating mobility of the Cr complexes.

Polysaccharides have been implicated in the uptake of metal ions by plant roots (Geraldes and Castro, 1990). Polyuronates, such as polygalacturonic acid and its methyl ester, pectin, originating in the root, allow the removal of metal ions from the soil. The outer part of the plant root, the cortex, is freely available to aqueous solutions of inorganic salts by diffusion, and polygalacturonates provide about 90% of the ion exchange capacity. Metal ions have been found to be associated with polyuronic acids in cell walls. It has also been suggested that in order to cross membranes into the vascular system of the plant, most metal ions are coordinated. It is also possible that other redox reactions can occur: reductions of Fe(III) to Fe(II), V(V) to V(IV), and Mo(VI) to Mo(V) take place on interaction of soil mineral species with polygalacturonic acids, followed by the complexation and solubilisation of the reduced ions (Deiana *et al.*, 1983).

Plant Cell Cultures

The development of a cell culture requires that the organisation of the plant, whereby cells perform specific functions and have different environments, be broken down. All the cells in a culture show primary metabolic features without morphological differentiation. Most studies have been conducted on finely divided suspension cultures. The uses of plant cell cultures have been described by Rhodes and Robins (1987). A number of the NMR studies described above has been carried out on plant cell cultures (for example Fox and Ratcliffe, 1990; Folsom *et al.*, 1986; Sillerud and Heyser, 1985).

The heterogeneity of cells in a suspension culture can be exploited and variant cell lines can be isolated that show resistance to toxic substances or specific nutritional requirements, or the ability to grow in environmental stressed conditions. It had been shown (Smith 1975, 1980) that there is enhanced uptake of SO_4^{2-} by *Nicotiana tabacum* cells when the intracellular sulphate levels are low, which can be suppressed by S-containing metabolites including L-cystine. Using *Daucus carota* cells and a selenocystine analogue, Furner and Sung (1982, 1983) isolated two variants: one showed enhanced sulphate uptake even in the presence of cystine while the other showed hypersuppression of sulphate uptake by cystine. Sulphate metabolism and tissue culture studies are discussed in detail by Rhodes and Robins (1987).

The use of cell tissue culture methods in the study of metal toxicity and tolerance would appear to offer great potential scope.

Mass Spectrometry

The mass spectrometer has a number of applications in plant studies (Raven 1987). It can measure mass/charge ratios and thus has been of importance in the identification and quantitation of a number of physiologically important organic molecules. Thus MS coupled with separation techniques has been used to identify metal organic compounds extracted from plant tissues (eg, Kersten *et al.*, 1980; Homer *et al.*, 1991). A second major use has been the study of biochemical pathways and mechanisms where no radioactive isotope is available, by the use of stable isotopes as tracers. Lastly mass spectrometry can be used to study isotopic discrimination and isotopic ratios. Those elements where isotopic ratio studies have yielded interesting information include, hydrogen, carbon, nitrogen, oxygen and sulphur.

Isotope ratios are generally expressed as delta values (δ 'del') in parts per thousand ('per mil'):

$$\delta H(‰) = \left(\frac{(H/L)_{\text{sample}}}{(H/L)_{\text{standard}}} - 1\right) \times 1000$$

where $(H/L)_{\text{sample}}$ and $(H/L)_{\text{standard}}$ are the ratios of the heavy to the light isotope of the element in the sample and the standard, respectively. For example the natural abundance of ^{15}N is referred to atmospheric nitrogen, $^{15}N_{\text{atmos}} = 0‰$. The data on $\delta\,^{15}N$ of NH_4^+ and NO_3^- in soils have been reviewed by de Niro and Hastorf (1985). The values have a wide range with mean around 5–7‰. The $\delta\,^{15}N$ associated with uptake and assimilation of NH_4^+ or NO_3^- by terrestrial and aquatic plants is usually small (Minaguwa and Wada, 1984; Schoeninger and de Niro, 1984). However in some studies, substantial discrimination against ^{15}N has been observed. Mariotti *et al.* (1980) found that delta ^{15}N in *Pennisetum americanum* taken up from 12 mol m^{-3} nitrate solution were $-18‰$ at 3 days from germination and $-3‰$ after 25 days. At lower, more 'natural' levels of nitrate, the depletion is much smaller. The questions raised by this apparent discrimination have been discussed by Raven (1987).

Variations in the $\delta\,^{34}S$ values for plants have been summarised by Krouse *et al.* (1991), these range from -20 to $+34‰$ and are comparable to those reported in sedemintary rocks. It has been found that the sulphur uptake by the plant is mainly from soil sulphate where ambient SO_2 levels are low. In some cases ambient SO_2 dominates and the $\delta\,^{34}S$ values are close to those of the SO_2. These two uptake mechanisms are a factor in the differing isotope ratios in different parts of an individual plant (Chukhrov *et al.*, 1978). Thus sulphur isotope ratios are helpful in the determination of sulphur sources in the environment.

References

Ambler, J. E., Brown, J. C., Gauch, H. G. (1971). *Agron J, 63,* 95–97.

Armstrong, E. M., Calviou, L. J., Charnock, J. M. Collinson, D., Ertok, N., Garner, C. D., Mabbs, F. E., Naismith, J. H. (1991). *J. Inorg Biochem, 43* (2–3), 413.

Baker, A. J. M., Brooks, R. R. (1989). *Biorecovery, 1,* 81–126.

Baxter, M. J., Burrell, J. A., Crews, H. M., Massey, R. C. (1989). In: *Aluminium in Food and the Environment:* Massey, R. C., Taylor, D. (eds.) Cambridge: Royal Society of Chemistry.

Bayer, E., Kneifel, H. (1972). *Z Naturforsch, 27B,* 207–213.

Bell, P. F., Chaney, R. L., Angle, J. S. (1988). *J Plant Nutrition, 11,* 1237–1252.

Belton, P. S., Lee, R. B., Ratcliffe, R. G. (1985). *J Exp Bot 36,* 199–210.

Berliner, L. J., Reubin, J. (eds.) (1985). *Biological Magnetic Resonance, Vol. 6:* New York: Plenum.

Bertrand, D. (1943). *Bull Soc Chim Biol, 25,* 194–197.

Bowling, D. J. F. (1976). *Uptake of Ions by Plant Roots:* London: Chapman and Hall.

Bowen, H. J. M., Cause, P. A., Thick, J. (1962). *J Exp Bot, 13,* 257–267.

Brooks, R. R., Shaw, S., Asensi Marfil, A. (1981). *Physiol Plant, 51,* 167–170.

Brown, J. C., Ambler, J. E. (1974). *Physiol Plant, 31,* 221–224.

Buckout, J. T., Bell, P. F., Luster, D. G., Chaney, R. L. (1989). *Plant Physiol, 90,* 151–156.

Carrondo, M. A., Duarte, M. T. L. Pessoa, J. C., Frausto da Silva, J. J. R., Vaz, M. C. T. A., Vilas-Boas, L. F. (1988). *Chem Com,* 1158–1159.

Deiana, S., Dessi, A., Micera, G. Gessa, C., de Cherchi, M. L. (1983). *Inorg Chim Acta, 79,* 231–232.

Chaney, R. L., Bell, P. F. (1987). *J Plant Nutr, 10,* 963–994.

Chukhrov, F. V., Yermilova, L. P., Churikov, V. S., Nosik, L. P. (1978). *Geochemistry International,* 489–491.

De Niro, M. J., Hastorf, C. A. (1985). *Geochim Cosmochim Acta, 49,* 97–115.

Etherton, B. (1968). *Plant Physiol, 43,* 838–840.

Farago, M. E. (1981). *Coord Chem Revs, 36,* 155–182.

Farago, M. E., Mahmoud, I. E.D.A.W. (1983). *Min Environ, 5,* 113–121.

Farago, M. E., Mullen, W. A. (1979). *Inorg Chim Acta, 32,* L93–L94.

Farago, M. E., Mullen, W. A. (1981). *Inorg Nucl Chem Lett, 17,* 275–277.

Farago, M. E., Parsons, P. J. (1983a). *Platinum, Gold and Other Chemotherapeutic Agents, ACS Symposium Series, 209:* Lippard, S. J. (ed.) Washington DC: ACS; pp. 297–311.

Farago, M. E., Parsons, P. J. (1983b). *Inorg Chim Acta, 79,* 233.

Farago, M. E., Mullen, W. A., Cole, M. M., Smith, R. F. (1980). *Environ Pollut Ser A, 21,* 225–244.

Folsom, B. R., Popescu, A., Kingsley-Hickmann, P. B., Wood, J. M. (1986). In: *Frontiers in Bioinorganic Chemistry:* Xavier, A. N. (ed.) Weinheim: VCH.

Fox, G. G., Ratcliffe, R. G. (1990). *Plant Physiol, 93,* 512–521.

Frausto da Silva, J. J. R. (1989). *Chem Speciation Bioavail, 1,* 139–150.

Furner, I. J., Sung, Z. R. (1982). *Proc Natl Acad Sci USA, 79,* 1149–1153.

Furner, I. J., Sung, Z. R. (1983). *Plant Physiol, 71,* 547–550.

Geraldes, F. C. G., Castro, M.M. C. A. (1990). In: *Metal Speciation in the Environment, NATO ASI Series, Vol. 23:* Broekeart, J. A. C., Gucer, S., Adams F. (eds.) Berlin: Springer-Verlag; p. 105.

Gerasimowicz, W. V., Tu, S. I., Pfeffer, P. E. (1986). *Plant Physiol, 81,* 925–928.

Gerson, D. F., Poole, R. J. (1972). *Plant Pysiol, 50,* 603–607.

Gillard, R. D., Lancashire R. J. (1983). *Phytochem, 23,* 179–180.

Gordon, B. M., Jones, K. W. (1991). In: *Biological Trace Element Research. Multidisciplinary Perspectives, ACS Symposium Series 445:* Subramanian, K. S., Iyengar, G. V., Okamoto, K. (eds.) Washington DC: American Chemical Society.

Grill, E., Winnacker, E. L., Zenk, M. H. (1985 a). *Science, 230,* 674–676.

Grill, E., Zenk, M. H., Winnacker, E. L. (1985 b). *Naturwissenschaften*, 72, 432–433.

Grill, E., Winnacker, E. L., Zenk, M. H. (1988). *Experientia, 44,* 539–540.

Grill, E., Gekeler, W. Winnacker, E. L., Zenk, M. H. (1986). *FEBS Letters, 205,* 47–50.

Gurr, E. (1958). *Methods of Analysis, Histology and Histochemistry:* London: Leonard Hill.

Hewitt, E. J., Notton, B. A. (1960). *J Sci Food Agric,* 653–658.

Homer, F. A., Reeves, R. D., Brooks, R. R., Baker, A. J. M. (1991). *Phytochem, 30,* 2141–2145.

Huang, J. W., Shaff, J. E., Grunes, D. L., Kochian, L. V. (1992). *Plant Physiol, 98,* 230–237.

Karez, C. S., Bonotto, S., Puiseux-Dao, S. (1991). In: *Carcinogenic and Mutagenic Metal Compounds 3. (Interrelation between Chemistry and Biology):* Merian, E., Bronzetti G. L., Haerdi, W. (eds.) New York: Gordon and Breach.

Kersten, W. J., Brooks, R. R., Reeves, R. D., Jaffre, T. (1980). *Phytochem, 19,* 1963–1965.

Kime, M. J., Ratcliffe, R. G., Loughman, B. C. (1982). *J. Exp Bot, 33,* 670–681.

Kneifel, H., Bayer, E. (1973). *Angew Chem Int Ed, 12,* 408–513.

Krauss, P., Bayer, E., Kneifel, H. (1984). *Z Naturforsch, 39 b,* 829–832.

Krouse, H. R., Stewart, J. W. B., Grinenko, V. A. (1991). In: *Stable Isotopes: Natural and Anthropogenic Sulphur in the Environment SCOPE 43:* Krouse, H. R., Grinenko, V. A. (eds.) Chichester: John Wiley.

Krull, I. S. (ed.) (1991). *Trace Metal Analysis and Speciation, Journal of Chromatography Library Series,* Vol 47; Amsterdam: Elsevier.

Kunoh, H. (1990). *Ann Rev Phytopath, 28,* 93–111.

Kwan, K. H. M., Smith, S. (1990 a). *Chem Speciation Bioavail, 2,* 77–84.

Kwan, K. H. M., Smith, S. (1990 b). *Chem Speciation Bioavail, 2,* 97–103.

Lauchli, A., Swarder, H. (1966). *Experientia, 22,* 503–505.

Lauchli, A., Spurr, R. A., Epstein, E. (1971). *Plant Physiol, 48,* 118–124.

Lee, J., Reeves, R. D., Brooks, R. R., Jaffe, T. (1977). *Phytochem, 16,* 1503–1505.

Mariotti, A., Mariotti, F., Champigny, M.-L., Amanger, N., Moyse, A. (1980). *Plant Physiol, 69,* 880–884.

Martin, F. (1985). *Physiol Veg, 23,* 463–490.

Micera, G., Dessi, A. (1988). *J Inorg Biochem, 35,* 157–166.

Millard, M. M., Foy, C. D., Coradetti, C. A., Reinsal, M. D. (1990). *Plant Physiol, 93,* 578–583.

Minaguwa, M., Wada, E. (1984). *Geochim Cosmochim Acta, 48,* 1135–1140.

Morrell, B. G., Lepp, N. W. Philipps D. A. (1986). *Env Geochem Health, 8,* 14–18.

Morrison, R. S., Brooks, R. R., Reeves, R. D., Malaisse, F., Horowitz, P., Aronson, M., Merriam, G. R. (1981). *Phytochem, 20,* 445–458.

Nagata, T., Hayatsu, M., Kosuge, N. (1992). *Phytochem, 31,* 1215–1218.

Newman, I. A., Kochian. L. V., Grusak, M. A., Lucas, W. J. (1987). *Plant Physiol, 84,* 1177–1184.

Nolting, H.-F., Hermes, C., Strasdeit, H., Dahme, A.-K., Kneer, R., Zenke, M. H. (1991). *J Inorg Biochem, 43,* L011.

Parsons, P. J., Farago, M. E. (1987). *J Electron Micros Tech, 5,* 291–301.

Peterson, P. J. (1969). *J Exp Bot, 20,* 863.

Perry, C. C. (1987). *Recueil, 106,* 399.

Perry, C. C. (1989). In: *Biomineralisation, Chemical and Biochemical Perspectives:* Mann, S., Webb, J., Williams, R. J. P. (eds.) Weinheim: VCH.

Perry, C. C., Lu Yun. (1991). *J Inorg Biochem, 43* (2–3), 673.

Pfeffer, P. E., Tu, S. I., Gerasimowicz, W. V., Cavanaugh, J. R. (1986). *Plant Physiol, 80,* 77–84.

Raven, J. A. (1987). In: *The Biochemistry of Plants, Vol. 13. Methodology:* Davies, D. D. (ed.) London: Academic Press.

Reilly C. (1969). *New Phytol, 68,* 1081–1087.

Reilly, C. (1972). *Z Pflanzenphysiol, 66,* 993–997.

Richards, R. E., Thomas, N. A. (1974). *Perkin Trans, 2,* 368–374.

Rhodes, M. J. C., Robins, R. (1987). *The Biochemistry of Plants, Vol. 13. Methodology:* Davies, D. D. (ed.) London: Academic Press.

Roberts, J. K. M. (1984). *Ann Rev Plant Physiol, 35,* 375–386.

Roberts, J. K. M. (1986). In: *Modern Methods of Plant Analysis:* Linskens, H. F., Jackson, A. F. (eds.) Berlin: Springer-Verlag.

Roberts, J. K. M. (1987). In: *The Biochemistry of Plants, Vol. 13. Methodology:* Davies, D. D. (ed.) London: Academic Press.

Roberts, J. K. M., Jardetzky, O. (1985). In: *Modern Physical Methods in Biochemistry:* Neuberger, A., van Deenen, L. L. M. (eds.) Amsterdam: Elsevier.

Roberts, J. K. M., Wade-Jardetzky, N., Jardetzky, O. (1981). *Biochemistry, 20,* 5389–5394.

Schoeninger, M. J., De Niro, M. J. (1984). *Geochim Cosmochim Acta, 48,* 625–640.

Sijmons, P. C., Bienfait, H. F. (1983). *Physiol Plant, 59,* 409–415.

Sillerud, L. O., Heyser, J. W. (1984). *Plant Physiol, 65,* 269–272.

Smith, I. K. (1975). *Plant Physiol, 55,* 303–307.

Smith, I. K. (1980). *Plant Physiol, 66,* 877–883.

Smith, S., Peterson, P. J., Kwan, K. H. M. (1989). *Toxicol Environ Chem, 24,* 241–251.

Smith, S., Peterson, P. J., Kwan, K. H. M. (1991). In: *Carcinogenic and Mutagenic Metal Compounds 3. (Interrelation between Chemistry and Biology):* Merian, E., Bronzetti, G. L., Haerdi, W. (eds.) New York: Gordon and Breach.

Spry, B., Bochem, H.-P. (1981). *Fresenius Z Anal Chem, 308,* 239–245.

Tu, Shu-I., Cavanaugh, R. R., Boswell, R. T. (1990). *Plant Physiol, 93,* 778–784.

Turner, R. G., Marshall, C. (1972). *New Phytol, 71,* 671–676.

Wergin, W. P., Bell, P. F., Chaney, R. L. (1988). In: *Proceedings of the 46th Meeting of the Microscopy Society of America:* Bailey, G. W. (ed.) San Francisco: San Francisco Press; pp. 392–393.

Williams, R. J. P. (1986). In: *Frontiers in Bioinorganic Chemistry:* Xavier, A. V. (ed.) Weinheim: VCH.

Winick, H., Doniach, S. (eds.) (1980). *Synchrotron Radiation Research,* New York: Plenum Press.

Index